SPIRIT OF THE WEB

A L S O   B Y   W A D E   R O W L A N D

Ockham's Razor:
A Search for Wonder in an Age of Doubt (1999)

Polar Passage (with Jeff MacInnis) (1989)

Nobody Calls Me Mr. Kirck (with Harvey Kirck) (1985)

Making Connections (1979)

Fuelling Canada's Future (1974)

The Plot to Save the World (1973)

Spirit of the Web

THE AGE OF INFORMATION

FROM TELEGRAPH TO INTERNET

Wade Rowland

PATRICK CREAN EDITIONS

An Imprint of

KEY PORTER BOOKS

Canadian Cataloguing in Publication Data

Rowland, Wade, 1944-
Spirit of the web: the age of information from telegraph to Internet

ISBN: 1-894433-02-5

1. Telecommunications – History. I. Title.

TK5102.2.R69 1999 384′.09 C99-931325-8

The publisher gratefully acknowledges the support of the Canada Council for the Arts and the Ontario Arts Council for its publishing program.

Canadä
We acknowledge the financial support of the Government of Canada through the Book Publishing Industry Development Program (BPIDP) for our publishing activities.

Key Porter Books Limited
70 The Esplanade
Toronto, Ontario
Canada M5E 1R2

www.keyporter.com

Electronic formatting: Heidi Palfrey
Design: Peter Maher

Printed and bound in Canada

99 00 01 02 03 6 5 4 3 2 1

ACKNOWLEDGEMENTS

The preparation of this book has been a congenial journey for me, from the Dantesque land of network television management, back to the sources of a lifelong passion for journalism, scholarship and writing. A number of people have been enormously helpful along the way, though they may not realize it, like the roadside bystander who offers good directions and keeps the wayfarer out of the brambles and on the trail to his destination. First among them is Professor Abraham Rotstein of the University of Toronto's faculty of economics. He gave me initial encouragement when it was needed most, and introduced me to Professor Derrick de Kerckhove, Director of the University of Toronto's McLuhan Program in Culture and Technology, with characteristic generosity and amiability embraced both me and my project and drew me into the McLuhan Program's continuing dialogues. I was reacquainted with my publisher, Patrick Crean (we worked together on another book, years ago), through Professor de Kerckhove. Patrick's enthusiasm, integrity and professionalism have been as refreshing as mountain air. His cheerful factotum, Janice Zawerbny, has been indispensable.

My thanks to Douglas Rowland and Everest Munro for reading the manuscript and making useful suggestions. And to the indefatigable Bernice

Eisenstein for saving me from many embarrassing gaffe with her impeccable copy editing. To my daughter Hilary and son Simon I owe much of what I know about computers and networks, and I have benefited from their advice throughout the project.

During the period in which I was mainly occupied with this book, my wife Christine Collie Rowland, a former television network art director, was busily establishing a successful website development firm which she named with characteristic whimsy, Blue Cat Design. Her prodigious design talent and endless capacity for adapting it to new technologies has been both a revelation and inspiration to me. I owe her much more than I can acknowledge here. Without her, there would be no book. Together, and with our children, we have become accidental pioneers on the electronic frontier of which I write in the concluding chapters of this work. It is an exciting and rewarding place to be, but I doubt that I would have had the courage to find my way here without her.

For Laura and Frank Collie
with love and gratitude.

Some Milestones in Communications Technology

c. 3300 B.C.	Earliest known Sumerian clay tablets
c. 3100 B.C.	Earliest Egyptian hieroglyphic inscriptions
c. 730 B.C.	Appearance of phonetic alphabet in Greece
c. 1450 A.D.	Gutenberg's printing press
1753	Electric telegraph proposed in Scots' magazine
1794	First messages carried on Chappé visual telegraph (France)
1822	Babbage begins work on Difference Engine, computer precursor (Britain)
1837	Cooke and Wheatstone patent electric needle telegraph (Britain)
1840	Morse patents electric telegraph (United States)
1857	First attempt to lay transatlantic telegraph cable (Britain, United States)
1861	Completion of transcontinental telegraph in United States
1866	Completion of transatlantic cable, Ireland to Newfoundland
1876	Bell patents telephone (United States)
1884	Disk-scanning television (Nipkow)
1888	Hertz publishes discovery of radio waves
1896	Marconi patents wireless telegraphy*radio
1899	Fleming invents vacuum tube diode
1900	Marconi patents radio-tuned circuit
1901	Experimental voice transmission by radio (Fessenden) Marconi's first transatlantic radio transmission
1902	Trans-Pacific telegraph cable, Vancouver to Australia
1905	First practical photo-facsimile machine (Korn)
1906	Successful broadcast of voice and music on radio (Fessenden)
1907	De Forest patents vacuum tube (triode) amplifier
1919	Radio Corporation of America (RCA) incorporated
1920	Westinghouse establishes first commercial radio station, KDKA
1922	AT&T sells first radio commercials BBC begins radio broadcasting in Britain
1924	First radio-facsimile transmission of photograph
1925	AT&T establishes first radio network via telephone lines
1929	BBC begins experimental television broadcasts
1932	Fully electronic television demonstrated by NBC

1936	BBC begins regularly scheduled television broadcasts Turing invents modern digital computer
1939	NBC begins regularly-scheduled television broadcasts in New York City
1942	First fully automatic digital computer (Harvard University)
1943	Colossus, first electronic digital computer (Britain)
1945	ENIAC, first electronic digital computer in United States
1946	First cable television systems in United States
1948	Shannon publishes first work on Information Theory Transistor invented at Bell Labs
1949	Pilot Ace, first computer using Random Access Memory (Britain)
1951	Remington Rand UNIVAC, first commercial computer using RAM
1957	Sputnik I, first artificial earth satellite, placed in orbit (USSR)
1959	First integrated circuits (Fairchild Semiconductor)
1962	Distributed communications network proposed (Barand)
1965	Early Bird, Syncom, geosynchronous communications satellites First commercial digital telephone installations (United States)
1968–9	ARPANET launched using packet technology to link computers
1971	First microprocessors (Intel Corporation)
1972	Internet proposed at First International Conference on Computer Communication, Washington, D.C.
1974	TCP/IP protocols introduced, placed in public domain CompuServe incorporated
1975	Intel 8080 microprocessor Altair 8800, first personal computer (United States) Apple Computer incorporated
1976	First electronic telephone switching systems (United States) Microsoft incorporated
1977	Apple II personal computer introduced
1980	IBM PC introduced: Microsoft Disk Operating System (MS DOS)
1982	TCP/IP become uniform Internet protocols
1984	Macintosh computer introduced by Apple
1993	World Wide Web protocols (Berners-Lee) Mosaic, graphical-interface WWW browser (Andreesen)
1994	Netscape incorporated (Andreesen, Clark)
1995	Small dish satellite television in United States
1998	Iridium global satellite personal communications network

CONTENTS

Part Two

The Digital Era / 207

PROLOGUE

Extensions of Man

The pages that follow contain the story of a family of inventions: how they were conceived and then coaxed into existence and finally offered to an unsuspecting world. Linking together in unforseen ways, they have created an entirely new and awesomely powerful medium of communication, the impact of which provides a running subtext and a collection of concluding thoughts and speculations to this story. The uses and effects, past, present and future, of these technologies are also on the menu, as are the stories of the lives, motivation and inspiration of some of the geniuses who bore them into the world. And because, as a group, these technologies have been instrumental in changing in the most profound ways our view of the world we live in, we will be looking into some of the philosophical ideas that have influenced their development, and which have in turn been shaped by those technologies.

The new medium of communication is in fact a *meta*medium—digital, interactive, multimedia—delivered by switched networks, the most visible current application of which is the Internet. It can be ranked among the handful of most important technical advances in history. Many predict that it will fuel the next great postindustrial economic transformation as it matures and realizes its full potential; it has already changed the way business is done in the world. It will be—indeed, is now—the most sophisticated, engaging, all-embracing medium of communication ever seen on the planet. What does it communicate? Information—which is an assemblage of data (literally, "the given"). Information, through the application of intelligence and experience, produces knowledge. And knowledge is a modern economy's most valuable asset.[1]

Not everyone wishes to be on the welcoming committee for this new metamedium. Educator and social critic Neil Postman is one of a number of commentators who complain that modern communications technologies have conjured up an information deluge of scriptural proportions, one that has drowned or is drowning the last vestiges of coherent culture and intelligent discourse. He fears that what is most valuable in Western civilization will be swallowed up before we realize what is at risk, until,

> years from now, . . . it will be noticed that the massive collection and speed-of-light retrieval of data have been of great value to large-scale organizations but have solved very little of importance to most people and have created at least as many problems for them as they may have solved.[2]

In a broader sense, he says,

> the uncontrolled growth of [communications] technology destroys the vital sources of our humanity. It creates a culture without moral foundation. It undermines certain mental processes and social relations that make human life worth living.[3]

Critics have completed their job when they have pointed out a problem, and should not be expected to provide solutions. But Postman does make some suggestions in the area of education. He proposes that all subjects be taught as history, so as to give context and meaning to information, as opposed to passing it on to students as a mere consumer product:

> To teach about the atom without Democritus, to teach about electricity without Faraday, to teach about political science without Aristotle or Machiavelli, to teach about music without Hayden, is to refuse our students access to The Great Conversation. It is to deny them access to their roots, about which no other social

> institution is at present concerned. For to know about your roots is not merely to know where your grandfather came from and what he had to endure. It is also to know where your ideas come from and why you happen to believe them; to know where your moral and aesthetic sensibilities come from. It is to know where your world, not just your family, comes from.[4]

It is in this spirit that this book is an exploration of the roots of modern communications technologies. Its goal is to help readers to understand where their sometimes bewildering world comes from. While it acknowledges the attitudes of Postman and other media critics, it is not a pessimistic work. There are two reasons for this. Despite the manifest problems they have created (one need look no further than television's cultural depredations for examples), the technologies we will be dealing with have undeniably brought great and sweeping benefits, which are daily felt by virtually every inhabitant of the planet. Only a misanthrope could find this surprising: they are, in the phrase of McLuhan and others, "extensions of man" in a very direct sense; as such it is difficult to see how they could be wholly or even mainly negative developments.

A more direct and immediate reason for the book's ultimately hopeful outlook is that it takes the historical narrative one step beyond where Postman left off, to the new era of networked computers and the Internet. If earlier critics have been harsh in their judgments of the impact of computers, it is perhaps because the true nature of these marvelous machines as gateways to networks had not yet fully emerged and was therefore not generally understood. Postman says, for instance: "Although I believe the computer to be a vastly overrated technology, I mention it here because, clearly, Americans have accorded it their customary mindless inattention; which means they will use it as they are told, without a whimper."[5]

Here, Postman is wrong on both counts. The biggest surprise in the tumultuous, astonishing history of the personal computer industry has in fact been the way people have *chosen* to use their machines—for record keeping and computation, certainly, but also and primarily as communications tools. On-line publishing, e-mail, newsgroups, Internet telephony and video conferencing, the World Wide Web itself, were all unanticipated, user-driven applications of the personal computer. They came from the bottom up. It would be difficult to appreciate the value of an automobile or a subway train in the absence of a network of roads and rail beds; so we may forgive the Postmans of the world their shortsightedness, since it has been difficult to appreciate the true potential of the computer as a communications tool prior to the very recent advent of national and global networks.

Furthermore, Americans, no less than Canadians or Britons or people elsewhere in the world, historically have been anything but submissive in the presence of technical advances in the area of communications. Computer engineers and marketers were not alone in being caught off guard by a public demand for applications they'd scarcely considered for their product: as we shall see, the same thing happened with the telephone and radio, even, to some extent, the telegraph.

It is likely that this ability to appropriate for personal communication, what began as essentially industrial or institutional technologies, goes back even further, back to the printing press and even to the phonetic alphabet. People have always felt the compulsion to communicate with one another over both distance and time. Communication with those remote from us in time has been accomplished by recording techniques and devices ranging from the charcoal sticks and mineral paints of cave dwellers to the compact disk. Its success is determined by both the durability of the storage medium and the quality of the content being preserved. The inherent quality of content helps to preserve it because it increases its perceived value, deterring casual destruction. Communication over distance, on the other hand, is concerned mainly with speed, and quality of the content is usually less important than its currency or timeliness. In general, the more immediate the message, the more valuable it is. Thus, the ideal distance-communication system operates in real time as does, for example, the telephone.

The quest for better distance communication has been the quest for technologies that extend our principal "distance-capable" senses, vision and hearing, so that they may be employed at ever greater distance and with ever smaller time delays in communicating thought and data. It has led us from the megaphone to the drum, bugle and other musical instruments, to smoke signals and signal fires, to the optical telegraph and penny post, to the electrical telegraph, to telephony, to radio and television and finally to today's digital multimedia delivered via switched networks, satellites and wireless cellular systems.

This raises the question: What is all of this information used for? One analyst who has closely examined the issue, G. J. Mulgan, states that information in industrial societies is used principally for purposes of control. This may be the straightforward control of a machine that is regulated by feedback (e.g., a thermostat), or it may be the complex cultural controls, from laws to customs exercised by the various estates of society over a population in the name of social order. "Information technology," he argues, is really a misnomer: the phrase should be "control technologies." The "information revolution" is really a shift toward more rapid, comprehensive and efficient methods of control of objects, processes and people.

Seen from this perspective, the burgeoning development of communications technologies, beginning with the telegraph and continuing through to the Internet, is a necessary response to the natural growth in size and complexity of industrial society. The difficulties of control, and thus the need for information, grow disproportionately with the increasing complexity of any system: as the scale of a system doubles, the number of possible control points is squared. As well, improved communication *permits* increased complexity and the two feed on one another to further accelerate demands for yet more information.

Information and the networks that distribute it have both positive and negative potentials, just as "control" has positive and negative connotations. They can mean the liberty that arises out of "peace, order and good government," as Canadians say, or they can be used for autocratic and repressive purposes. Information may be a neutral commodity, but the technologies that make it useful by giving it a context are not: they have social and political implications that need careful examination. Mulgan writes:

> When governments introduced compulsory universal education, public libraries and later broadcasting it was clear to many that beneath the rhetoric of benevolent liberal reform these were also deliberate attacks on the autonomous realm of working-class activity, with its own material base, its own networks of exchange

> and its own meanings. The self-education movements and subversive literary cultures of subordinate classes were effectively preempted and coopted, as governments and ruling classes learned to use access, rather than the denial of access, as a tool of power.[6]

We will argue here that broadcast or unilateral communications technologies ("unilateral" because they move information in one direction only, i.e., from the top down), by their very essence, are authoritarian, and that bilateral communications technologies ("bilateral" because information moves, or may be moved, in equal volumes in both directions), like the telephone and the Internet, are intrinsically favorable to the advancement of democracy. The product of all communications technologies is organization: communication is what allows the functioning of bureaucracies (in the benign sense of organizational structures). Broadcast communications technologies tend to promote vertically oriented, hierarchical organizational structures due to the fact that they originate information at one point and disseminate the same information to all those on the receiving end. Bilateral technologies, on the other hand, of necessity organize laterally, through conversation and cooperation.

The technical advances that created the Information Age occurred during two distinct periods of great creativity. The first extends roughly from the middle of the last century through to the outbreak of World War I. It is an era in which the hitherto jealously distinct fields of science and what we would now call engineering, but was then called "craft," came together in cooperation, with astounding results. Then a chasm occurred. The nightmare carnage of The Great War wiped out a generation of the best and brightest idealists and intellectuals and shattered the prewar dream of human perfectibility through material progress and invention. In one of the truly pathetic ironies of history, one of the reasons for what seems an otherwise lunatic war was the technological optimism that was so pervasive in the prewar period. It was believed on all sides that technology had brought the arts of war to such a pinnacle of sublime perfection that offensive military action would be devastating and irresistible, making a protracted conflict impossible. The dreadful truth was that it led to unbreakable deadlock and a sustained level of bloodletting unparalleled in history.

The period between World War I and World War II is justifiably treated by most historians as a time of retrenchment, economic confusion, moral cynicism, and technical stagnation in many fields. It amounted to an uneasy bridge spanning the thirty years between two periods of global warfare and was devoted to a regrouping of forces in preparation for a continuation of conflict. Following World War II, even after the worst of the physical devastation that had been inflicted on Europe and east Asia had been repaired, a sense of optimism was slow to rebuild. The forty-year Cold War, with its truly terrifying nuclear arms race, gave good cause for despair over the uses to which advanced technologies were being put. The world had arrived at a state in which it had become possible, with the deliberate or accidental touch of a button, to exterminate virtually the whole of the human race and much of the natural ecosystem in which it had flourished. The materialist, technology-centered optimism of the late nineteenth century seemed laughable in retrospect. It is not until very recently, with the end of the Cold War and the rapid winding down of the arms race and its deeply psychotic strategy of "mutually assured destruction," that it has become possible once again to see a future in which new technologies might actually advance humanist goals.

For a time, immediately after World War II, technological development continued to build on the great advances of the late nineteenth century, which in turn had capitalized on the scientific discoveries of the previous two hundred years and more. But, increasingly, it was the science of the early twentieth century that began to drive technological innovation. Whereas earlier science had focused on the macrocosmic issues of the visible worlds of astronomy, physics and chemistry, this new science was involved with the unseeable, and (as it would develop) literally unknowable, submicroscopic world of atomic and subatomic activity.

The science of quanta or elementary particles has led to technologies of a radically different nature from those of the machine age, particularly in the field of communications. From a historical perspective, these new communications technologies can be seen as the outgrowth of the convergence of mid- to late-twentieth-century digital computer technology with the analogue electronic technologies of the late nineteenth and early twentieth centuries, including the telephone and radio. This has led to the gradual dominance of digital over analogue techniques and devices. (Analogue technologies work by measuring; digital technologies work by counting. It is a distinction which may seem tenuous and obscure but which has nevertheless had profound effects on everyday life, as we shall see.) The impact of the digital revolution has been greatest of all in the area of communications (or, more broadly, information) technologies, which in turn are fomenting social change by placing unprecedented power in the hands of ordinary people around the world. The political ramifications have been startling and will be discussed at some length in the concluding chapters.

The trends sketched here take us to a point where it is necessary to question the dominant historical viewpoint of the latter part of our own century, as expressed by technology historian Gene I. Rochlin, writing in 1974:

> We seem no closer to the dream of a social structure in which industry is the servant of humanity rather than the forger of its chains. The chains may be more ephemeral now than they once were, but they still bind us to our machines.[7]

A more hopeful outlook is called for in light of the way in which the latest communications technologies are placing more and more political and economic power in the hands of ordinary citizens, by giving them both greater access to, and greatly improved ability to make use of, information of all kinds. Monopoly control of information and the techniques of employing it by organized religion, corporations, government bureaucracies and the media operating in various patterns of alliances has been the occasion and the agency of the subservience of individual human beings throughout history. That monopoly may well have been shattered in our own time by the very technologies it created. Powerful new bilateral forms of communication appear to be in the process of engulfing and subsuming older broadcast media. Whether the crumbling of authoritarian instruments of control continues, or is reversed, is the central issue for the politics of the Age of Information.

PART ONE

The Analogue Era

CHAPTER ONE

The Meaning of the Age of Information

For better or worse, we live in interesting times. The computer is still in its early adolescence and the potential of the global computer networks is just beginning to be explored. Already, however, we can see a smudging of the boundary between human and machine by the notion of the brain as an elaborate, biological computing device and intelligence as an emergent, perhaps generic property of complexity in natural systems. As well, we have the Internet, a network of digital computers, proliferating like an organic creature. Whether in the end substantial or illusory, this strange convergence between the animate and the inanimate, the organic and the inorganic, seems likely to mark humanity as profoundly as did Copernicus's momentous conclusion that the earth orbits the sun.

When he published *On the Revolution of the Celestial Spheres* (1543) in the last year of his life, Copernicus changed the world. By observing that the movements in the heavens as they had been seen and understood from the dawn of history were an illusion created by the earth's motion around its star, he threw into question every other accepted truth about nature and the universe. If so obvious and fundamental a precept as the earth-centeredness of the universe could be challenged and disproved, what other astounding revelations might be possible? Where else was the received wisdom of the ages in error? The Medieval notion that all knowledge was contained within the Scriptures (as later complemented by the writings of the ancients of classical Greece) could no longer be sustained.

One might reasonably have expected the relegating of the earth to the status of just another celestial body—quite a step down from being the hub of the universe!—would have been a demoralizing experience for Western man. But the discovery came at a time when the return from Byzantium of many of the original texts of ancient Greece had produced intense new interest in the ancients[8]. A resurgence of interest in the science and philosophy of classical Greece and Rome fueled the growth of a new humanism, with which the Copernican revolution was in sympathy. Western civilization's values were set adrift with the loss of their moorings in the immutable, ordained truths of an unchanging universe, gradually to find a new anchorage in a renewed faith in values arising out of human wisdom. Experience, rather than zealous acceptance of received dogma, became the guide to understanding. And this had an inevitable impact on social values and the way in which they were expressed in politics. Petrarch, the chief humanist of the age, set the tone with his famous aphorism: "It is better to will the good than to know the truth." Explorers, making practical navigational use of the new cosmology and of new technologies of measurement, discovered and began exploring the Western Hemisphere. It was the time of Dante, Boccaccio, Rembrandt, the Medicis, da Vinci and Machiavelli.

The humanism based on rhetoric, the rediscovered philosophy of the Greek Sophists, flourished until about the mid-seventeenth century as a worldview in which the truth was arrived at by observation and discussion of observed reality, in all its untidiness. "Man is the measure of all things" is the famous watchword of the movement. It meant, man is a center of values; that values are not external to him or in some way "given" or predetermined. The process of establishing truth was much like a courtroom trial in which material evidence is brought forward and examined according to established rules and conventions. As in law, once a decision has been arrived at, it becomes a precedent on which new "reality" can be built. The Truth was more invention than discovery and therefore was not immutable. It could change if new evidence was presented. It is not difficult to see how humanism was a comfortable cradle for the modern, empirical science brought into the world through the successive ministrations of Copernicus, Kepler, Galileo, Descartes and, ultimately, Newton.

The fact that Newton was born in the year Galileo died (1642) is a coincidence of history made all the more striking by the inescapable metaphorical notion that Galileo died giving birth to Newton. The well-known story of Galileo's persecution by the Roman Catholic Church, which insisted he recant his published conviction that the earth revolves around the sun, marks the climax of a long struggle by the Church to accommodate new physical knowledge of the universe without abandoning the metaphysical structures which defined its values. With that incident, though by no means entirely because of it, came an historic parting of the ways between Religion and Science, between the spiritual and the material. Science flourished; the Church withered. Increasingly, Western societies looked to the victorious and robustly practical Science as a source of moral and ethical values, the sort of humanist values that had played such an important role at its birth.

Gradually, though, the accumulation of scientific knowledge led to a new determinism, more rational but, in retrospect, in some ways no less confining than the dogmatic religious faith it had replaced. In the new scientific worldview, the universe was complete and perfect, a closed system whose mysteries could be discovered by rational analysis. Newton's physics epitomize this conception: Newton's universe was a well-oiled machine in which the whole was exactly equal to the sum of its parts and in which there were no surprises, no unexpected events once one understood completely the machine's workings. It was a linear world in which equations all plotted in straight lines or regular curves. The values inherent in such a system could only be as fixed and invariable as the humanist's were mutable and subject to evolution, and over time, in its insistence on material, quantifiable answers to all phenomena, science would become as rigidly dogmatic as the medieval Church, until

> [t]he world that people had thought themselves living in—a world rich with color and sound, redolent with fragrance, filled with gladness, love and beauty, speaking everywhere of purposive harmony and creative ideals—was crowded into minute corners in the brains of scattered organic beings.[9]

And yet, the one willful blindness made no more sense than the other, since

> science deals with but a partial aspect of reality, and . . . there is no faintest reason for supposing that everything science ignores is less real than what it accepts. . . . Why is it that science forms a closed system? Why is it that the elements of reality it ignores never come in to disturb it? The reason is that all the terms of physics are defined in terms of one another. The abstractions with which physics begins are all it ever has to do with. . . .[10]

The role of a Creator and lawgiver in the universe devolved step-by-step from engaged participant, to detached observer, to Newton's Prime Mover, to a mere embodiment of the mathematical principles underlying material reality. God became a kind of constitutional monarch, a figurehead held over from a more naive era, held on to for nostalgic reasons.

The ascendance of the scientific rationalist worldview was considerably accelerated by the first great broadcast-style communications technologies, the printing press and the mechanical clock, products of the sixteenth and fourteenth centuries respectively. Learning and literacy left the close confines of the monasteries and were made widely accessible, as were the techniques of organization: the two technologies combined to create what was, in its own way, an Age of Information.

The printing press, by fostering literacy, shaped our minds, our very thought processes. The mechanical clock, audible and visible in every prosperous town, day and night, taught people that the "river" of time was made up of quanta or bits: it taught them, some historians argue, to quantify. More concretely, by making it possible and practicable to organize time, it made possible the organizing of production processes, with potent results, as Lewis Mumford has pointed out:

> The gain in mechanical efficiency through coordination and through the closer articulation of the day's events cannot be overestimated. . . . The modern industrial regime could do without coal and iron and steam easier than it could do without the clock.[11]

Although it was probably invented simply as a means of regulating monastic routines, the clock was, in late twentieth-century terms, a potent piece of information technology that did its job by reducing uncertainty and increasing predictability. (Information may be defined, in fact, as "the elimination of uncertainty.") The information the clock created made it a powerful agent of control. It did no work on its own, but it came to govern those processes and machines that did, in much the same way as computers control processes today. However, it not only created information, it disseminated it, in broadcast fashion, to the four corners of the earth: at first, monastery bells ringing the seven daily calls to prayer became important to surrounding villagers, and soon clocks and watches everywhere were synchronized, giving life a universal and precise, though synthetic, regularity. Clock-created information, clock time, became virtually inescapable and invaded cultures worldwide at every level.

But the clock did much more than that. By dissociating time from human activity, the mechanical clock created a new model of reality; more efficient but in some ways impoverished. In bequeathing us the benefits of the ability to organize time, it also further distanced us from an essential part of our humanity, namely, the subjective, the emotional, the nonrational. It did this by promoting a method of thought whereby entities could be subdivided, the segment standing in for the whole as an abstraction of it. And this was a decidedly mixed blessing, as Mumford has noted:

> [I]solation and abstraction, while important to orderly research and refined symbolic representation, are likewise conditions under which real organisms die, or at least cease to function effectively. . . . In short, the accuracy and simplicity of science, though they were responsible for its colossal practical achievements, were not an approach to objective reality but a departure from it. In their desire to achieve exact results the physical sciences scorned true objectivity: individually, one side of the personality was paralyzed; collectively, one side of experience was ignored. To substitute mechanical or two-way time for history, the dissected corpse for a living body, dismantled units called "individuals" for men-in-groups, or in general the mechanically measurable or reproducible for the

> inaccessible and the complicated and the organically whole, is to achieve a limited practical mastery at the expense of truth and of the larger efficiency that depends on truth.[12]

Society and its values had drifted from the faith of the medieval Roman Catholic Church to the renaissance of humanist thought to a new faith in the demonstrable truths of science, which more and more would be written in the supremely logical language of mathematics.

It was the determinist, rationalist position of Newton that was being advocated as recently as the early years of our own century by Bertrand Russell and Alfred North Whitehead in their master work, *Principia Mathematica*, a heroic attempt to integrate logic and mathematics as the pinnacle of human knowledge. And then, along came Einstein with Relativity, and Gödel with his proof that mathematics and logic are inescapably incomplete and, in an important sense, "untrue," and Heisenberg with proof that there are some things under the sun that simply cannot be known, ever—and the very foundations of the mechanistic, determinist worldview were fatally undermined. Pulverized, in fact. Rationalist civilization's values—if they were indeed inherent in natural systems—could decidedly no longer be fixed and invariable, since the systems themselves were now known to be indeterminate, incomplete and relativistic. The center no longer held.

Once again, in our own century, we are left to fend for ourselves after having been set adrift from a secure-seeming moral anchorage. We are currently in the midst of a revolution of outlook and perception no less disorienting and no less pregnant with possibilities than any experienced by our ancestors. It is being broadened and accelerated by the insights provided by our computer-mediated information networks.

The impulse to spread the net of science to take in more and more of life as it is experienced and observed led to quantum theory, and more recently to chaos theory. Reality in the Newtonian universe had been reduced to an affair of linear equations about static perfection because that is what was manageable with the technology (and the mathematics) of the day. That the world was not a predictable place, that the sum of systems was often greater than their parts, that systems sometimes were dynamic and seemed to wobble out of control or have a mind of their own, that response sometimes preceded the action said to have caused it, were all accounted for as problems of information deficiency. (Deviations of planetary orbits from the perfectly circular had been explained away in similar terms in the pre-Galilean Middle Ages.) Observations inconsistent with classical physics were set aside and ignored as irrelevant. "If we had more information, we'd be able to explain what is going on." Computers made that argument increasingly untenable because they removed many of the roadblocks to information acquisition and processing, allowing more and more of the real world to be brought in to the closed system of physics and mathematics.

We are developing a conception of the world in which systems are best understood by examining their "bottom-up" behavior rather than by looking at them "top down." This shift in perspective is a response to an understanding that natural systems were not created complete and fully mature, like Athena springing from Zeus's forehead, but rather evolved over time according to laws which we are only just beginning to examine with the help of our most powerful computers. We now understand that even very complex, seemingly chaotic systems obey rules, though they are not the sort of linear rules set out by Newton. Thanks to quantum mechanics, we understand that the various systems that make up our universe are tightly interconnected and that what happens in one affects in some way all the others. Our understanding of the world is much more biological, evolutionary, Darwinian—rhetorical, in fact—than Newton's. Our world exists on the edge of chaos, constantly in flux, forever changing, never quite exploding into complete anarchy but, because of its dynamism, alive in a way that Newton's was not.

The rhetorical-humanist worldview, Richard Lanham says, has always been an evolutionary view, one which fits well with a modern understanding of the world we live in. Such a system, he says, "does not aim at *stasis*, as does the utopian tradition that descends from Plato, at finding an ideal pattern of life and then *shutting the developmental process down*. The developmental process *is* life. To shut it down is to kill it. . . . [T]he evolutionary niche that favours life poises itself in the interface between the two."[13] That is, between the static and the chaotic, in dynamic flux.

> The rhetoricians, the "Greek Liberals" . . . built their world on the *logos*, on the word, or more largely on what we would call "information." It was there that they found their natural home, not in the "real" world beyond language, either in Platonic Forms or Newtonian bodies. It was words that constituted life, not embodiment. This assumption has now returned to the sciences as well. Life is information; life is the *logos*. It is an evolutionary system, dynamic, perpetually emergent. It creates new meanings, as does poetry, rather than simply communicating preexisting knowledge in a transparent capsule.[14]

The Information Age, we might then argue, is also the age of the return of the humanist or rhetorical worldview, which is essentially a democratic outlook. Where determinist thought spawned utopian schemes of all descriptions, top-down societies all of them, rhetoric has been concerned with the give-and-take of real politics, real life and incremental, bottom-up progress. The Information Age brings with it a worldview that maps our current understanding of physical reality and our highest aspirations for human progress and happiness much better than the machine-age paradigms it replaces.

It could be said that, in the Age of Information, we are returning to our roots as literate people, roots that reach back nearly two millennia, to the Greece of the ancients so venerated by our medieval ancestors.

CHAPTER TWO

The Need to Communicate

Human communication of course began with gestures, vocalizations and finally the spoken word. Communications technology, though, begins with writing. The earliest writing known is found on Sumerian clay tablets dated to 3300 B.C.; hieroglyphic inscriptions began appearing in Egypt around 3100 B.C. The Greek alphabet, to which we owe our own, was not invented until about 730 B.C. And "invented" may be the appropriate verb here: judging from archeological evidence, it appeared rather suddenly, and it was used from the beginning for what we would now call literary purposes, rather than for accounting, list keeping and other trade and business functions. This has led some scholars to believe that it was the fruit of a spontaneous act of creation by some brilliant Greek thinker, who sought a means of making a permanent record of Homer's great epics, the *Iliad* and the *Odyssey*.[15]

As unlikely as that may seem, it is certainly true that the Greek alphabet made possible great refinement of the art of writing and of composition because it was so flexible and adaptable, two virtues which derived from the fact that it was composed entirely of phonetic elements that, when assembled into a word, could be "sounded out" to extract their meaning. Other written languages of the period, like Egyptian hieroglyphics, were typically extensive and hugely complex, containing both phonetic content and symbols which represented complete words. Some symbols did multiple duty as ideas, names or words with more than one meaning. The

Egyptian sign [hieroglyph] means "open," "hurry," "mistake" or "light" depending on the sign following (the determinative); it can also mean "becoming bald," and, as well, stands for the word *wn* (wen). The intended meaning could only be determined by the context, and if the reader had not previously memorized the symbol, it was next to impossible to decipher.

That meant that until the advent of early Greek writing, literacy was of necessity confined to small, highly trained elites. A ruler receiving a message recorded on clay would have to have it translated by his scribe, who would then compose the reply as well. At the receiving end, it would be decoded by another scribe and spoken aloud to the intended recipient. As a communications technology, this kind of writing had somewhat limited impact on society at large because of its failure to penetrate beyond the ranks of the professional.

By virtue of their literacy skills, the scribes of Egypt became quite influential in government, since their records allowed them to predict the size of the harvest, estimate government revenue and allocate appropriations to government departments based on those estimates. In Babylon, the literate unfailingly mentioned this accomplishment on their seal-stones (which were not dissimilar in content to today's extended e-mail signatures), in the same way that a present-day academic might list his degrees on his business card. It was the key to high religious, commercial or government office.

It is a situation rather reminiscent of the nascent days of personal computer communications, when the all-but-impenetrable MS DOS reigned as the dominant operating system and only the highly skilled "computer whiz" was able to make the technology work. Or the time of the telegraph, when messages were transmitted in Morse code, which could only be deciphered by trained telegraphers. Pre-Greek writing, like many of our present communications technologies, was *mediated* communication.

The Greek alphabet removed the need for mediation or interpretation—in the current jargon, it disintermediated literacy. By making widespread literacy possible, the alphabet also made it an object of desire as a social goal. But the spread of literacy would have profound and surprising effects upon civilization, which did not go unnoticed at the time. Plato, and perhaps Socrates too, well understood that the medium is the message.

In his brilliantly persuasive *Preface to Plato,* Eric Havelock follows in the footsteps of Harold Innes and Marshall McLuhan with an engaging interpretation of Plato's *Republic* as an impassioned rejection of the oral culture which had previously dominated Greek life. Until writing was introduced, Greek culture was preserved and passed down through the generations by means of epic poetry which, using mnemonic devices such as storytelling and rhythmic meter, conveyed social, legal, ethical, historical and religious information essential to the perpetuation of the civilization. McLuhan called it "the tribal encyclopedia . . . the specific operational wisdom for all contingencies of life . . . Ann Landers in verse." But oral culture, Plato asserted, stood in the way of intellectual growth, the growth of a true philosophy, because it trapped men's minds in a poetical or "musical" state in which understanding and learning are emotional and not rational, subjective and not objective, concrete and never abstract, based on opinion rather than fact.

> [Plato] asks of men that . . . they should examine this experience and rearrange it, that they should think about what they say, instead of just saying it. And they should become the "subject" who stands apart from the "object" and reconsiders it and analyses it and evaluates it, instead of just "imitating" it. . . . This amounts to accepting the premise that there is a "me," a "self" a "soul" or consciousness which is self-governing and which discovers the reason for action in itself rather than in imitation of the poetic experience. The doctrine of autonomous psyche is the counterpart of the rejection of oral culture.[16]

And it was the alphabet, Havelock asserts, that made this epochal change possible. He asks how the Greeks were able to shake themselves out of their poetic somnambulance and become a thinking, analytical people:

> The fundamental answer must lie in the changing technology of communication. Refreshment of memory through written signs enabled a reader to dispense with most of that emotional

> identification by which alone the acoustical record was sure of recall. . . . No longer . . . was it a matter of "this corpse on the battlefield" but of "body" anywhere and everywhere. No longer was it "this basket which happens to be empty but will be full in a moment": it is the cosmos which is empty or has emptiness always and everywhere.

In other words, by making possible the purely objective communication of information through written words capable of unambiguous understanding by any reader, the alphabet made possible the evolution of the detached intellectual state which makes philosophy or abstract thought possible, and which allows rational inquiry leading eventually to the scientific method. No less a role is claimed for it by Havelock.[17]

Marshall McLuhan extended this thought in *Understanding Media:*

> [A] single generation of alphabetic literacy suffices in Africa today, as in Gaul two thousand years ago, to release the individual, initially at least, from the tribal web. This fact has nothing to do with the content of the alphabetized words; it is the result of the sudden breach between the auditory and the visual experience of man. Only the phonetic alphabet makes such a sharp division in experience, giving to its user an eye for an ear, and freeing him from the tribal trance of resonating word magic and the web of kinship.[18]

The alphabet allowed information to be removed from its human, emotional, tribal context for examination and evaluation, transmission and transformation. McLuhan himself had serious reservations as to whether this was an entirely positive development. When it comes time for us to examine digital media, which take this abstraction to its ultimate limits of undifferentiated 1s and 0s, it would be well to keep his reservations in mind.

The printing press, an invention of the fifteenth century ascribed to Johannes Gutenberg, is generally recognized as the next great advance in communications technologies, one that is often alluded to by authors searching for adequate comparisons for today's computer network revolution. In making possible the mechanized duplication of manuscripts by using movable type, it did indeed transform the world, though not as quickly as is often suggested, and often in surprising and instructive ways. The printing press is credited with having loosened the grip of Catholicism on Europe; it also led to rapid advances in learning in almost every discipline due to the widespread adoption of indexing, alphabetization, encyclopedic collections and other methods of organizing data. It arrested linguistic drift (fragmention or corruption of languages) and paved the way for uniformity within European languages, thereby encouraging the evolution of the nation-state. It was responsible for the ideas of copyright, plagiarism and patents. With it came the whole modern notion of authorship, virtually unknown in earlier eras of scribes and anonymous translators and interpreters. But perhaps the most telling change wrought by the printing press is one that gets little attention in the libraries of tomes devoted to the subject. It is nicely summed up in a quotation that has been unearthed by Elizabeth Eisenstein in *The Printing Press as an Agent of Change*: "Why should old men be preferred to their juniors now that it is possible for the young by diligent study to acquire the same knowledge?" The question is asked by Jacob Filippo Foresti, in his newly minted world history, published in Venice in 1483.[19] Communications technologies—whether the printing press or the World Wide Web—tend, in the modern vernacular, to level the playing field.

The printing press was a decisive step in the long process of abstracting and codifying information in discrete symbolic units, which made it reproducible with minimal erosion of content, and easily transportable, but which involved losses as well, as Frances A. Yates observed in the following passage from his *Art of Memory*:

> In Victor Hugo's *Notre-Dame de Paris*, a scholar, deep in meditation in his study . . . gazes at the first printed book which has come to disturb his collection of manuscripts. Then . . . he gazes at the vast cathedral, silhouetted against the starry sky . . . "*Ceci tuera cela,*" he says. The printed book will destroy the building. The parable which Hugo develops out of the comparison of the building,

> crowded with images, with the arrival in his library of a printed book might be applied to the effect on the invisible cathedrals of memory of the past of the spread of printing. The printed book will make such huge built-up memories, crowded with images, unnecessary. It will do away with habits of immemorial antiquity whereby a "thing" is immediately invested with an image and stored in the places of memory.

To recapitulate, communications media carry with them their own unique "message" in the form of their impact upon the societies they serve. As techniques for transmitting or distributing information, they are not neutral, but rather color and shade that information to greater or lesser degrees. More important, they can shape the very consciousness of their users, by their emphasis on a single sense or set of senses to the exclusion of the others. The message of the Greek alphabet, a static, visual medium, was thus the autonomous personality, the objective fact, the analytical mind-set—a profound message, indeed. The printing press helped to make it a paradigm for Western culture.

A medium, however, is more than merely a method of using symbols to encode, record and preserve human thought. As the word is used today, it also includes the means of transmitting thought thus recorded. The phrase "the medium of television," for instance, takes in not just the aural, visual and textual matter that is presented to us, but also the technological infrastructure that packages this information in a way that allows it to be moved from place to place, that is, to be *delivered* to its audience. A book as a medium of communication includes not just the words it contains, but the paper and binding that make it portable and durable and allow the delivery of the message it contains. We might say, then, in a more abstract sense, the "medium" includes both the coding of the message and the delivery of the message. It is the "delivery" that makes the information accessible. As we pursue this examination of modern communications technologies, it will become evident that both aspects of media can have an impact on their users: coding, as we've seen in the example of the alphabet, has received much more scholarly attention than delivery.

Until the era of the telegraph, communication and transportation were inextricably linked. There could be no reliable communication over distance without involving some means of transportation: letters, for instance, had to be shipped from place to place on horseback or by stagecoach or by packet steamer. The early empires of Europe and Asia each developed sophisticated postal courier systems, which relied on a network of good roads for their efficiency. The Persian postal system was much admired by the classical Greeks, who were unable to establish a similar network due to their fragmented political condition. By far the most efficient of these early imperial systems was the Roman *cursus publicus*, a staged relay system of postal delivery whose speed was not to be duplicated in Europe until the eighteenth century. The importance of the Roman road network throughout Europe is normally thought of in terms of its usefulness in hurrying troops from place to place, but it served an equally important function in speeding mail on its way. It was thus a communications network in the modern sense, as well as a transportation network. The *cursus publicus* would outlast the Roman Empire, adopted and maintained by the various barbarian rulers. It died a slow death only with the deterioration of the Roman road system on which it ran.

The modern national postal system saw its genesis in Europe of the seventeenth century, in the era of the emerging nation-state with its strong central government based on efficient command and control structures. So important was postal communication to government that what had been a collection of private services in most countries was everywhere taken over by the state and monopolized. Like the Romans, the European rulers thought the ability to communicate rapidly and reliably a valuable strategic asset which ought to be in the exclusive hands of government. The regularity and security of these state postal systems was such, however, that there was immediate and ever-increasing pressure to open them up for nonofficial communication. Though this was resisted initially, governments everywhere were not long in recognizing the opportunity to make money by carrying unofficial correspondence, and this became common practice throughout Europe by the mid-seventeenth century. The issue of state security, which had been the main concern of those arguing against public carriage and which is a recurring theme throughout the history of communications technology, was dealt with in legislation which permitted state postal inspectors to open suspicious mail. In Britain, a warrant was required; in other states, this was seen as an unquestioned prerogative of the ruling regime. Interestingly, the continuing state monopolies on postal services were justified largely on the grounds that the public, through their government, had built and improved the network of roads on which the postal systems ran and was therefore entitled to share in the income earned through its use. In Britain, private postal concerns were permitted to continue doing business, but only over roads not used by the state Post Office.

Before long, public use of British postal services superseded the government's in volume and per-unit costs fell steadily with increasing usage. By the 1820s, mail was averaging more than ten miles an hour around the clock in the British Isles and a letter mailed in Manchester at 4:00 P.M. would reach London next day at 10:00 A.M. Jane Austen paid famous tribute to the British postal service in her novel *Emma,* through the character Jane Fairfax:

> "The Post Office is a wonderful establishment!" said she. "The regularity and dispatch of it! If one thinks of all that it has to do, and all that it does so well, it is really astonishing! It is certainly well regulated. So seldom that any negligence or blunder appears! So seldom that a letter, among the thousands that are constantly passing about the kingdom, is even carried wrong and not one in a million, I suppose, actually lost! And when one considers the variety of hands, and bad hands too, that are to be deciphered, it increases the wonder."[20]

Mail was nowhere else nearly so rapid or efficient, but despite the fact that it was not uncommon for an international letter to take half a year to reach its destination, people corresponded in remarkable volume and frequency. "Even in France," notes historian Paul Johnson, "regarded as backward in this respect, the number of letters leaving Paris each day had risen to 36,000 by 1826. . . . In London, outgoing mail had long ago passed the 100,000 mark."[21] The very eagerness of the public everywhere to use even the slow and sometimes inefficient public post pointed to a great untapped demand for any innovation which would speed communication.

It was a British tax reformer and educator named Rowland Hill who first sorted out the true relationships in any communications network, from the Roman *cursus publicus* to the modern Internet. In an 1837 report entitled "Post Office Reform: Its Importance and Practicability," Hill demonstrated conclusively that transport costs were an insignificant portion of the overall cost of delivering a letter. In fact, he found that by failing to understand this, postal authorities were needlessly and very significantly increasing the cost of mail. This was because the complex distance-based system of postal tariffs then in place required armies of accountants and paper-pushers to administer. Hill proposed a single tariff for all mail delivered in Britain, regardless of distance, and he suggested it be a penny, which he had calculated was slightly higher than the actual average cost of delivery was likely to be under such a regime. Ignored by officialdom, the penny post idea nevertheless found many champions among the general public and politicians were soon persuaded to adopt it. It was, of course, a resounding success, greatly increasing both mail volumes and government revenues. Like the Internet, it was a radical simplification of the technology of communication; the profound changes it heralded for both commerce and society are currently being mirrored in our own era of the World Wide Web.

In finally agreeing to carry nonofficial mail, governments had implicitly accepted the principle that the administrator of the delivery network should stay clear of the business of content, except for official correspondence of its own: in other words, they had accepted the principle of the common carrier, which was later extended to the telephone and telegraph. Hill, in turn, had discovered that in a communications system the cost of establishing and maintaining the delivery network is trivial when compared with the revenue to be gained from the information traffic it will carry. This is particularly true, of course, when the network has been built to serve, or is able to serve, other purposes. The early mail services did not have to bear the entire expense of building and maintaining the road systems they used, just as the Internet is greatly subsidized by the billions of dollars of telephone infrastructure built up and paid for by earlier voice communication services. In this sense the Internet and the penny post can be thought of as value-added services, though in the case of the Internet the new service appears to be in the process of swallowing up all previous uses. But that is a story best told later on.

CHAPTER THREE

Prelude to the Telegraph

In all of human history there have been no discoveries more important than those that have enabled us to control fire and electricity. The former gave the human race the power it needed to tame and cultivate the natural environment; the latter gave us the ability to extend our senses far beyond our bodies and to communicate information and ideas with others of our species, worldwide. We have lately discovered through quantum physics that these two sets of tools have much in common; they are in fact identical at the subatomic level where reality has its foundation and where information in the form of what we call "energy" is all there is.

Fire was, in a sense, a bridge to electricity in the long history of technology—a tool to sustain humanity while it struggled to uncover the initial secrets of the material world. Fire, of course, but electricity as well were abroad in the world and known to man from prehistoric times. Indeed, electricity in the form of lightning was the father to fire. Both fire and electricity would eventually provide motive power for people with work to do, survival to see to. As well, they both afforded means of projecting thought over distance. In that sense, they might be considered sequential tools in the fulfillment of the unique evolutionary destiny of Homo sapiens, the only species we know of that has taken control of its own evolution, through the communication of information.

The signal fire has been used for rapid communication from time immemorial. "Oh children of Benjamin," the Old Testament admonishes in Jeremiah 6:1, "gather yourselves to flee out of the midst of Jerusalem, and blow the trumpet in Tekoa, and set up a sign of fire in Beth-haccerem: for evil appeareth out of the north, and great destruction."

When the Spanish Armada threatened England in 1588, watchmen stationed all along the English coast gave the alarm, painting a picture described by Macaulay in *The Armada*:

> From Eddystone to Berwick bounds, from Lynn to Milford Bay,
> That time of slumber was bright and busy as the day.
> For swift to east and swift to west the ghastly war-flame spread.
> High on St. Michael's Mount it shone; it shown on Beachy Head.
> Far on the deep the Spaniard saw, along each southern shire,
> Cape beyond cape, in endless range, those twinkling points of fire.

Out of the signal fire developed the idea of more manageable and expressive means of long-range communication. Visual telegraphy using flags and lanterns evolved in the mid-seventeenth century and achieved its greatest fame a century later with the exploits of the Chappe brothers of France.

Claude Chappe began experimenting with machines for long-distance communication in 1790 with the idea that such a device might help protect the Revolution, threatened as it was on all sides by hostile monarchist armies. He erected a test device in Paris the following year, hoping to demonstrate it to members of the Assembly, but it was torn down by a mob who took it to be a machine for communicating with enemies of the state. Undaunted, he set up at another site outside the city, only to have this apparatus, too, torn down by citizens who, this time, thought he must be signaling to royal prisoners in the Temple Tower. Finally, in 1793, the National Assembly's Committee of Public Safety agreed to finance an officially sanctioned trial, which was to be overseen by an eminent scientist and a leading mathematician. The test took place over the thirty-five miles between a Paris park and Saint-Martin-du-Tertre, with a midway post at Ecouen. Each of the three installations consisted of a tall mast with two semaphore arms which could be moved by means of ropes and pulleys. The position of the arms indicated numerals or letters of the alphabet. Messages were passed successfully in both directions and the device was hailed as miraculous. In its report to the committee, the evaluators concluded with the encomium: "What brilliant destiny do science and the arts not reserve for a republic which, by its immense population and the genius of its inhabitants, is called to become the nation to instruct Europe!" This set the pattern of euphoric enthusiasm with which each succeeding generation of communications technology would be received.

Chappe's telegraph was understood immediately to be of great strategic value. The scientist who had observed the test believed it was the best answer to those critics of the revolutionary regime who thought France was too big to be a republic. The Committee of Public Safety decided to erect lines from Paris to Montmartre to Lille. The first message the line carried brought word of the capture of Quesnoy from the Austrians, August 15, 1794. It took an hour for word to reach Paris: a mounted courier would have taken eleven. By 1844, France had three thousand miles of telegraph line, involving 533 signaling stations, and messages could be flashed from Paris to Calais in 4 minutes, to Brest in 6 minutes 50 seconds, and to Bayonne in 14 minutes.

Chappe was disappointed that his invention was being exploited solely for military purposes and, as reported by historian Geoffrey Wilson in *The Old Telegraphs,* he

> proposed to Bonaparte a pan-European commercial system stretching from Amsterdam to Cadiz and even taking in London, as he claimed to be able to correspond between Calais and Dover. He also proposed to relay stock exchange news daily. Yet another of his ideas was an official journal to be sent from Paris by post to all Departments and supplemented by a telegraphed summary of the news of the day. . . . Unfortunately all these schemes were rejected as impracticable, but Bonaparte at least consented to the weekly transmission of the numbers of the winners of the national lottery.[22]

The lot of an important inventor is seldom an easy one: in our own era, siege-length lawsuits challenging patents are the norm and often they are enough to drive the principals to distraction. In Chappe's day, challenges were more public and direct and were carried out in the atmosphere of the fervid politicking surrounding the guillotine-happy revolutionary Convention. While exhausting himself building telegraph lines, Chappe was forced to fight off challenges from several sources which claimed his inventions as their own. By 1804 he was becoming increasingly despondent: he suspected an attack of food poisoning which had laid him low was a deliberate attempt to kill him. On January 23, 1804, he committed suicide by throwing himself down a well outside a telegraph station in Paris. The note he left behind said: "I give myself to death to avoid life's worries that weigh me down; I'll have no reproaches to make myself." He was just forty-two.

The success of Chappe's telegraph lines spurred the introduction of similar systems throughout much of Europe. Sweden was an early adopter, throwing up lines beginning as early as 1794. The visual telegraph remained in operation in that country, often alongside or supplementary to the electrical telegraph, until 1891, longer than in any other European nation.

In every European country there was a nationalistic reluctance to adopt the proven French design without indigenous "improvements." Napoleon, on being shown a German technique that would improve the operations of his own telegraph, is said to have rejected it out of hand with the objection: "It's a German idea." When news of Chappe's telegraph arrived in Britain in 1794, it prompted this cheeky lyric which appeared in a London musical production:

> If you'll only just promise you'll none of you laugh
> I'll be after explaining the French Telegraphe!
> A machine that's endowed with such wonderful pow'r
> It writes, reads and sends news 50 miles in an hour.
> Then there's watchwords, a spy-glass, an index on hand
> And many things more none of us understand,
> But which, like the nose on your face, will be clear
> When we have as usual improved on them here.

Adieu, penny posts! mails and coaches, adieu!
Your Occupation's gone, 'tis all over wid you.
In your place telegraphs on our houses we'll see
To tell time, conduct lightning, dry shirts and send news.

The "improved" system predicted in the song and in fact adopted by the British Admiralty was to be of little use as a clothesline: it was a radical departure from the Chappe model, consisting of six large wooden shutters mounted three-by-three in a massive frame. The shutters could be pivoted to a horizontal position with the pull of a rope, in which case they would be invisible at a distance. Combinations of visible and invisible shutters constituted a code for numbers and letters. To pass an average message between London and Portsmouth took about fifteen minutes: words were spelled out but often contracted by leaving out vowels. The brief preparatory message or protocol that preceded transmissions could be sent from Portsmouth to London and back, a distance of five hundred miles, in three minutes!

During the British naval blockade of France that followed the Revolution, the first international telegraph network was established with a link between the systems in Denmark and Sweden, with a view to sharing intelligence on the movements of the British fleet. Following the Napoleonic Wars, Britain itself reverted to a telegraph system similar to the Chappe model, having discovered that its extended arms could be seen at a greater distance in adverse weather than shutters enclosed in their bulky frame. (Smog was a persistent problem in London: messages often had to be carried by courier to the city's outskirts for transmission.) Chappe's telegraph had the added advantage of being easily rotated to accommodate transmissions to stations at right angles to the main line, whereas the shutter telegraph was fixed in position.

Though in England, as in France, the visual telegraph was used almost exclusively for military purposes, the very fact that it facilitated rapid communication over great distances kindled the imaginations of contemporary writers who sensed the wider implications of such technologies. Alexandre Dumas, in *The Count of Monte Cristo,* described the machine in darkly ominous terms as "the insect with the black claws and the terrible name." Leigh Hunt wrote more optimistically in *The Town*:

> Telegraphs now ply their dumb and far-seen discourses, like spirits in the guise of mechanism, and tell the news of the spread of liberty and knowledge all over the world.[23]

The transition to the early electric telegraph began in Britain in 1842 with the conversion of the old London-Portsmouth visual line to an electric system buried along the London and South West Railway right-of-way. The switch was not universally applauded for reasons that sound to a modern ear both quaint and very contemporary. Fears were expressed that the electrical line was prone to error and, worse, to sabotage: it would be no trouble for anyone intent on interrupting vital communications to simply cut the wire. This, of course, could not happen with the visual telegraph.

The *Times* of London, in its coverage of the switch, deplored the loss of jobs and attendant distress that would be involved. The newspaper noted that visual telegraph stations had long been operated by retired naval officers, who had no other prospect of employment, and listed a dozen such men by way of illustration: "Four of the above are [Battle of] Trafalgar men; one was a mate in Sir Richard Strachan's action in 1805; one was a lieutenant of the *Denmark* in the Walcheren expedition; one lost a leg at Navarino and all the others have distinguished themselves in their country's service." In the event, operators were given three months' notice of termination and were granted permission to stay on rent-free in their stations. On the whole, it seems a more humane response to the fallout of technological change than we have come to expect in our own era of economic restructuring.

In the Americas, Canada has the distinction of having introduced the visual telegraph. In 1794, Prince Edward, Victoria's father and commander-in-chief of Nova Scotia, organized a system of flags, wickerwork balls and drums for signaling by day, and lanterns by night. Telegraph posts were built near Chebucto Head, at York Redoubt and on Citadel Hill in Halifax, and at the Naval Dockyard. When Edward moved his mistress, Madame St. Laurent, to more secluded quarters outside Halifax on Bedford Basin, he had a telegraph installed there as well, on a hill behind the lodge. In 1799, newly promoted to commander-in-chief of all of British North America, he extended the telegraph to the Annapolis Valley, around the Bay of Fundy to Saint John, New Brunswick, and up the Saint John River valley to Fredericton. Another line, to give early warning of the arrival of shipping, was set up between L'Isle-Verte, below Rivière-du-Loup, and Quebec City, 120 miles farther up the St. Lawrence.

Perhaps we should not be surprised that the first visual telegraph to operate in the United States was not a military, but a commercial venture. Jonathan Grout, Jr., a lawyer from Belchertown, Massachusetts, built a line modeled on descriptions he had seen of European systems in a magazine. It was designed to carry news of the arrival of shipping off Martha's Vineyard to Boston. Grout's line was superseded by the famous line run from Boston's Exchange Coffee House, a merchants' meeting place, all the way to Long Island Head and Boston Light. It made a profit from the outset, signaling the arrival of 799 vessels in 1825, a number which climbed to 2,104 ships eight years later. An early system between New York and Philadelphia was used mainly to report stock prices until, like others of its kind everywhere, it was made obsolete in the mid-1840s by electric telegraphy, the first practical application of European breakthroughs in the generation, control and application of electricity.

By then, the visual telegraph had played out its crucial historical role in establishing the value of rapid, long-distance communication for both military and commercial ends, and had performed the invaluable task of creating a market for the electric telegraph, the device which heralded the twentieth-century era of light-speed electronic communication by wire and through the air. Paul Johnson succinctly sums up the reason why the visual telegraph was entirely superseded and not merely complemented by the electric telegraph: "It [the visual telegraph] was one of those good ideas not susceptible to fundamental improvement. In April, 1829, the system [in France] was nearly 40 years old but when news of the election of Pius VIII reached Toulon from Rome, at 4 a.m., it still took till noon to get to Paris."[24]

Nevertheless, it would be a mistake to think of this as a "primitive" means of communication: in its way, it was surprisingly sophisticated. All of the systems eventually developed complex transmission protocols. These included elaborate dictionaries of abbreviations which were an early form of compression; numerical codes which contained complex stock or generic messages; initialization sequences used on initiation of contact; error control protocols and codes for controlling the speed of transmission. The system was quite secure, since there were no wires to be cut as in a modern telegraph or telephone system. In times of unrest or war, the telegraph towers could be armed and defended.

But its speed could not be significantly improved. Nor could the volume of traffic it was capable of carrying be increased in any other way. It could pass only one message at a time. That meant that, in practice, it would be forever restricted to serving the exclusive needs of government and the military: there simply was not enough "bandwidth" to carry private or business messages. And without a commercial market, it was doomed to stagnation, and ripe for replacement.

The stage was set for the advent of electronic communication: it only required the invention of the electron.

CHAPTER FOUR

The "Invention" of the Electron

Modern communications technology begins with a brilliant British experimenter's coaxing to a climax several hundred years of slowly accumulating knowledge of electricity and magnetism. His name was Michael Faraday and he posed himself a simple proposition: If electricity flowing in a wire can act on a magnet to deflect it (a phenomenon that had been observed by the Danish scientist Hans Christian Oersted), it ought to be possible to reverse that process, to *create* electricity using magnetism. It was the experimental discovery of the nature of electric current by Faraday and the subsequent formalizing of the theory of electrical and magnetic fields by James Clerk Maxwell (called by many the greatest scientist since Newton) that made possible light-speed communication beyond line of sight.

The impact of their discoveries would be difficult to overstate. Biographer Bern Dibner said correctly of Faraday that, in discovering electrical induction, he did nothing less than "transform society into an ever-growing, integrated network."[25] Much of the remainder of this book is a narrative in support of that assertion.

Faraday is one of those figures whom editors of juvenile books and encyclopedias used to make much of, as a wholesome and altogether admirable role model for budding young scientists. The classic 1911 edition of the *Book of Knowledge* has several references to him, including a biography entitled "Michael Faraday, the Blacksmith's Son Who Helped to Change the World." Faraday, the article says,

> was born in 1791, the son of a poor London blacksmith. After very little schooling he was apprenticed to a bookbinder, and after working hard all day he would study science at night. One day a gentleman, on entering the shop, found the boy at work binding an encyclopedia, and studying hard at the article in it on electricity. The gentleman was surprised to see a boy so interested in a subject of such difficulty, and questioned him. He found that Faraday, working late at night, had already been making experiments of his own, though he was too poor to possess anything but an old bottle for his battery. The visitor was so pleased that he gave him four tickets for the lectures which Sir Humphry Davy was then delivering at the Royal Institution. . . . [Faraday] made notes on what he heard, and then at the end of the lectures he went, in fear and trembling, to the great man and showed him his notes. . . .[26]

Readers of a skeptical bent appropriate to the late twentieth century may be surprised to learn that the story of Faraday's rise to prominence through hard work and native intelligence, his personal integrity and scrupulous honesty, not to mention his lifelong loyalty and devotion to his wife Sarah, are facts so well documented as to be unimpeachable. Not only that, he was by all accounts a superb platform speaker, and he initiated Christmas holiday lectures on science for the children of London that were nothing short of wonderful. His scientific writing is a model of clarity. Throughout his working life, he gave much of his modest earnings to charity. When he felt his intellectual powers waning late in life, he gracefully resigned his position as director of the Royal Institution, despite flattering pleas to stay on from those around him.

His early life was in fact harder than is suggested in the passage quoted above. His father was in poor health and unable to work regularly. Faraday, with his mother, older sister and younger brother, moved from lodging to lodging in West London. When his father died, his mother took in boarders. "My education," Faraday recalled in later life, "was of the most ordinary description, consisting of little more than the rudiments of reading, writing and arithmetic at a common day school. My hours out of school were passed at home and in the streets." The benefactor of the lecture tickets is reliably identified as a Mr. Dance. In fact, though, Faraday on his own initiative (and with a shilling borrowed from his blacksmith brother) had earlier signed up for regular lectures on "natural philosophy" given by the City Philosophical Society in the evenings. During the two years of his attendance, he heard lectures on astronomy, geology, hydrostatics, chemistry and electricity, among other subjects. He took detailed notes and later wrote out each lecture in full and illustrated it with a natural draughtsman's skill and precision. He was eventually invited to address the society on the subject of the nature of electricity.

Faraday was twenty-one when he attended Humphry Davy's lectures at the Royal Institution, where Davy, a surgeon by trade, was employed as a professor in chemistry and geology. Continuing his City Philosophical Society practice, Faraday wrote out the lectures and bound them in book form. His apprenticeship ended at about this time and he was engaged as a journeyman bookbinder with a Mr. De La Roche, who seems to have been a hard taskmaster. Faraday desperately wanted a job in science. As fate would have it, Davy was temporarily blinded by a laboratory explosion; Faraday, perhaps through the good offices of Mr. Dance, was taken on as Davy's secretary while the scientist recovered his sight. Later, Faraday sent Davy the bound and illustrated book of his lectures and asked for a job. Nothing was available at the time, but before long a young laboratory assistant at the Institution was fired for brawling; Davy sent for Faraday and offered him the job. He was formally hired by the board of governors in 1813 at a guinea a week, and given two rooms under the eaves to live in, with fuel and candles supplied. He was to remain with the Institution for the rest of his working life, making scores of significant contributions to chemistry, metallurgy and electrical theory and practice, none greater, however, than his discovery that passing a wire through a magnetic field will cause an electrical current to be generated in it. On that foundation, as elaborated and formalized by the great Cambridge physicist James Clerk Maxwell, has risen the modern edifices of both electrical engineering and electronics.

Nowadays, we're pretty blasé about electricity, most of us believing we have a reasonable understanding of how it works: electrons running through wires, setting up electromagnetic fields, lighting light bulbs and generating radio waves. The truth is, in the 150-odd years since Faraday's experiments, the mysteries of the electron have only gotten deeper and deeper. The more science learns, the more magical it all seems. In exploring the world discovered experimentally by Faraday, science left behind the realm of sensory experience and entered a looking-glass universe of subatomic particles where logic and common sense can no longer be counted on, where paradoxes abound, where insights into the true nature of the universe demand the abstract, intuitive understanding of the mystic in equal measure with the discipline of the scientist.

The universe as it was understood by Faraday was the three-dimensional world described by Newton, in which matter was composed of indestructible particles, space was an absolute and time was a river in which all change occurred, flowing immutably from the past, through the present, into the future. God, Newton believed, created the material particles, the forces between them and the laws of motion; He set the whole construct running and it is running still, a perfectly self-regulating machine.

But when Faraday induced an electric current in a copper wire by moving a magnet close to it, he did something Newtonian physics said was impossible: he converted mechanical energy required to move the magnet into electrical energy in the wire, even though there was no physical contact between the two objects. The imaginations of Faraday and Maxwell were thus captured not so much by the result or action of the electromagnetic force they had observed, but by *the force itself,* an area of inquiry which Newtonian physicists had all but ignored. It proved the key to a Pandora's box of revolutionary scientific insight.

In translating Faraday's experimental discoveries into the precise language of mathematics, Maxwell found it necessary to describe the interaction between positive and negative electrical charges in a new way. It was not enough to state simply that unlike charges attract one another in some way analagous to the way masses attract one another (by gravity) in Newtonian physics. For the mathematics to work, he found he had to replace the idea of a force with a *force field,* a notion Faraday had arrived at intuitively from his laboratory observations. In Maxwell's conception, an electrical charge creates a "condition" or "disturbance" around it which is felt as a force by another charge introduced to its area of influence. The area of disturbance is a "force field" and it exists whether or not another charge is brought into the picture.

It was a subtle but ultimately shattering change in our understanding of physical reality. In the Newtonian world, forces were firmly rooted in the physical entities they acted upon. In the new world of electrodynamics, forces had their own reality and could be studied independently of the objects they acted upon. Very soon it was understood that light is another form of electromagnetism, and that there was in fact an entire universe or spectrum of such disturbances, a continuum ranging from radio waves through visible light to ultraviolet, X-rays, cosmic rays and beyond. Maxwell's Rainbow, it has been called.

Maxwell's equations were to provide the necessary engineering data for the development of the telegraph, radio, television and computers, all very concrete examples of natural forces being harnessed to perform practical functions—to do work in the world. All electric communication stems from the notion that if electrons can be induced to flow in a wire or radiate through space, then it ought to be possible to communicate electronically by manipulating that electron flow in such a way as to carry intelligence—by dots and dashes, for example, generated by switching the electron flow on and off. The only trick is to develop suitable electron-generating devices at the transmitting end and appropriate electron-detecting devices at the receiving end. This proved simpler in the case of a wired connection than a wireless link, and wired telegraphy thus preceded wireless telegraphy or radio.

The picture of orderly ranks of electrons marching through wires and streaming through space as electromagnetic waves at the touch of a telegraph key is an easy one to visualize, and it sufficed to carry us through the spate of invention that followed Faraday's discoveries. But it has little to do with what really goes on in electrodynamic processes. It is probably just as well that the early inventors were innocent of the insights that were to grow out of Faraday's and Maxwell's work: they would have thrown up their hands in despair at the idea of ever being able to harness so utterly bewildering a phenomenon as the electron. On the other hand, in the light of present-day understanding of electrical phenomena, their achievements in doing exactly that seem all the more amazing, justifying in every way the awe those accomplishments inspired among the public.

While Maxwell had made a great leap forward from Newton's conception of the universe, he was unable to disengage himself completely from Newtonian mechanics. He felt there needed to be a medium in which electromagnetic force fields existed, some substance in which these disturbances were created, and he proposed several under the rubric of "ether." Thus, the comfortable familiarity of Newton's universe of action and reaction, cause and effect, logic and predictability were salvaged and preserved for the next fifty years.

Enter Dr. Einstein. In two articles published in 1905, Albert Einstein sought to resolve the discrepancies between Faraday's and Maxwell's electrodynamics and Newtonian mechanics. After all, both seemed to work in the real world, despite their apparent mutual contradictions: there must be a way to bring them together into a unified theory of broad application. The first of those papers, on "special relativity," remains among the most stunning intellectual achievements of the century. Quickly verified in detail through experimentation, it demonstrated that space is not three-dimensional, and that time does not exist outside space. There is a fourth dimension called "space-time," in which the two are inextricably bound together. Time is not a river; it is not inexorable. Events that seem to take place simultaneously to one observer, may be separated in time from the vantage point of another observer traveling at a different speed. Thus, all measurements involving time and space lose any absolute significance; they become *relative to the point of view of the observer.* Space and time are removed from their central position in the physical universe, to become merely two more elements of a description of physical phenomena.

As Einstein was demolishing time and space as discrete theoretical phenomena, the new communications technologies arising out of electrodynamics were drastically altering the impact of time and space on day-to-day life and changing their meaning in popular perception. The telegraph, telephone and radio—even though their invention preceded Einstein's insights—were each spontaneously greeted in the popular media as having the miraculous ability to "annihilate time and space." On the day following Samuel Morse's first successful demonstration of the news-carrying capacity of his electric telegraph (he brought the results of the Democratic party presidential nominating convention from Baltimore to Washington), the New York *Herald* observed: "Professor Morse's telegraph is not only an era in the transmission of intelligence, but it has originated in the mind . . . a new species of consciousness."[27] Time and distance were no longer what they once had been.[28]

In the 1920s a polyglot group of scientists extended Einstein's insights in a sweep of great and sustained scientific creativity. They included Niels Bohr of Denmark, Werner Heisenberg of Germany, Louis de Broglie of France, Erwin Schrödinger and Wolfgang Pauli of Austria and Paul Dirac of Britain, and together they mapped the unexplored territory of the subatomic particle, a field called quantum mechanics.

It may help to pause here for a moment to get some idea of the scale of the world of subatomic particles which these scientists set out to explore. Travelers in southern Europe fortunate enough to find themselves among the ruins of the Minoan civilization on Crete, Thera and other islands often find it astonishing that, to the Romans, the Minoans were as remote in antiquity as the Romans are to us. John Gribben employs a similar kind of scale in groping for a way to convey the incredible tininess of quanta in his book *Schrödinger's Kittens*: "In very round numbers, the quantum world operates on a scale as much smaller than a sugar cube as a sugar cube is compared with the entire observable Universe. To put it another way, people are about midway in size, on this logarithmic scale, between the quantum world and the whole Universe."[29]

Early in the history of the new discipline of quantum mechanics, it became necessary to deal with a paradox that has haunted it ever since. Einstein had demonstrated conclusively that light is composed of particles, which he called photons. Confirmed experimentally to exist, they are exceedingly strange entities: they always travel at the speed of light, have no mass, and because of their speed of travel, time has no meaning to them. Like other quantum particles, they are able to communicate with one another *instantaneously*, if need be, from one side of the universe to the other. Unfortunately, about a hundred years before Einstein, an English scientist named Thomas Young had proved that light is composed of waves. This, too, was tested and confirmed beyond dispute in many experiments. So light had been "proved" to be both a wave and a particle. Quantum physicists struggled with that for some time, but there was no way out. The answer had to be that light was indeed both particle and wave and, by extension, other subatomic particles, including electrons, were also both waves and particles. In fact, it has recently been experimentally demonstrated that "objects" as large as atoms simultaneously exist in both states.

How can something be both a wave and a particle? The blithely outrageous answer agreed upon among the principal quantum mechanics investigators of the time was that subatomic entities have no concrete existence. They do not appear at specific times in specific places, but exhibit a "tendency" to exist. This tendency is expressed in the mathematical form of a probability wave. According to Heisenberg, the probability wave "meant a tendency for something. It was a quantitative version of the old concept of "potentia" in Aristotelian philosophy. It introduced something standing in the middle between the idea of an event and the actual event, a strange kind of physical reality just in the middle between possibility and reality."[30] The indeterminacy of quantum physics is not the result of shortcomings of test instruments or of the indifference of physicists to concrete definitions. As Niels Bohr put it, "in quantum mechanics, we are not dealing with an arbitrary renunciation of more detailed analysis of atomic phenomena, but with a recognition that such an analysis is *in principle* excluded."[31]

It gets stranger. A subatomic entity such as an electron can be said to exist only when its probability of being at a certain place at a certain time is 1; in other words, when it is certainly there. The only way to know whether it is certainly there, however, is for an intelligent observer to look and see it there. Until the observer observes, the probability of its existence is always less than 1: it may or may not exist. Until it is observed, it can have no concrete existence; it represents only a potential that may be described in terms of probability waves. It may or may not be there when the observer looks: if it is not there, it doesn't exist and the portion of the probability wave function indicating existence collapses; if it is there, it does exist, and it is the portion of the probability wave function indicating non-existence that collapses. Without the observer, there can be no collapse of the wave-function, and therefore neither existence nor non-existence—only potential. Thus, the observer plays an integral role in quantum theory, in the very existence of particles. In a very real sense, the observer *creates* the particle through the act of observing it. That makes him or her more than an observer, as pointed out by Princeton physicist John Wheeler:

> May the universe in some strange sense be "brought into being" by the participation of those who participate? . . . The vital act is the act of participation. "Participator" is the incontrovertible new concept given by quantum mechanics. It strikes down the term "observer" of classical theory, the man who stands safely behind the thick glass wall and watches what goes on without taking part. It can't be done, quantum mechanics says.[32]

If quantum mechanics is correct, and for seventy years it has passed every experimental test, then there is no substantive physical world, only a vast web of energy relationships. To try to extract discrete parts of the subatomic world for examination is futile, because there is no such thing as a "separate part." As Einstein said: "We may therefore regard matter as being constituted by the regions of space in which the field is extremely intense. . . . There is no place in this new kind of physics both for the field and matter, for the field is the only reality."[33]

Here is a more recent explication:

> The presence of matter (such as an electron) is merely a disturbance of the perfect site of the field at that place; something accidental, one could almost say, merely a "blemish." Accordingly, there are no simple laws describing the forces between elementary particles. . . . Order and symmetry must be sought in the underlying field. . . . The field exists always and everywhere; it can

> never be removed. It is the carrier of all material phenomena. It is the "void" out of which the proton creates the pi-mesons. Being and fading of particles are merely forms of motion of the field.[34]

Physicist David Bohm has written that, in the world of quantum physics,

> [p]arts are seen to be in immediate connection, in which their dynamical relationships depend, in an irreducible way, on the state of the whole system (and, indeed, on that of broader systems in which they are contained, extending ultimately and in principle to the entire universe). Thus one is led to a new notion of unbroken wholeness which denies the classical idea of analyzability of the world into separately and independently existent parts. . . .

Furthermore, Bohm states, "There is a similarity between thought and matter. All matter, including ourselves, is determined by 'information.'" In other words, Bohm asserts, " 'Information' is what determines space and time."[35] Which helps to make the notion of an "Information Age" one of compelling interest at unexpected levels.

If the movement of electrons through wires and across space, bearing human thought, seems a magical idea, as it certainly did to the contemporaries of the early electrical experimenters, it is only because it is indeed *magic*. How it happens in specific detail is in a sense beyond comprehension.[36] It is unknowable in a rational, scientific sense, because we are prevented in principle from having enough knowledge to make determinations that are not based on observation, due to the ambivalent wave/particle nature of these most basic of entities.

It is in this sense that the electron, the basis for all electronic communication, is an invention rather than a discovery.

CHAPTER FIVE

The Electric Telegraph

As the first of the practical applications to grow out of electrodynamic theory, the electric telegraph was slow to be accepted for the revolutionary breakthrough in communication that it was. Its value and potential were certainly not understood by the governments of Great Britain and the United States, each of which churlishly dismissed offers from its chief inventors to deed it to the state. The optical telegraphs then in use, along with the penny post, were deemed perfectly adequate for the job of handling necessary long-distance communication. The idea that a new medium might lead to the creation of novel services and products, or might alter the traditional pattern of doing things in useful or interesting ways, was not considered. Wheatstone in Britain and Morse in the United States faced uphill battles to have their inventions adopted. Indeed, if there is a great puzzle in the early history of communications technologies, it is in the inexplicable inability of even the great inventors themselves to predict the public appetite for the services they were to provide. Samuel Morse and Alexander Graham Bell saw limited applications for their creations, mainly in business and government. Cyrus Field, the builder of the transatlantic telegraph cable, thought there "might" be enough traffic to warrant the expense; Guglielmo Marconi at first thought of radio as primarily a ship-to-shore service, a niche product. And they were the optimists.

The almost universal failure to connect humanity's innate gregariousness with the application of these new communications technologies is a mystery, and one that remains endemic. The greatest surprise among industry savants observing the early development of the personal computer was that people adapted it for use as a communications device. And very few so-called experts foresaw the explosive development of the Internet as a medium of mass personal communication in the 1990s. If there is anything in common with these failures of imagination, perhaps it lies in thinking that the technologies were too complicated for the public to be much interested in them. But in each case the public in its wisdom saw the transcendent worth of the devices in maintaining contact with other people, and forced the market to begin serving that need. Modern communications technologies, beginning with the telegraph, have this in common: their growth has been a phenomenon of what we nowadays call "demand pull," as opposed to "supply push." Throughout the modern history of communications, the ordinary citizen has been out in front of industry, regulators and government in recognizing the potential of new media.

The idea of the electrical telegraph is an old one. Although Samuel Morse apparently believed he was first to conceive of transmitting intelligence over wires using electricity, he had many predecessors, and it was only his scant scientific training that prevented him from knowing as much. As early as 1753, an anonymous writer to the *Scots' Magazine* proposed a fully realized telegraph system that would use the movements of pith balls or the ringing of tiny bells activated by electrical currents to spell out messages. Each letter was to have its own wire. To prevent the "electric fire" from being dissipated in the atmosphere over long lengths of wire, the author proposed covering the strands with "a thin coating of jewellers' cement. This may be done for a trifle of additional expense; and as it is an *electric per se* [i.e., an insulator] will effectually secure any part of the fire from mixing with the atmosphere." The suggestion reflects the understanding of electricity at the time as having the mechanical properties of heat.

A more practical device using just one wire was invented in 1816 by Sir Francis Ronalds and tested over more than forty miles of wire coiled on a frame. An ingenious system of synchronized clockwork mechanisms at either end of the wire displayed each of the letters of the alphabet in turn, and when the correct letter appeared in the display window, the sending operator interrupted the circuit briefly to indicate to the receiving operator that he should copy that letter down. Ronalds sought an audience with the British admiralty to demonstrate his device but was informed by return mail that "telegraphs of any kind are now wholly unnecessary; and that no other than the one now in use [i.e., the visual telegraph] will be adopted."

Credit for the first working electromagnetic telegraph is generally given to Baron Schilling, who developed a system that used the deflections of a magnetized needle, and a code not unlike that attributed to Morse, to spell out words. Schilling built his telegraph in his hometown of St. Petersburg and demonstrated it to the czar in 1830 before traveling with it to the capitals of Europe. Czar Nicholas reacted with the instincts of a true autocrat: seeing in the telegraph an instrument of subversion, he forbade any mention of the device in the Russian press or scientific literature for the duration of his reign, with the result that Russia was among the last nations in Europe to adopt the new technology. Nicholas was in many ways a modern man, not the least in his understanding of information as both a means of control and, on the other side of the coin, a source of liberation.

Schilling's telegraph was seen in Heidelberg by William Cooke, a British scientific amateur with sharp entrepreneurial instincts. He at once saw commercial possibilities in the device and returned to England to develop it further. In search of sound scientific advice, he formed a partnership with Professor Charles Wheatstone, the chair of the King's College (Cambridge) experimental philosophy department. In 1837 they jointly patented a needle telegraph that used up to five wires and three to five needles which pointed to letters of the alphabet arranged on a grid; there was no need for the operator to know any code beyond the standard alphabet in order to receive a message. Problems with finding suitable insulating materials for the telegraph lines delayed deployment of the device in a commercial sense until 1846, when the two incorporated the Electric Telegraph Company and went into business.

It is an interesting and instructive fact that Wheatstone struggled for several years to find a solution to seemingly intractable problems with their telegraph—problems that had been solved a decade earlier by the Bavarian scientist Georg Simon Ohm. Ohm, for whom the unit of electrical resistance has been named, developed laws showing the relationship between electrical current, voltage and resistance in a circuit, and in those formulae was everything Wheatstone needed to know to iron out the bugs in his system. Clearly, communication among scientists and engineers was still at a primitive stage. Schilling, a military man by profession and scientist only by avocation, published almost nothing in the scientific literature, and although Ohm did publish, his work remained unknown in England until Wheatstone unearthed it in about 1840. It is evident that had communication been better, development of the telegraph could have been accelerated by several years, perhaps as much as a decade.[37]

Samuel Morse was a portrait painter of genuine talent, a politician of pro-slavery, anti-Catholic and anti-immigrant prejudice and a lifelong victim of poisonous pride who learned in mid-life of the electrical experiments of Oersted. His information came from fellow passengers on board the packet ship *Sully* carrying him back to America in 1832, from studying and painting in Europe. In a gush of jingoistic hyperbole, an early American biographer of Morse titled his book *Samuel Morse: An American Leonardo*, but it is clear that Morse was far from that, having discovered none of the scientific principles associated with the telegraph, nor invented any electromechanical devices that were not already known in Europe. Indeed, his fellow-countryman Joseph Henry, a man of milder disposition and scholarly modesty, had built a working telegraph at Albany in 1832 or perhaps even earlier. Henry published his work (and Morse saw it) but never sought a patent. There is even doubt that Morse invented the Morse code; that distinction is claimed for his associate Alfred Vail.

The code is an altogether ingenious conversion of the alphabet into binary, digital form based on dots, dashes and spaces. Vail is said to have worked out the details, perhaps at Morse's suggestion, by examining the numbers of various letters of the alphabet to be found in a printer's type drawer. The letters most frequently used, *e* and *t*, were assigned the simplest codings, one dot (·) and one dash (-), respectively. The letters *a* and *n* were assigned (·–) and (–·), and less frequently used letters such as *v* (···–) and *w* (·––) were given more complex combinations. Much effort was put into devising means of printing the code on paper tape, so that it could be translated into letters of the alphabet, until it was noticed that telegraphers were quickly able to do the decoding in their heads simply by listening to the clicks of the telegraph receiver as its circuit opened and closed. It was this code, and the rugged simplicity and cheapness of the system required for sending and receiving it (a telegraph key for transmitting and a simple electromagnetic sounder at the receiving end), that allowed Morse's system to eventually reign supreme over its competitors in the field.

Morse insisted until his death that he was the sole inventor of the electric telegraph and his compatriots have been inclined to believe him, despite the evidence. Britain, however, refused him a patent and the U.S. Supreme Court, in its comment on one of the many lawsuits he energetically pursued in defense of his American patent, averred that Morse's claims were far too broad for even a generous interpreter to accept.

What is clear is that Morse was captivated to the point of obsession by the idea of developing an electric telegraph from the moment he conceived it on board the *Sully*. Living in penury on income from teaching painting, Morse worked on his device fitfully until 1837, when he formed a partnership with the young engineer Alfred Vail, a wealthy New Jersey industrialist's son. In that year the U.S. House of Representatives asked the Secretary of the Treasury to report to it on "the propriety of establishing a system of telegraphs for the United States." Morse was among the many who replied to the circular distributed by the secretary asking for submissions. It was the beginning of a long and arduous process of convincing the government to finance testing of his device, which was patented in 1840, three years after Cooke and Wheatstone received a patent for their telegraph in England. Finally in 1843, by dint of persistence and a new partnership with the well-connected lobbyist and speculator F. O. J. Smith, Morse received approval from Congress to build an experimental line from Washington to Baltimore, some forty miles distant.

We tend to think in romantic terms of early technological triumphs, but what sold Congress on Morse's telegraph was mundane business logic. Though perennially improvident in his personal finances, Morse presented a soundly reasoned business case for the superiority of his electric telegraph over the visual telegraphs then in use. Taking the widely admired French network as his benchmark, he made some historically interesting comparisons:

> The French system of telegraphs is more extensive and perfect than that of any other nation. It consists, at present, of five great lines, extending from the capital to the extreme cities of the kingdom . . . making a total of 1,474 miles of telegraphic intercourse. These telegraphs are maintained by the French government at an annual expense of over 1,000,000 of francs, or $202,000.
>
> The whole extent, then, of the French lines of telegraph is 1,474 miles, with 519 stations . . . erected at a cost of at least $880 each—making a total of $456,720.
>
> The electro magnetic telegraph, at the rate [of] . . . $461 per mile (and which, it should be remembered will construct not *one* line only, but *six*) could be constructed the same distance for $619,514—not one-third more than the cost of the French telegraphs. Even supposing each line to be only as efficient as the French telegraph, still there would be six times the facilities, for not one-third more cost. But when it is considered that the French telegraph, like the English, is unavailable [due to inclement weather and darkness] the greater part of the time, the advantages in favor of the magnetic telegraph become more obvious.[38]

Morse estimated potential earnings from operations of the Washington–Baltimore line at $600,000 a year, and drove home his case with an argument that has a distinctly modern ring: "An important difference between the two systems is, that the foreign telegraphs are all a burden upon the treasury of their respective countries; while the magnetic telegraph proposes, and is alone capable of sustaining itself and producing a revenue."[39]

In his report to the U.S. Treasury Secretary on completion of the Washington–Baltimore line, Morse was of course enthusiastic about its potential, though the instances he gives of its usefulness seem, from this distance, to quaintly understate the case:

> An instance or two will best illustrate [the great utility] of the telegraph: A family in Washington was thrown into great distress by a rumor that one of its members had met with a violent death in Baltimore the evening before. Several hours must have elapsed ere their state of suspense could be relieved by the ordinary means of conveyance. A note was dispatched to the telegraph rooms at the Capitol, requesting to have inquiry made at Baltimore. The messenger had occasion to wait but *ten minutes* when . . . the answer returned that the rumor was without foundation. Thus was a worthy family relieved immediately from a state of distressing suspense.
>
> An inquiry from a person in Baltimore holding the check of a gentleman in Washington, upon the Bank of Washington, was sent by telegraph, to ascertain if the gentleman in question had funds in that bank. A messenger was instantly dispatched from the Capitol, who returned in a few minutes with an affirmative answer, which was returned to Baltimore instantly; thus establishing a confidence in the money arrangement, which might have affected unfavorably (for many hours at least) the business transactions of a man in good credit.[40]

The American Civil War provided the impetus needed to drive the telegraph out across the plains and deserts of the American heartland to the new settlements on the West Coast. President Abraham Lincoln deemed the communications link essential in keeping California loyal to the Union. The lines marched inland from both coasts and eventually met in Salt Lake City in October 1861. The challenges of crossing so much open, unsettled territory were formidable. Not the least of these was the fact that in much of Nebraska and Colorado territories, the Cheyenne and Arapahoe Indians were at war with the American government. Whenever the opportunity arose, the Indians tore down telegraph lines and attacked the isolated telegraph stations. But repairs were effected with surprising speed and energy. In one episode, repair crews protected by army howitzers rebuilt eight miles of telegraph line, replacing poles that had been burned or chopped down, and restrung another twenty-two miles of damaged wire in forty hours of continuous pick-and-shovel labor. In more normal circumstances, new line could typically be thrown up at a rate of between three and eight miles a day, depending on the terrain, with twenty-five poles per mile, each set in a five-foot-deep hole dug with special long-handled spades.

When the line was finished and signals flashed from Atlantic to Pacific for the first time, the *New York Times* intoned: "It is with almost an electric thrill that one reads the words of greeting yesterday flashed instantaneously over the wires from California. The magnificent idea of joining the Atlantic with the Pacific by the magnetic wire is today a realized fact. New York, Queen of the Atlantic, and San Francisco, Queen of the Pacific, are now united by the noblest symbol of our modern civilization."

National purpose played a large role in this mammoth construction feat, but perhaps not so great a role as the promise of equally mammoth profits. The eastern portion of the line from Omaha to Salt Lake City had been financed by a million-dollar stock sale: its actual cost was $147,000. The western line from Salt Lake to California is estimated to have cost just $500,000. Telegraph companies initially charged a dollar a word on the new transcontinental line, and even at that breathtaking rate there was plenty of demand for the service.[41] Once it became possible to communicate instantly from one coast to the other, it became *necessary* for business to do so.[42] Companies that did not take advantage of the new communications link could be quickly outmaneuvered by competitors who did. Commodities speculators, in particular, had to keep informed by the quickest means available if they were to stay in business, and that meant using the telegraph to check prices of gold, silver and other speculative resources. In the same way, government communication had to be carried out at the maximum speed available even in peacetime, if only because political courtesy demanded it.

Nor was owning a telegraph line the only way to get rich in telegraphy. Owning a glass factory would do just as well. Some lines used as many as two thousand insulators per mile to mount many converging wires. Manufacturers turned out millions of insulators on high-speed glass molding machines introduced in 1865.

In one of the best of modern media histories, Daniel Czitrom has observed that, in contemporary writing, "[s]erious considerations of the telegraph usually touched upon the other technological marvels of the age, the railroad and the stream-boat. Yet the inscrutable nature of the telegraph's driving force made it seem somehow more extraordinary."[43]

Mysterious electricity, an early historian of the telegraph wrote, "seems to connect the spiritual and the material." Another reflected: "The mighty power of electricity, sleeping latent in all forms of matter, in the earth, the air, the water, permeating every part and particle of the universe, carrying creation in its arms, is yet invisible and too subtle to be analyzed. . . . Its mighty triumphs are but half revealed, and the vast extent of its extraordinary powers but half understood."[44] The words chosen by a young woman of Samuel Morse's acquaintance[45] to be the first transmitted by telegraph were, "What hath God wrought!"[46] Historian and telegraph promoter T. P. Shaffner, in concluding a review of all previous forms of communication, said of the telegraph: "But what is all this to subjugating the lightnings, the mythological voice of Jehovah, the fearful omnipotence of the clouds, causing them in fine agony of chained submission to do the offices of a common messenger—to whisper to the four corners of the earth the lordly behests of lordly man!"[47]

The special significance accorded the telegraph, and other electrical means of communication, is wrapped up in the way "communication" was defined, according to Czitrom:

> Praisers of "universal communication" [made possible by the telegraph] no doubt had in minds the most archaic sense of the word: a noun of action meaning to make common to many (or the object thus made common). The notion of common participation suggested communion, and the two words shared the same Latin root, *communis*. . . . Those who celebrated the promise of universal communication stressed religious imagery and the sense of miracle in describing the telegraph. They subtly united the technological

> advance in communication with the ancient meaning of that word as common participation or communion. They presumed the triumph of certain [Christian] messages. . . .[48]

The telegraph, wrote one enthusiast, "gives the preponderance of power to the nations representing the highest elements in humanity. . . . It is the civilized and Christian nations who, though weak comparatively in numbers, are by these means of communication made more than a match for the hordes of barbarism. . . . [The telegraph] binds together by a vital cord all the nations of the earth. It is impossible that old prejudices and hostilities should longer exist, while such an instrument has been created for an exchange of thought between all the nations of the earth."[49]

For commentators like this, the telegraph, as an instrument of the ineffable, semispiritual stuff called electricity, was more than just an invention—it was a moral force in the world. It was a view shared in varying shades and degrees by many on both sides of the Atlantic: nothing else sufficiently accounts for the intensity of curiosity and unparalleled public enthusiasm generated by this invention. In Britain, it was seen as a providential tool which would assist in extending the influence of the empire to the far corners of the world; in America, it became an agent of the manifest destiny of the nation to achieve greatness and world moral leadership.

CHAPTER SIX

A Worldwide Web

For the business interests behind the great telegraph enterprises that were to develop in Europe and North America, it was not so much moral issues as the example of mountainous profits on the U.S. transcontinental line that provided the primary incentive for undertaking the seemingly impossible task of linking the two continents by wire. The story of the laying of the first transatlantic marine cables is remarkable in its own right and a magnificent tribute to nineteenth-century engineering prowess. But there was also a little-remembered, outrageously ambitious scheme to tie Europe to North America through the back door, via British Columbia, Alaska and Siberia. It was undertaken by the Western Union Company in 1865, with the cooperation of the governments of the United States, Canada and Russia.

With the transatlantic cable project suffering one failure after another (as we'll see in a moment) and dismissed in the conventional wisdom as a pipe dream, while the trans-America line was a roaring success, Western Union fell under the spell of a promoter named Perry D. Collins, who convinced the directors that the overland route to Europe via Russian Alaska and Siberia was the way to go. What he proposed seems in retrospect sheer lunacy, given the weather conditions and terrain involved, but the confidence of the age in its engineering capabilities seems to have been boundless. And the need for rapid communication with Europe was clear: when President Abraham Lincoln died from an assassin's bullet on the morning of April 15, 1865, word was received in San Francisco within the hour; Europe found out on April 26, eleven days later, when the steamer *Nova Scotian* docked in England.

The *New York Times* expressed the blithe opinion of many who had grown skeptical of the prospects of ever making a successful transatlantic link: "If there is ever to be electric communication with Europe, it will be by imitating the splendid example the United States has thus given in our transcontinental line. The bubble of the Atlantic submarine line has long ago burst, and it is now seen to be cheaper and more practicable to extend a wire over five-sixths of the globe on land, than one-sixth at the bottom of the sea."

Work was begun on the line within days of the end of the Civil War in the summer of 1865, with construction of sections linking San Francisco to New Westminster, British Columbia, and from there up the Fraser River to Quesnel in the B.C. Interior. Other crews were put to work on line stretching across the largely unexplored mountain ranges of Russian Alaska to the Bering Strait. A third work party sailed to Siberia to explore a route and construct line from the Bering Strait (which was to be spanned by a short underwater link) eighteen hundred miles southwest to Okhotsk at the mouth of the Amur River. A seven-thousand-mile Russian-built line from Okhotsk to St. Petersburg would provide the final connection to Europe and its growing web of telegraph communication.

The Russian party included a twenty-year-old telegrapher named George Kennan, who kept a diary thanks to which we have a detailed record of the events of the next two years. Kennan and the other Americans spent the summer of 1865 and the following winter mapping a route south. By early 1866, they had been out of touch with the rest of the world for nearly a year. It wasn't until August of that year that two supply ships arrived in a village on the Sea of Okhotsk, bearing sixty American construction workers and cargoes of building supplies and construction tools from San Francisco, two months away by sea. A third supply vessel arrived in September. Over the following winter six hundred Siberian laborers were hired and three hundred horses were purchased to cut and haul twenty thousand telegraph poles, in temperatures that often dropped to minus 60°F. In the spring of 1867, work on the line began in earnest, with confident predictions that it would be completed right to St. Petersburg by late 1869.

It was on the evening of May 31, 1867, that an American whaling ship appeared over the horizon and dropped anchor in the Siberian port settlement where the American expedition was headquartered. The following morning a party of Americans rowed out to the *Sea Breeze* to be greeted by the astonished captain.

"Have you been shipwrecked?" he asked them. What other explanation could there be for Americans to be stranded in such a remote corner of the globe?

Kennan explained that they were building a telegraph line, at which point the captain volunteered the devastating news that the transatlantic cable had been completed more than a year earlier and was functioning well. San Francisco newspapers, he said, were routinely publishing European news that was only a day old. Kennan and the other dispirited men were to learn from newspapers on board the whaler that construction had been halted on their own line through British Columbia seven months earlier. It wasn't until July that a second vessel arrived at the little port, with official news from Western Union that the entire project had been abandoned. The Americans were told to sell their tools and building materials as best they could and return home. It is said that glass insulators sold as tea mugs can still be found in Siberian farmhouses.

The line from the U.S. border to Quesnel was purchased by Canada in 1870. Abandoned poles and spools of wire left along the broad right-of-way cleared to the north of Quesnel were salvaged by local Indians, who used the copper for nails and fishing spears, and for binding logs together in footbridges and other construction projects. A famous native-built suspension bridge made almost entirely of telegraph wire spanned the river at Hagwilgaet for decades. Thirty years after the line was abandoned, the great Klondike gold rush saw the right-of-way north of Quesnel used as a highway to the gold fields of the Klondike and the Canadian government at last completed the telegraph link from Quesnel to Whitehorse.

Western Union and its shareholders had spent $3 million on the aborted project, a sum that gains significance when compared with the $7 million paid the following year by the United States to purchase all of Alaska from Russia. A traveler to the Skeena and Bulkley rivers in northern British Columbia in 1872 penned a haunting epitaph to the project:

> Crossing the wide Nacharcole River and continuing south for a few miles, we reached a broadly cut trail which bore curious traces of past civilization. Old telegraph poles stood at intervals along the forest-cleared opening, and rusted wires hung in loose festoons down from their tops, or lay tangled in the growing brushwood for the cleared space. A telegraph in the wilderness! What did it mean?
>
> When civilization once grasps the wild, lone spaces of the earth it seldom releases its hold; yet here civilization had once advanced her footsteps, and apparently shrunk back again, frightened at her boldness. . . .[50]

The Atlantic submarine cable that dashed Western Union's plans for a "back door" link to Europe was a high watermark of nineteenth century technological achievement. It bore all the hallmarks of Newtonian physics expressed in machines; it pushed the limits of engineering of all kinds to an extent that gives pause even to the modern observer. Everything about it was monumental, larger than life. Given the materials and technology then available, the laying of a wire cable right across the Atlantic Ocean seems every bit as impressive a feat as sending a man to the moon.

It took five attempts, beginning in 1857 and ending with success, finally, in 1866. One man, the American Cyrus W. Field, provided the bottomless supply of energy and optimism required to carry the project to completion through years of heartbreaking failures, and long delays caused by the Civil War. (He made more than fifty Atlantic crossings in the process.) Most of the financing came from British backers.

The first foray was launched with great fanfare and speech making, from Valentia on the southwest coast of Ireland. The British and American governments had each supplied a warship rigged for cable laying: the frigate USS *Niagara* was to lay the first half of the cable from Ireland, accompanied by its escort USS *Susquehanna*; HMS *Agamemnon* was to meet *Niagara* at mid-Atlantic, splice the cable and carry on to Trinity Bay in Newfoundland, escorted by HMS *Leopard*. The job went smoothly at first, until four hundred miles into the Atlantic a swell caught the stern of *Niagara*, lifting it high in the air. Before the men on board could react, the cable parted and dropped out of sight beneath the waves. The ships were forced to return to Britain, and Field had to raise new capital for a second attempt.

The squadron set out again with fresh supplies in June 1858. This time the plan was for *Niagara* and *Agamemnon* to meet and join the cable at midocean, *Agamemnon* then heading for Ireland and *Niagara* steaming slowly for Newfoundland. Several cable breaks in the first few days forced the vessels to return to Valentia to resupply before setting out once again for midocean and a fresh start. The vessels had steamed to within a few hundred miles of their mid-Atlantic rendezvous when they encountered one of the worst Atlantic storms on record. The heavily laden *Agamemnon* was nearly capsized, and saved only by superb seamanship. *Niagara* survived as well, although her cable had been thoroughly scrambled below decks, resembling a mass of cooked spaghetti. Six days after the storm the two ships met and the cable was spliced. They steamed off in opposite directions. Within an hour, the cable had broken as it was being paid off *Niagara*'s deck. A fresh splice was made, and the exercise began again. A day later the cable broke once again, but this time underwater. A third splice was made, and now the operation went smoothly while 146 miles of cable was laid. Then, the cable parted at *Agamemnon*'s stern and dropped into the ocean. There was no longer enough cable for another attempt, and the squadron returned to Britain, thoroughly disheartened.

There were those on shore who counseled abandoning the project as impossible, but Field and other optimists prevailed, and on July 17, another attempt was mounted with the same ships. There were no celebrations this time when the vessels departed for midocean, only faint hope and foreboding. Tension on board was palpable as they spliced the cable and crept east and west toward their respective destinations. On

August 4, *Niagara* entered Trinity Bay, Newfoundland, while lookouts on *Agamemnon* raised the Irish coast. The ocean had finally been bridged, and the link worked, though there were disturbing problems with weak signals and long gaps when nothing could be heard.

The success triggered rapturous celebrations on both sides of the ocean, and Field was hailed as a hero. In his hometown of New York, festivities included a church service, two parades, a banquet and speeches. The streets were alive until well after midnight. So enthusiastic was the fireworks display that night that it set the city hall alight and burned it down. And in the state capital, Albany,

> [c]rowds of persons flocked to the newspaper offices and Telegraph offices for confirmation of the news, which most at first doubted, but when the conviction of the truth of the report forced itself on the public mind, the scene in the street was as though each person had received some intelligence of strong personal interest. . . . The people are wild with excitement.[51]

In reflecting on the spontaneous exhilaration that greeted the news, the *New York Times* referred to the telegraph as a "divine boon," and added: "From some such source must the deep joy that seizes all minds at the thought of this unapproachable triumph spring. It is the thought that it has metaphysical roots and relations that makes it sublime."[52]

But Field could not enjoy the celebrations because he and a few others shared the awful secret that all was not well with the cable. It continued to work only intermittently, and messages took hours to transmit when they should have taken minutes. Later investigations indicated that the rubberlike gutta-percha insulation had broken down on portions of cable that had been stored in sunlight. Further damage was done experimenting with high voltages to boost the signal strength. The link lasted only two months: a total of seven hundred messages were transmitted before it died forever.

It was a terrible blow for Field, who became, as dethroned heroes often do, an object of scorn and ridicule, excoriated in the press that had so recently lionized him. Remarkably, he persevered, arguing that the brief success of the 1858 cable had proved beyond doubt that the idea was feasible. So enormous was the potential for profit, and so great was the importance assigned to the project by the governments of the United States and Britain, that Field was eventually able to raise enough capital and government guarantees for yet another attempt. By then, cable-making methods had improved, as had telegraphic instruments and cable-laying technology.

Ultimate victory was assured by putting into service as a cable-layer the British vessel *Great Eastern,* which was, at the time of her launch in 1858 (the year of Field's first, partial success), the biggest ship afloat, five times larger than her nearest competition at nearly 700 feet long, with a displacement of 27,000 tons. Her twin, 60-foot side paddle wheels weighed 90 tons each, and gave her superb maneuverability. Her single cast-iron screw propeller was 24 feet high and weighed 36 tons. Her coal-fired engines developed eleven thousand horsepower. Only a vessel of this size could manage the great weight of the more than two thousand miles of armored and insulated copper cable that needed to be paid out on the voyage from Ireland to Newfoundland: though only a little bigger around than a thumb, it weighed in at about 7,000 tons.

But, once again, success was not to come without bitter setbacks. On the first attempt in 1865 there were four signal failures on the cable in the first four days out from Ireland. Each time the wire had to be rolled back on board the ship, inspected for damage and repaired. Sabotage was suspected, since the failures had been caused by tiny needles of wire shorting the cable to its outer casing of wire armor, and all had occurred when the same crew was on duty. Eventually, however, it was determined that the fault had been in the brittle composition of the outer shielding, which had a tendency to splinter. Redoubling their caution and signaling constantly through the line back and forth to Ireland, the crews continued paying out cable until *Great Eastern* was within 660 miles of Newfoundland. There, another fault was detected, and while the line was being hauled on board for repair, it snapped and dropped back into the sea.

Plundering the huge ship's stores for enough rope, the captain rigged a grapple and steamed back and forth across the cable's track. Twice the crew snagged it and raised it from the ocean floor, but both times the grapple line parted under the strain. Finally there was not enough line left to try again. *Great Eastern* sailed back home.

A new company was organized by Field and his backers, and financing was arranged. New cable was manufactured over the winter at the rate of twenty miles a day. On July 12, 1866, *Great Eastern* sailed from Valentia with more than 2,300 miles of cable stored in her ample holds. On July 27, at 5:00 P.M., the ship's crew dragged the cable ashore through the surf at Heart's Content, Newfoundland, and it was soon operating perfectly. Two weeks later, with Field aboard, the ship returned to the site of the previous year's break, and dragging with new, stouter grappling gear, raised the broken cable. The crew cleaned it up, attached telegraph equipment, and tried signaling Ireland. It worked! Field strode manfully to his cabin, closed the door and collapsed in tears. There were now two functioning transatlantic cables.

Despite the awesome logistics of these early transatlantic undertakings, by 1892 there were ten such telegraph links spanning the ocean, where the indefatigable Field and the backers of the original success had wondered whether there would be enough demand to pay for one. The economics tell the story of the rapid expansion: a message on the first cables cost about five dollars a word. Working at speeds of up to seventeen words per minute, the potential revenue from each cable was calculated at upwards of $2 million a year. The total investment for all five original cables, three unsuccessful and abandoned, one broken but repaired and one completely successful, plus ancillary lines across Newfoundland to Nova Scotia and on to New York, was about $12 million over twelve years, much of that guaranteed by long-term contracts with the governments of the United States and Great Britain. By 1867, Field had paid off all his company's creditors with an added 7 percent interest. Cable stock that had sold for 30 guineas per £1,000 (about 30 cents on the dollar) was paying £160 in annual dividends.[53] The bottomless, pent-up demand for rapid communication made it a sound business proposition despite the enormous costs and risks.

In 1930, with the world sliding calamitously into economic depression, there was a global web of about 350,000 miles of telegraph line.

About 265,000 miles of that was long-distance trunks, including transoceanic links. British private and government interests were by far the largest owners of long-distance cables, with the United States a distant second. Britain's Eastern Telegraph Company had long-established cable links with India, Australasia, the Far East, Africa and South America, making it the largest cable operation in the world. The system amounted to 136,000 miles of cable, and it was carrying about seventy million words a year. The Pacific was spanned by two cables linking Vancouver with Australia, jointly developed by the British, Canadian, Australian and New Zealand governments. A section between Vancouver and the Fanning Island coral group in the Pacific was the longest in the world, at 3,500 miles. The Fanning relay station was opened in 1902.

As long as the telegraph held its monopoly on long-distance communication, ample profit was a foregone conclusion for any cable-laying endeavor. The industry, as one might expect, became technically conservative and administratively hidebound. It took radio's competition to prod cable into making renewed technical advances in the 1920s and 1930s. The baffling and debilitating problem of capacitance on the line was at long last overcome. Undersea cables behave electrically like capacitors, with the insulated central conductor as one plate and the outer armor of wire as the other. A feature of capacitors is that they store electricity until fully charged, and then allow current to flow as a discharge. The effect was that code had to be sent very slowly in the early cables, in order to be received distinctly at the other end. If it was sent too quickly, characters would be lost in the charging period; words would pile up and the backlog would never reach the other end in intelligible form. Early transatlantic cables were limited to speeds of about fifteen to seventeen words per minute.

This was a problem that could be safely ignored so long as the cables held on to their monopoly on transoceanic communication, but when radio arrived, telegraph engineers got busy. It was discovered that if the cable were "loaded" (i.e., if the inductance to ground were increased) by wrapping the center conductor with a ribbon of nickel-iron alloy (Permalloy), the capacitance problem could be all but eliminated. The loading or increased inductance had the effect of lowering the cable's capacitance or storage capacity. Speeds on these new loaded cables reached four hundred words per minute, with dramatic consequences for the cable operator's bottom line. This and other technical advances made it possible, at last, to transmit telephone signals. As well, techniques for laying undersea cables were developed to the point where, in 1928, a high-capacity cable was laid and in operation between Newfoundland and the Azores, a distance of 1,341 miles, in just eight days.

The last of the scores of transatlantic telegraph cables that were laid in the hundred years following Field's first successes was abandoned in 1966, with telegraph traffic switching over to newer, more efficient telephone cables and to satellite transponders. Fiber optic cables of undreamed-of capacity now span both the Atlantic and Pacific oceans, carrying everything from live television pictures to e-mail.

As the global cable web grew, so did the reality of a global market, now that shipping movements, prices, supply and demand could be learned instantly for stocks and bonds and currencies and for staple commodities like minerals and foodstuffs. The effect on the grain market was typical: the telegraph, combined with a parallel improvement in ground and marine transport technologies, involved North America directly in what had been a European-dominated world market for grain. In 1874 the cost of shipping a bushel of grain across the Atlantic was twenty cents. In 1904, it was two cents. Grain exchanges in Canada, Britain, the United States and Australia were in constant communication, and the buying and selling of grain became transoceanic, at world prices. Turn-of-the-century proponents of globalization lauded the development as lowering prices and making crop specialization feasible; others worried about the loss of diversity in agriculture in individual nations, as "inefficient" sectors were threatened by foreign competition. Ironically, the era of free trade that had begun early in the nineteenth century, and which British Prime Minister Benjamin Disraeli and other European leaders assumed in 1860 was a permanent fixture in world markets, died as rapidly as the expiry of trade treaties would allow in the 1880s, and high tariff walls were a feature of international trade right up to World War II. The telegraph was a major factor behind the change: it instantly created powerful lobbies in commodities sectors and in industry for protection from "too much" foreign competition, pressure which politicians in Europe and North America were unable to resist.

While the telegraph enabled business to operate more efficiently, the invention's impact in transforming the public's patterns of information consumption was perhaps even more important. For one thing, the telegraph changed the definition of "news." Until the telegraph, newspapers, apart from strictly local coverage, had consisted mainly of analysis and interpretation of events that were often days, even weeks or months, old. With the telegraph, newspapers got into the national and international spot-news business and freshness supplanted relevance as the most important defining characteristic of the product. The telegraph made relevance in news less relevant. Thoreau put his finger on it when he commented in *Walden*: "We are in great haste to construct a magnetic telegraph from Maine to Texas; but Maine and Texas, it may be, have nothing important to communicate. . . . We are eager to tunnel under the Atlantic and bring the old world some weeks nearer to the new; but perchance the first news that will leak through into the broad flapping American ear will be that Princess Adelaide has the whooping cough."

As instantaneous communication via the telegraph affected business, so it would affect news, which is, after all, merely a business defined by business considerations particular to the news industry. Once it became possible to bring newspaper readers word of events from around the world on the very day they occurred, it became *necessary* to do so. Simple imperatives of competition forced a redefining of "news." The notion of currency or freshness had always had a place in the definition of the word; now it had a dominant role.[54] News moved away from being analysis and interpretation of events of interest and importance to readers—useful or functional information—toward a simple recitation of happenings. What kind of "happenings" met the definition of "news"? To some degree, the fact that events occurred at all made them news or news*worthy:* the more dramatic or unexpected the occurrence, and the farther from home, the more weight its exoticism carried in the balance with its possible relevance to readers. The fact that "Princess Adelaide has the whooping cough" became news of the highest order, as was made clear in the first news bulletin ever carried by transatlantic cable. Transmitted from England and addressed to the Associated Press, August 27, 1858, the complete, unedited text read:

> EMPEROR OF FRANCE RETURNED TO PARIS SATURDAY. KING OF PRUSSIA TOO ILL TO VISIT QUEEN VICTORIA. HER MAJESTY RETURNS TO ENGLAND 31ST AUGUST. SETTLEMENT OF CHINESE QUESTION: CHINESE EMPIRE OPENS TO TRADE; CHRISTIAN RELIGION ALLOWED. MUTINY BEING QUELLED, ALL INDIA BECOMING TRANQUIL.[55]

The great news wire services were formed to satisfy the newly created demand for news of the world and put its supply into the hands of professionals. Before the Associated Press in the United States and Reuter in Britain, Agence France Presse in France and the forerunners to the Canadian Press in Canada, most telegraph news had been provided by telegraph operators in the employ of the telegraph companies. In Canada, that meant that Canadian Pacific Telegraphs controlled the flow of news across the continent, and the company was not averse to turning off the spigot to newspapers who dared to criticize CP in their pages. In the United States, the Associated Press newspapers fought a long and expensive battle with the Morse interests and Western Union to gain unrestricted, uncensored access to the wires for news. It ended in a settlement so cozy that the AP and Western Union—which had grown to become America's biggest corporation—were the objects of mounting criticism as two-headed monopoly controlling what should have been a public utility. The collusion between the two was the focus of intense disappointment among many of those who had been most enthusiastic in greeting the telegraph's arrival, disappointment that the bright promise of the telegraph as a divinely given instrument of moral force and a common carrier of intelligence should have fallen into the hands of crass commercial monopoly.

The news agencies, like some of the more responsible newspapers, tended to subscribe to more or less thoughtful standards for the definition of news, including in their characterization the notion of information needed by citizens to organize their lives and make responsible choices in a democratic society. However, they were all dependent for their survival on the satisfaction of the maximum number of subscribers and idealism was inevitably forced to find a modus vivendi with coarse commercial imperatives.

Media critic Neil Postman says of the telegraph that it ". . . gave a form of legitimacy to the idea of context-free information; that is, to the idea that the value of information need not be tied to any function it might serve in social and political decision-making and action, but may attach merely to its novelty, interest, and curiosity."[56] The news organization became the buyer and seller of the generic commodity called "news."

When broadcast news was still young enough to be idealistic, CBS News writing guidelines forbade calling the newscast a "show" in order to preserve, to the extent possible, the distinction between news and entertainment. In the nineties, after network television had fallen into the hands of multinational entertainment conglomerates and Harvard-trained management specialists, this quaint tradition was relegated to history and the coffee-shop conversation of the few aging news people who had somehow survived the multiphased corporate downsizings. Today, it is not unusual in newsrooms to hear the word "content," used in preference to any of the more descriptive terms for news and information programming. Plane crash, budget speech, obituary, flood: it's all "content," nicely demonstrating Postman's reminder that news has become a commodity, a "thing" to be bought and sold.

Postman argues correctly that the value of any information (as distinct from entertainment) depends on its usefulness to the recipient, which can be further defined as the possibilities for action it presents. News of happenings or circumstances that will never impinge on the recipient's life, and which allows no action by the recipient in response to it, is of very little, if any, use to anyone.[57] With the coming of the telegraph, the proportion of the information flooding into people's lives that had any relevance to them—that presented any possibility for action, and could thereby be said to be "useful" to them—dwindled dramatically. Furthermore, says Postman:

> Prior to the age of telegraphy, the information-action ratio was sufficiently close so that most people had a sense of being able to control some of the contingencies in their lives. What people knew had action-value. In the information world created by telegraphy, this sense of potency was lost, precisely because the whole world became the context for news. Everything became everyone's business. For the first time we were sent information which answered no question we had asked, and which, in any case, did not permit the right of reply.
>
> Thus, the telegraph's contribution was "to dignify irrelevance and amplify impotence."[58]

It is not a new idea. The London *Spectator* had put the argument even more eloquently and succinctly nearly a century earlier: "[With] the recording of every event, and especially every crime, everywhere without perceptible interval of time the world is for purposes of intelligence reduced to a village. . . . All men are compelled to think of all things, at the same time, on imperfect information, and with too little interval for reflection. . . . The constant diffusion of statements in snippets, the constant excitements of feeling unjustified by fact, the constant formation of hasty or erroneous opinions, must, in the end, one would think, deteriorate the intelligence of all to whom the telegraph appeals."[59]

The nineteenth-century American critic W. J. Stillman accused the telegraph of having ". . . transformed journalism from what it once was, the periodical expression of the thought of the time, the opportune record of the questions and answers of contemporary life, into an agency for collecting, condensing, and assimilating the trivialities of the entire human existence. In this chase for the day's accidents we still keep the lead, as in consequent neglect and oversight of what is permanent and therefore vital in its importance to the intellectual character."[60]

The early, uncritical enthusiasm for the telegraph had begun to seriously erode by the end of the century. "In place of the enormous faith invested in the telegraph by the earliest observers," notes Daniel Czitrom, "late nineteenth-century thinkers increasingly identified the telegraph and the modern newspaper as both symptom and cause of the frantic pace of industrial life." Eventually, "[t]he disturbing challenge of the periodical press to classical notions of culture began to elicit troubling doubts about the ultimate cultural import of modern communication."[61]

Late in our own century, the laying of cables beneath vast stretches of ocean and across continents has once again inspired extravagant hope and ebullient predictions. These modern cables are fiber optic lines of such vast capacity that they are making satellites obsolete for data transmission. Whereas copper submarine cables can carry only a few dozen telephone circuits, or about 2,500 kilobits of data per second (kbps), the new cables can carry 120,000 circuits or about eight gigabits per second (gbps), or about 3,200 times more data. Several of these high-capacity cables are either nearing completion or are well off the drawing board. FLAG (Fiber-optic Link Around the Globe), completed in 1997, stretches 28,000 kilometers from England to Japan, via Gibraltar and the Mediterranean Sea, Egypt, the Red Sea, India, Malaysia, China and Korea. A new fiber optic cable across the Atlantic is planned, along with a link called SEA-ME-WE 3 (Southeast Asia-Middle East-Western Europe #3). APCN (Asia-Pacific Cable Network) will link Japan, Malaysia, Thailand, Indonesia, Korea, Hong Kong and Taiwan. Older, second-generation fiber optic cables tie these new lines to North and South America.

The wiring of the world continues apace, and the hope that it will lead to better lives for the planet's inhabitants has not died.

CHAPTER SEVEN

The Invention of the Modern Inventor

The period of European and North American history that coincides roughly with the reign of the Morse telegraph from about 1860 to the outbreak of World War I in 1914, is one of exceptional interest to students of social history and technology. They have variously dubbed it the Age of Materialism, the Age of Invention and the Second Industrial Revolution. It is essentially a period in which the scientific discoveries of the previous two hundred years were exploited to produce practical devices for everyday use. These included an array of electric and electronic communications devices, as well as the electric incandescent light and the internal combustion engine (a lightweight, portable power plant which made the automobile possible). As well, enormous advances were made in chemistry and pharmaceuticals, steel-frame construction demonstrated in the Eiffel Tower made possible the first skyscrapers, iron bridges spanned impossible distances and railways transformed commerce and created nations.

It was an era of breathtaking developments in geopolitics as well. The great European imperial expansions were in full swing: between 1870 and 1900 Great Britain alone acquired about five million square miles of territory in the Eastern and Southern hemispheres. It was not the land the imperialists were after so much as the markets represented by the people who inhabited the land, as noted by J. D. Bernal in his *Science in History*:

> Already towards the end of the [eighteen] sixties the first, simple, optimistic phase of early capitalism was beginning to draw to an end. . . . The enormous productive forces liberated by the Industrial Revolution were by then beginning to present their owners with the problem of an ever larger disposable surplus. This could not, under capitalism, be returned to the workers who made it. When invested at home it led to even greater production and to a more hectic search all over the world for markets that were soon filled. The result was colonial expansion, minor wars, and preparation for the larger wars which were to come in the next century.[62]

The telegraph, like the telephone and many other technical achievements of the era, is a "modern" invention in that it owes its genesis to scientific research. Prior to Michael Faraday's work, virtually all technology had been the result of empirical discoveries by practical men searching for solutions to practical problems. Motives for innovation were thus as varied as the human condition, and not exclusively or even primarily economic in nature. It is argued, for instance, that the primary motivation for the widespread adoption of wind and water mills in the Middle Ages was the Christian view that menial labor requiring brute force to the exclusion of the intellect was inhumane, and inconsistent with the doctrine of the intrinsic value of all human beings. In other cases, innovation was suppressed despite economic logic: the Moslem world, for example, refused to adopt the printing press for several centuries after its introduction in Christian Europe. The reasons for this are obscure, but are presumed to have been rooted in a (justified) fear that widespread availability of books would undermine existing power structures.

From the pre-Christian era through to the mid-nineteenth century, technology and science had been distinct traditions. The intelligentsia of classical Greece shared a well-known prejudice against manual labor in its most literal interpretation: they relegated to lower social strata the farmer and miner, but also the artist and artisan and even the musician, all of whose work, being manual, was stigmatized by the taint of slavery. Indeed, it is a prejudice which has proved remarkably resilient, despite Jewish rabbinical and Christian teachings to the contrary, and the monastic idea of *laborare est orare,* work is worship. Historian Sir Desmond Lee has remarked:

> That the upper and controlling classes in Greece and Rome thought poorly of manual labour in the sense that they regarded it as a lower class occupation is undoubted. The texts speak clearly enough. . . . But I would ask in reply at what time in the world's history has the attitude of the upper and controlling classes been different? When have they commended manual labour as a way of life, except for other people or in sentimental pastoral?[63]

However, beginning in the nineteenth century with Faraday and men like him, a new relationship between theory and practice was forged: technology provided problems for science and science provided laws and rules that allowed technology to progress. A new community of practitioners evolved, people who had scientific and mathematical knowledge and who also had an intimate knowledge of technology. They came to be called engineers, not a new coinage but the revival of a term used sporadically since it had first appeared in twelfth-century Catalonia, in connection with specialists in siege engines and fortifications.

The age-old class structure that had relegated technology to the trades while stereotyping practitioners as trained, but not educated, broke down. By the mid-twentieth century, Arthur Koestler was able to write of the person who did not understand and appreciate technology as being every bit as much a "barbarian" as one who had no interest in or appreciation of art. For technology, he observed, is after all a human product, growing not only out of humanity's need to improve its material lot but also out of its love of play and adventure.[64] Morse and Vail, Marconi, Edison, Bell and Fleming are all exemplars of Koestler's insight into the soul of the gifted engineer.

Nevertheless, experience in our own century, particularly in its second half, has made it possible to question the extent to which modern technology does in fact represent humanity's highest aspirations and most sublime qualities, as opposed to our baser economic drives. In fact a convincing argument can be made that the conditions under which much late twentieth-century technical innovation has taken place has effectively eliminated all facets of motivation except the economic from the invention equation. Since this is an issue that will be raised explicitly and implicitly a number of times in succeeding chapters, it is worth a brief exploration here.

With the possible exceptions of Fleming, the inventor-entrepreneurs mentioned above also exemplify, particularly in their later careers, an important social and political phenomenon that is described eloquently in David F. Noble's *America by Design*:

> From the outset the engineer was in the service of capital and, not surprisingly, its laws to him were as natural as the laws of science. . . . The technical world of the engineer was little more than the scientific extension of capitalist enterprise; it was through his efforts that science was transformed into capital.[65]

Noble goes on to quote A. A. Potter, the founding dean of the faculty of engineering at Purdue University, as summarizing the relationship this way: "Whatever the numerator is in an engineering equation, the denominator is always a dollar mark."[66]

The rise of the modern engineer was fostered by and concurrent with the rise of the great industrial corporations which, by exploiting the fecundity of the wedding of science to craft, would come to dominate economic and, thus, social existence in the industrial nations of the world. Prior to this, practical arts had been the helpmate of science; following the marriage, the reverse was true: science came to be more and more in the service of the practical arts, or engineering. This was a crucially important turning point in the history of technology, bringing into intimate contact, as it did, two quite different sets of motives and ideals.

Initially, the distinction in values was subtle: on the one hand science wished to reach an understanding of the universe for its own sake; on the other hand, engineering sought to apply that understanding for the happiness and well-being of humanity. Who could be blamed for seeing it as a marriage made in heaven? However, as the nineteenth century wore into the twentieth and the power of the technology-based corporation grew, a disturbing gulf arose between the objectives of the partners. Science's goals remained essentially the same (though engineering increasingly encroached on its territory), but engineering's goals merged with those of the corporation, which were, by definition, to create a profit through the provision of goods and services. It can be argued, and often is, that this science-engineering-corporate nexus is the most effective system yet devised for providing widespread material well-being; nevertheless, it has the effect of making social goals subservient to corporate goals. Throughout most of history prior to World War I, this would have been regarded as an intolerable affront to civilized values. Today, it is widely acquiesced to by a society conditioned to accept it as the natural and inevitable order of things.

The rise of the industrial corporation with its phalanxes of professional engineers also brought with it the evolution of the great industrial laboratories, to which so many of this century's technological advances are owed. The General Electric Lab opened in 1901 with 8 staff; in 1902 there were 102, and by 1932, when Irving Langmuir brought it its first Nobel prize[67] it employed more than 600 researchers. In 1907 several Bell Telephone research establishments were amalgamated with Western Electric; in 1911 the famous Bell Labs were established, with a mandate to do fundamental scientific research. By 1925 more than 3,600 people were employed there, and it was the largest and best-equipped research facility in the United States, including those of the best universities. It won the first in a long series of Nobel prizes in 1937. Similar research centers were established in the oil, automotive, steel and chemical industries. In 1920 there were 526 American companies boasting research facilities; in 1983 there were more than 11,000.

The indispensable David Noble describes the difference between the new industrial and the traditional university laboratory this way:

> Whereas the university researcher was relatively free to chart his own paths and define his own problems (however meager his resources), the industrial researcher was more commonly a soldier under management command, participating with others in a collective attack on scientific truth. [The role of the scientist

> within a large industrial lab] came more and more to resemble that of the workmen on the production line and science became essentially a management problem.[68]

In addition, corporations increasingly played a leading role in financing and in other ways molding and shaping the engineering faculties at major universities, especially in the United States. From the industrial corporation's point of view, engineering schools were providers of what we have lately come to refer to as "human resources," and corporations were at pains to see that students were trained according to corporate requirements. The very nature of university education was shifted in the direction of corporate priorities.

The corporatization of technical innovation did not eliminate the freelance inventor. A study of seventy key inventions of the first half of the twentieth century shows that about half of them came from independent inventors.[69] These include bakelite plastic, the automatic transmission, the ballpoint pen, cellophane, the cyclotron, insulin, the gyrocompass, the jet engine, color photographic film, xerography, the zipper and the safety razor. But the momentum in favor of the corporate lab had nevertheless been firmly established.

Paradoxically, perhaps, technology throughout this era of materialism and the rise of corporate capitalism was seen increasingly as holding the key to the realization of the humanist aspirations of Victorian civilization: modern technology, it was argued, had ended slavery (by making industrial processes using wage labor cheaper than maintaining slaves); technology had elevated the status of women and children, rescuing them from the "dark Satanic mills" of early industrialism. Technology had provided the means to banish hunger, made social welfare a reality, linked the peoples of the world in instantaneous communication. Might it not be expected to eliminate such remaining evils as warfare and prejudice and inequality as well?

But there was a dry rot at work. In the United States, Britain and most other industrial countries, the patent system had long been the key to encouraging innovation. It was designed to guarantee that the inventor of a new product or process would be the first to benefit financially from its production, and it was available only to individuals: corporations need not apply. It also allowed the inventor to release details of his or her creation to the public without fear of having it stolen, thus, in theory, further accelerating the process of innovation. It had grown out of the granting of royal "letters patent" or monopolies on trade and industrial processes, initially with no foundation in law other than the authority of the monarch. As early as the seventeenth century in England, however, it had been regularized in statutes passed by Parliament, granting seventeen years of exclusivity to inventors who met the legislation's criteria. Abraham Lincoln described the patent process, in words later engraved over the entrance to the U.S. Patent Office in Washington, as adding "the fuel of interest to the fire of genius."

But as corporations got involved, they naturally acted to make the system work on their behalf; this inevitably meant that it worked increasingly against the interests of the lone inventor, who was the corporation's main competition in the field. Corporate managers were to develop a number of tactics, including the pooling of patents which they controlled through purchase from inventors or by virtue of the fact that the patent holders were corporate employees. In these cartel-like arrangements, companies that had been competitors but found themselves hobbled by constant patent litigation agreed to divide up the industrial landscape under complex and often secret contracts which eliminated strife between them and barred the door against new entrants. Corporations sought out and purchased any and all patents related to their field of interest, and then ferociously defended their patent rights in the courts. Here is how the first president of Bell Telephone candidly described the tactic:

> It appears to me that the policy of bringing suit for infringement on apparatus patents is an excellent one because it keeps the concerns which attempt opposition in a nervous and excited condition since they never know where the next attack may be made, and since it keeps them all the time changing their machines and causes them ultimately, in order that they may not be sued, to adopt inefficient forms of apparatus.[70]

L. H. Baekeland, one of the founders of the plastics industry and the inventor of bakelite, was equally blunt, if more judgmental, in his observation of the patents scene in 1909:

> [B]efore the courts, the poor inventor is entirely at the mercy of a legalized system of piracy. . . . This game is so successfully played that I know of rich corporations here in the U.S. whose main method of procedure is to frighten, bulldoze, and ruin financially the unfortunate inventor who happens to have a patent which he is not willing to concede to them on their own terms, which is to say, for next to nothing. . . . Thus has it come about that an otherwise liberal patent law intended for the protection of the poor inventor has become a drastic method for building up powerful privileges in the interest of big capitalistic combinations.[71]

Corporations also established "industrial research" departments, whose task was to follow closely patent applications and news of industrial developments, so that patents could be purchased or, if that were not possible, in-house inventors could quickly secure related patents with which the original inventor could be harassed in court. The Bell Telephone Company began its life in 1875 with two patents; in 1935 it held 9,225, acquired through purchase, mergers and in-house research.

As independent inventors found it more and more difficult to patent and commercially exploit their innovations, more and more of them chose to accept the proffered job opportunities with corporate R&D departments. There, they would have security and the satisfaction of working out their ideas with the best equipment and facilities, and help from corporate colleagues. But policy within AT&T, General Electric, Western Electric and most other research-based corporations forced them to sign over to the employer patent rights to their inventions, without compensation beyond their salaries.

Such patent reform as was from time to time adopted by the U.S. Congress tended to serve corporate interests rather than those of the lone inventor. According to Noble,

> [t]he successful reform efforts between 1900 and 1929 . . . brought the American patent system more closely into line with the needs of corporate industry. They set the basis for a "formalism" in the handling of patents which progressively eliminated the individual inventor who, unlike the large corporations with their well-staffed legal departments, was not equipped to cope with its intricacies and complexities.[72]

Once corporations became deeply involved in the process, invention was no longer strictly something that grew out of human needs. New consumer needs were instead created by inventions to promote corporate growth. Edison, for example, did not invent the incandescent bulb and then merely speculate on how it might be used. He built on a range of earlier scientific discoveries and technical advances to fabricate a practical bulb: that meant a high-voltage, low-current bulb, one which would meet the commercial requirements of a system of widespread electrical distribution as he envisaged it. The bulb had to be cheap, it had to last for many hours before burning out, and it had to operate in such a way as to minimize the cost of the electrical generation and distribution system behind it. Edison did not invent the electric light so much as he invented *electric lighting*. And he did it not so much to fill an obvious existing need as to create a product which would, through its very existence, *create* a need.

Of course, there would continue to be all manner of serendipitous by-products of this process, but the focus and direction of invention had been redefined and, one might say, rationalized. It was an enormously successful technique and it led to the creation of entire new industries. These

> arose in and helped to perpetuate a social climate where the possession of material goods was becoming a sign of status, where desires were replacing needs as a reason for the acquisition of products. The search for market expansion in the automobile industry was to lead Henry Ford first to adopt mass production (previously used in stockyards and at Sears Roebuck), then mass

> advertising, and then credit buying to broaden mass consumption. The auto industry is the archetype of the twentieth-century consumer industries: it was among the first to create a necessity out of a luxury and subsequently to alter the very fabric of society to perpetuate its own market.[73]

Once progress in technological invention became independent of the inspired individual and came to reside in the group processes of the industrial laboratory, incremental change became more or less continuous, and finally even predictable. The current state of the art in any technology pointed clearly to the direction of expected development, and, increasingly, time lines could be assigned in advance to future "breakthroughs" made by teams of engineers and scientists working in industrially-financed R&D labs. For this reason, it has been said that the real significance of the Age of Invention lies in the invention of the modern process of invention.

The negative side to all of this "progress" is too well known to bear much discussion: the depredations of the automobile, for instance, are many and well understood. The daily toll in traffic fatalities alone would have stunned the most jaded nineteenth-century industrialist. Television's damage to the social fabric seems indisputable, if unquantifiable. That the social costs of modern technologies have been, on occasion, great, is beyond argument. What may be puzzling is the degree to which society at large has allowed itself to be burdened with destructive side effects or "externalities" arising out of these technologies, such as the destruction of farmlands, pollution, violent social turmoil, chemical-induced illness, family breakdown and so on. The simple reason appears to be that the side effects were never the subject of public debate; until very recently, the process of technical development had managed to side-step the ordinary democratic processes on which all other social institutions were organized. As a creature of the market economy, technology was deemed to be free of any need for social supervision; like the rest of the economy, it was "democratized" through the agency of the market itself and the choices made by individual citizens acting en masse. It was a flawed argument: every other aspect of the market economy is in fact supervised to some greater or lesser degree by government, through legislation and regulatory bodies and the courts. The great mainstream of technological development had, however, managed to escape any sort of social scrutiny, thanks mainly to its impenetrable fortifications of scientific and engineering jargon and exclusive professional priesthoods of practitioners.

It seemed an unassailable power structure, liable to persist like the pyramids. But we'll see how social rather than corporate ideals did come to be served by the flurry of recent invention surrounding computer-mediated communications technologies—in spite of, rather than because of, the role played by corporations and their R&D labs and patent research offices. And we'll see how narrow corporate interests were effectively subverted by massive government financing of early Internet development, along with the anarchic behavior of the early software hackers who, as a matter of principle, made their most revolutionary products available to the world free of charge. Partly because of the power of new communications and information technologies, the corporate-engineering grip on the direction of technological development has been substantially weakened in recent years. It was a grasp that had maintained its stranglehold through the power of exclusive access to information and knowledge, and, like all such power structures, it is being eroded by the Internet's apparently irresistible capacity to distribute information equitably and to shatter artificial barriers to knowledge.

CHAPTER EIGHT

The Telephone

Everybody knows who invented the telephone, but there is a sense in which the invention of the telephone *system* might be attributed at least in part to somebody few people have heard of—Theodore Vail, who was neither scientist nor engineer, but a professional manager, the first general manager of the Bell Telephone Company. Alexander Graham Bell himself fit the traditional mold of the bohemian inventor who was content to leave further development of his new device to others. It was an archetype he helped make obsolete through the success of his company and the leading role it would play in the corporatization of innovation. As we shall see, Vail provided the second half of the modern invention equation, the commercial half, by understanding that the telephone was of commercial value only insofar as it was part of a network.

History, though, has granted to Bell the undisputed title of "inventor" of the telephone itself—if only by a whisker. As if in a photo finish at an Olympic event, Bell's agent had hardly had time to gather his papers and leave the U.S. Patent Office on February 14, 1876, when a law clerk for Elisha Gray puffed up the stone steps, clutching the portly, graying inventor's sketches and specifications for an almost identical device which Gray had concocted independently while employed as chief engineer for Western Electric, the research and manufacturing arm of the Western Union telegraph company. Gray was philosophical: he wrote to his patent attorney, "[t]he talking telegraph is a beautiful thing in a scientific point of view. . . . But if you look at it in a business light it is of no importance. We can do more . . . with a wire now than with that method. . . . This is the verdict of practical telegraph men."[74] Having missed out on "the most valuable patent ever issued" by an excruciating two hours, Gray went on to take out fifty others during his long career, while Bell had the dubious pleasure of defending his own through some sixty lawsuits between 1879 and 1897. Most significant among these would be the epic patent fight with Western Union.

The two experimenters had the insight that led to the telephone at virtually the same time. Both were trying to solve the commercially important question of how to squeeze more than one signal down a telegraph wire, Bell as a freelance inventor and Gray as an acknowledged authority in the practical applications of electricity. Both were testing transmission techniques which allowed vibrations of different frequencies to be sent down the line simultaneously and separated at the receiving end by devices tuned to respond to those frequencies. It occurred to each of them that it ought to be possible to transmit enough different frequencies to replicate the human voice. What was needed was a device that would increase or decrease the amount of electricity sent down the transmitting wire in exact correspondence with the variations in volume and tone of the voice. Bell won the development race, but only just, and probably because of his background in speech pathology and a trainer of the deaf. Gray, an established inventor and electrical experimenter of long standing, approached the problem as one of finding a way to get electricity to carry sounds imitating speech; he saw the telephone as an extension of telegraphy. Bell, from his background in teaching the deaf to speak, saw the problem as getting an electrical apparatus to imitate human physiology. He had in fact done his most productive work on the telephone receiver only after a medical friend had given him a complete human ear, removed from a cadaver, and he was able to observe how sensitive was its operation. He once said, "Had I known more about electricity, and less about sound, I would never have invented the telephone." Whereas Gray saw the telephone in terms of its potential for improving commercial and industrial productivity, Bell saw it as an extension of man.

It took a year of solid work after his patent was issued for Bell to get his invention to work. His first success was achieved using a copy of Elisha Gray's device, which used the variable resistance of an acid solution to convert sound vibrations into electrical impulses.[75] His famous words to collaborator Thomas Watson on March 10, 1876, "Watson, come here. I want you," were uttered because he had spilled some of the acid on his clothes and worktable. It would be several more months before he could get his own invention, which used a vibrating metal diaphragm and an electromagnet, to work. But it proved to be a far superior instrument. The microphone, or sender, and receiver, or earpiece, were identical; in fact, early telephones used the same piece of equipment for both talking and listening and subscribers were alerted not to "talk with the ear and listen with the mouth." What made it possible was Bell's discovery of how little electrical energy was required for the system to transmit speech. The device used a thin metal disk placed close to an electromagnet: the vibrations of the plate when struck by sound waves were enough to induce a tiny electric current in the wire coils of the magnet, which was then transmitted along wires to the receiving device. The receiver's electromagnet, reacting to the fluctuating current flowing through it, acted on the metal disk causing it to duplicate the vibrations of the sending disk. The telephone, thus, "spoke."

Just two months after his first laboratory successes, Bell demonstrated his invention at the Philadelphia Centennial Exposition. There, the prestigious judging committee would likely have overlooked his modest booth among the scores of scientific displays (including the first electric light bulb, the first grain-binder and Elisha Gray's multiwavelength telegraph) had not Dom Pedro de Alcatrana, the second emperor of Brazil, appeared on the scene with his wife and entourage in tow. Dom Pedro recognized Bell, with whom he shared an interest in teaching the deaf, and paused to greet him. He was persuaded to listen to a demonstration, and was volubly astonished. "My God, it speaks!" he exclaimed. The judging committee members, caught up in the eddy of public and press interest created by the emperor, had little choice but to politely follow suit and try Bell's invention. Among its members were such luminaries as Joseph Henry, the foremost American physicist, and William Thomson, later to be Lord Kelvin, perhaps the world's leading authority on electricity and the engineer for the first transatlantic telegraph cable. Said Thomson: "It does speak! It is the most wonderful thing I have seen in America!" And Henry said: "This comes nearer to overthrowing the doctrine of the conservation of energy than anything I ever saw!" Reporters furiously scribbled notes.

For the remaining weeks of the exhibition, the telephone was its star attraction. As with the telegraph, people found it impossible to believe the telephone could work until they had actually tried it. So remote was it from their everyday experience that they simply could not take it in. Some who saw it work thought there must be a hole through the wire along which sound waves passed.

Buoyed by the Philadelphia success, Bell and his associate and collaborator Thomas Watson formed the Bell Telephone Association in partnership with two financial backers, Gardner Hubbard and Thomas Sanders. The group had great difficulty raising capital and in 1877, in desperation, they tried to sell rights to the invention to the Western Union telegraph company for $100,000. Just a decade after the great Alaska–Siberia telegraph debacle, Western Union president William Orton made the stupendous blunder of turning the offer down. Other potential investors shared Orton's apparently low opinion of the commercial promise of Bell's invention[76]: the fledgling company was starved for capital and in dire straits. Sanders, a shoe manufacturer and the only member of the quartet with any money to speak of, pushed his credit to the limit to keep the venture afloat, more out of gratitude for Bell's having taught his deaf son how to speak than out of any business sense. Hubbard, the group's untiring promotion maestro, was also attached to the project as much by sentiment as anything else: his daughter, also deaf, was Bell's fiancée and future wife. Fewer than eight hundred telephones had been sold, mostly in pairs to businessmen with special communications needs.

To make matters worse, Western Union chose this moment to change its mind about the telephone's commercial potential. The company had been supplying business customers with sophisticated printing telegraphs and automated sending devices which could transmit sixty words a minute. They were seen as the ultimate in business communications technology. But a Western Union subsidiary reported to head office that, despite these improvements, several of its customers had recently switched to Bell's telephone. That galvanized the telegraph giant into action. It formed a telephone company of its own, seeded it with $300,000 in capital and appointed Thomas Edison, Elisha Gray and a third inventor, Professor Amos Dolbear, to its engineering staff. It announced that it had "the only original telephone," invented by its three engineers, and that it was ready to supply "superior telephones with all the latest improvements." It was a thinly veiled threat that it was prepared to mount a protracted and expensive court challenge to Bell's patent if need be, a tactic typical of the freewheeling, brawling practices of the robber-baron era in American business. Western Union was at the time controlled by one of the legends of the breed, William H. Vanderbilt, who summed up the prevailing ethic with the famous quip: "The public be damned. I am working for my stockholders." Vastly experienced in the takeover or elimination of competing businesses, Western Union expected a quick success against the puny Bell Association.

Edison had in fact invented a single-purpose transmitter that was markedly superior to the dual-purpose transmitter-receiver used in the tiny Bell system. It was essentially the same device as is used in today's telephones: a microphone containing a cylindrical "button" of carbon particles which, when compressed by sound waves, provide a reduced resistance to an electric current flowing through them. The electric current varies, in other words, with the sound waves produced by the speaker's voice; this variable current is then applied to the electromagnet within the metal-disk receiver at the other end of the conversation, causing the disk to vibrate in sympathy with the speaker's voice.

Edison's carbon microphone thus required something Bell's telephone did not: electric current from a battery. The strength or amplitude of the electrical current flowing through the wires to the receiver was not dependent on the loudness of the voice using the transmitter as it was in Bell's telephone, but on the battery current. The sound waves produced by the voice simply modulated the battery current as it flowed through the carbon particles.[77] In compensation for this added complexity, Edison's microphone provided the enormous advantage of acting as an amplifier of sound waves, which meant that voices could be transmitted more clearly over greater distances. Western Electric's first test of the device took place over 160 miles of existing telegraph line and was a complete success.

The Edison microphone was an enormous competitive asset for Western Union. Bell's customers soon began demanding a transmitter as effective as Edison's. The situation facing the beleaguered company was now so bleak as to seem hopeless.

Then, the unexpected happened. The Western Union strategy boomeranged. The very fact that the great telegraph conglomerate was jumping into the telephone business legitimized the industry overnight and gave retrospective credence to all the promotion work done by Bell and Hubbard over the past year, demonstrating the invention in auditoriums and church basements all over the eastern United States. Very soon, the delighted Sanders found himself leasing telephones at the rate of a thousand a month. Half a dozen wealthy businessmen put up $50,000 in capital. And, in a moment of inspiration, the company hired Theodore Vail as its general manager.

Vail would become one of the legends of American business management, both for his masterful piloting of Bell Telephone and for his management philosophy. He was exactly the right man for the job. His family had operated the historic Speedwell Iron Works in Morristown, New Jersey, for generations. And, small world! Alfred Vail, the indispensable associate of Samuel Morse, was his cousin—young Theodore had grown up amid telegraph lore on the Vail estate where Morse had lived in genteel poverty for several years. Most importantly, Theodore Vail had been a department head in the U.S. Postal Service, where he was responsible for implementing notable efficiencies. Telephone historian Herbert Casson noted that "by virtue of his position [in the Postal Service] he was the one man in the United States who had a comprehensive view of all railways and telegraphs. He was much more apt, consequently, than other men to develop the idea of a national telephone system"[78]—which is exactly what he did. What made Vail a uniquely valuable leader was his understanding of the importance of networks. As a consequence, he focused on making the company a nationwide institution, with standardized equipment and practices, and rates designed to encourage system expansion: the value of the telephone, he knew, resided in the network rather than the appliance. It may seem a commonplace observation today, but it was a rare insight at the time, when communications networks were in their infancy.

With a conventional economic product—for example, gasoline—if the customer base expands by ten percent, then sales can be expected to increase by ten percent. The relationship is linear. But in a network, if the customer base expands by ten percent, the value of the network *to each and every subscriber* increases at the same time, and network traffic can be expected to increase accordingly. Revenues expand at a rate much greater than in a linear relationship.

It was with the network in mind that Vail continued the early Bell policy of leasing rather than selling telephone equipment. For the network to function, technical standards needed to be uniform across the system; incompatible equipment would lead to inefficiencies and breakdowns. Bell had a choice: it could publish its technical criteria for network equipment and allow competing companies to sell appliances that would meet those standards, or it could restrict use of the network to Bell equipment and preserve the standard in that way. The first option would have been, in modern terms, an *open* network architecture; the second one, the one chosen, was a *closed* architecture. These two approaches to maintaining uniform standards across a network are much in evidence in the modern world of computer communication. Each has its advocates, although proponents of open systems can point to indisputable evidence of increased innovation and more rapid technical development due to the benefits of competition. The explosive development of the Internet has been assisted by the fact that it is a system with open technical specifications.

Prior to Vail's network insights becoming widely accepted, the manifest destiny of the telephone as a home appliance was not immediately clear, even to many of its backers. At first it was believed its main market was limited to business and commerce, it being too expensive for the average householder. This was particularly true in most of Europe, where Sweden was the sole exception that proved the rule. Britain, France, Germany, Russia and Austria-Hungary all lagged far behind the United

States and Canada in adopting the telephone, and suffered the consequences in business efficiency and competitiveness as a result.

For a time, telephone promoters experimented with various broadcast applications for the device. As late as 1890 a vice-president of AT&T described "a scheme which we now have on foot, which looks to providing music on tap at certain times every day, especially at meal times. The scheme is to have a fine band perform the choicest music, gather up the sound waves, and distribute them to any number of subscribers."[79] In London, Paris and Budapest, telephone technology was actually used for wired broadcasting from the 1890s until displaced by radio broadcasting. In Budapest, Theodore Puskas's six-thousand-subscriber "telephonic newspaper" network broadcast to paying subscribers fourteen hours a day with a mixture of stock market reports, news, music and drama, which sounds remarkably familiar to a modern ear. It remained in business until it was merged into the new Hungarian radio broadcasting organization in 1925 and became merely a "cable" delivery system for radio broadcasts. Historian Carolyn Marvin notes: "Photographs and illustrated advertising posters show that subscribers listened to the [telephone broadcasting service] through two small round earpieces hanging from a diamond-shaped board mounted on the wall. The audience for which the service was intended apparently possessed wealth, education and leisure. Its cultural relaxations were those of the opera and the theater. Its attachment to sport was aristocratic."[80] A copycat service set up in Newark, New Jersey, in 1911 appears to have been a victim of its own success: demand for subscriptions far outstretched the company's financial abilities to supply equipment, and it soon went bankrupt.

A number of telephone companies on both sides of the Atlantic offered either news or scheduled musical entertainment as an incentive to new subscribers, and there were many experiments in broadcasting concerts, particularly operas, over the wires. Church services were another frequent subject of experimental broadcasts. In Woodstock, Ontario, in 1890, an enterprising pub owner anticipated Court TV of one hundred years later by placing telephone microphones in a local courtroom for a notorious murder trial of the day. There were twenty receivers in the tavern, which patrons could rent for twenty-five cents an hour; four more were available in a private room for ladies.

The failure of wired broadcasting to put down roots may be attributed at least in part to technical reasons. An early problem with the system was the steady weakening of the audio signal as more and more subscribers signed on. By the time suitable amplifiers were developed, radio was a competing reality which afforded the key advantages of reception beyond wire lines and true portability. Nevertheless, it would be an oversight not to list lack of vision among telephone company executives as one of the reasons as well. They seemed wholly preoccupied with enlisting subscribers and fighting their court battles.

It was in January 1878 that the first automated telephone switchboard went into operation, in New Haven, Connecticut. The facts of its origins are somewhat comical: it was started by an irate undertaker who was convinced the local telephone operator was turning his business away to a competitor. His automated service made her redundant. Its success was an event of great significance, for the switchboard provided the means by which the telephone could be fully exploited. It meant that the telephone network would become an *addressable system*, that every telephone could be linked to any other telephone, that users could choose where they wanted to be when they made a call. It was the switchboard that made the realization of Vail's network vision possible and which granted users an entirely new freedom to move from place to place while still remaining in social contact. The telephone switchboard was the system's linchpin that made the world a noticeably more secure and predictable place.

Late in the same year of 1878, Western Union went to court claiming Bell had infringed Elisha Gray's patent rights to the telephone, and the Bell Telephone Company, newly reorganized and recapitalized and with Vail at the helm, met the telegraph giant head-on with the country's best patent attorneys. The case carried on for a year, before ending with stunning suddenness when Western Union withdrew its suit and sought a negotiated settlement. It had become clear it could not win in court. Under the settlement, Western Union agreed to admit that Alexander Graham Bell was the sole inventor, that his patents were valid and to retire from the telephone business. Bell, for its part, agreed to buy the Western Union telephone system, pay Western Union a royalty on revenue from telephone rentals for several years and keep out of the telegraph business. A more brilliant victory for Bell could scarcely have been scripted.

An engineer who worked at Edison's Menlo Park research establishment during the period, Francis Jehl, summed up the consequences succinctly in his memoirs: "The Bell instrument is a delicate hearer, while the Edison transmitter possesses a good lung and speaks loudly, so that each does its share of the work, and, together, they have formed an ideal combination and have made the telephone a universal success."[81]

With the settlement, Bell shares soared to $1,000. The American Bell Telephone company was organized, capitalized at $6 million. In 1882, the company doubled in size and gross earnings reached $1 million. Bell, Watson, Sanders and Hubbard cashed in their chips to become wealthy men, leaving further commercial development to others. Bell went on to explorations in aircraft and hydrofoils and a dozen other areas at his Cape Breton retreat; Watson founded a successful shipbuilding venture; Sanders lost most of his money in a Colorado gold mine venture; Hubbard became a founder of the National Geographic Society.

Three years after the settlement, in 1882, American Bell purchased Western Union's research and manufacturing arm, Western Electric. In 1907, Bell's new corporate parent, American Telephone and Telegraph Company (AT&T) coolly wrote a check for $30 million to purchase control of Western Union itself. AT&T would go on to become the world's largest corporation. By 1900, it already had twice as many miles of line as Western Union (five times as many by 1905), and it had installed well over a million telephones in the United States and Canada. Another million had been installed by independent phone companies. So numerous were telephones that the proliferation of overhead wires in cities was becoming a serious blight, with dense forests of eighty- to ninety-foot poles carrying thirty or more cross arms and three hundred wires, and major engineering efforts were devoted to finding the right kind of cable for underground burial. For decades, many North American cities looked as if they'd grown a grotesque system of nervous ganglia on the outside of their skins.

With local urban markets quickly reaching saturation, long-distance telephony became a new growth area, though it took some imaginative marketing to get the public used to the idea; the telephone had not been seen initially as direct competition to the telegraph for intercity communication. In New York, a special long-distance *salon* was set up at Bell headquarters: customers were picked up in cabs and escorted over Oriental carpets to a silk-draped booth, where their calls were placed. Transcontinental calls would have to await the development of the vacuum-tube amplifier as a by-product of radio; transoceanic telephony by wire rather than radio would become possible only at mid-century with further improvements in electronics.

Transmission over even moderate distances involved two additional problems beyond the straightforward amplification hurdle: distortion and interference on the one hand and, on the other, the need to multiplex signals, put more of them down the same wire, so as to get maximum commercial value from very expensive long-distance cables. Telegraphy shared these difficulties, though on a less complex level due to the relative simplicity of the signals being transmitted.

Answers did not come easily. The early grounded-circuit telephones were plagued with baffling noises. Casson provides a colorful description:

> This was, perhaps, the most weird and mystifying of all telephone problems. . . . Such a jangle of meaningless noises had never been heard by human ears. There were spluttering and bubbling, jerking and rasping, whistling and screaming. There was the rustling of leaves, the croaking of frogs, the hissing of steam, the flapping of birds' wings. There were clicks from telegraph wires, scraps of talk from other telephones, and curious little squeals that were unlike any known sound. The lines running east and west were noisier than the lines running north and south. The night was noisier than the day, and at the ghostly hour of midnight, for what strange reason no one knows, the babel was at its height. Watson, who had a fanciful mind, suggested that perhaps these sounds were signals from the inhabitants of Mars or some other sociable planet.[82]. . . "We were ashamed to present our bills," said A. A. Adee one of the first agents, "for no matter how plainly a man talked into his telephone, his language was apt to sound like Choctaw at the other end of the line."[83]

The answer to the worst of the early noise problem turned out to be getting rid of the ground return system and instead using a two-wire connection. This (very expensive) conversion was begun in 1883. To help defray some of the cost of doubling up on circuits, a method of "ghosting" a third conversation on two lines was developed.[84]

The real solutions, however, would have to wait for the middle of our own century and the introduction of electronic amplification and pulse code modulation. Pulse code modulation reduced analogue (wave) signals to digital, binary code, the simplest possible (and therefore least error-prone) signal configuration—off-on, or 0-1. This was done with elegant simplicity by sampling the analogue signal wave at discrete intervals, assigning a binary code to the amplitude of the wave at that point and transmitting that code as a stream of 1s and 0s. At the receiving end the binary information was decoded and translated back into an analogue waveform duplicating the original. It was a relatively simple matter to identify and eliminate electrical interference from a binary message, and replication of the signal by the many amplifier/repeater stations along the route of telegraph lines no longer involved the introduction of error or distortion as it had with the more complex analogue waveform. Binary information could be replicated ad infinitum with little or no error.

Pulse code modulation meant that several telephone conversations could be carried simultaneously on a single line. This was accomplished by interleaving packets of binary information in the interstices between wave samplings and in pauses in conversation. The first digital telephone transmission was recorded by Bell Labs in 1956, and the first commercial installations were made by Illinois Bell in Chicago in 1962. The system was called T1 and could carry twenty-four voice signals or 1.5 megabits of information over a standard pair of wires.

Now operating in digital mode, telegraph and telephone communication were ready for wholesale adoption of the ultimate digital machine, the electronic computer. Introduced initially for improved automatic error (distortion) correction, buffering and switching, the digital computer would swiftly become the mainstay and workhorse of both systems. Increasingly, the telephone and telegraph systems which shared so much history and so many of the same technologies were being melded together as the telecommunications industry.

The first entirely electronic telephone switching systems were installed in the United States in 1976. Special-purpose computers, they could handle half a million calls an hour, as well as special services including toll-free 1-800 numbers and call forwarding. By 1980, more than half of all calls in North America were being switched electronically. A decade later the AT&T system could boast of handling sixty-one billion calls a year on its 2.75 billion circuit miles of facilities, a system that was now entirely digital from end to end.

CHAPTER NINE

A Convivial Technology

Of all the significant new technologies of the late nineteenth and twentieth centuries—the telegraph, the automobile, the jet aircraft, nuclear energy, television (the list could go on)—the telephone shares with electric lighting the unusual distinction of near-universal public approval. Large numbers of people devote vast amounts of time and energy to criticizing almost every other highly visible technology as perverse and dangerous in one way or another. There are remarkably few jeremiads of the telephone.[85] Why is this?

There is a clue to this mystery in the fact that as an interactive, point-to-point communication device, the telephone had two powerful advantages over that other destroyer of time and space, the telegraph. First, the telephone enabled users to reach one another directly. To send a telegram, by contrast, one had to rely on a series of intermediaries from telephone operators and order clerks to telegraph operators and messengers. The second, and more profound, difference was that the telephone put the electrical encoding device—the microphone—in the hands of the user. You did not have to know Morse code to use the phone. You did not need an operator's license, or special training, or special equipment, or "official" permission to use it. Ordinary speech did the trick.

Philosopher and social critic Ivan Illich includes the telephone in his list of modern technologies which foster self-realization, by enhancing "the ability of people to pursue their own goals in their unique way." Illich calls these "convivial" tools:

> Tools foster conviviality to the extent to which they can be easily used, by anybody, as often or as seldom as desired, for the accomplishment of a purpose chosen by the user. The use of such tools by one person does not restrain another from using them equally. They do not require previous certification [licensing] of the user. Their existence does not impose any obligation to use them. They allow the user to express his meaning in action.[86]

Most hand tools and simple mechanical devices, for instance, meet this definition, whereas nuclear power plants and modern, computerized automobiles do not. Convivial tools (and institutions) foster self-reliance and self-realization; "manipulative" tools foster dependency, passivity and alienation. But even simple tools can be made nonconvivial, or "manipulative" in Illich's terms, if they are restricted "through some institutional arrangements. They can be restricted by becoming the monopoly of one profession, as happens with dentist drills through the requirement of a license and with libraries or laboratories by placing them within schools."[87]

The conviviality of a tool really has nothing to do with its level of technological complexity. As we'll discover, radio fits the definition of a convivial tool that was made "manipulative" by institutional arrangements. The telephone, however, with its complex hi-tech network, has fit Illich's definition of a "convivial" or people-friendly technology from its introduction, and continues to do so. "Anybody can dial the person of his choice if he can afford a coin," notes Illich. "The telephone lets anybody say what he wants to the person of his choice; he can conduct business, express love, or pick a quarrel. It is impossible for bureaucrats to define what people say to each other on the phone, even though they can interfere with—or protect—the privacy of their exchange."[88]

If one were to rate communications technologies according to their conviviality, the telephone would rank near the top of the list, right after the alphabet, and the printing press, which Illich has called "almost ideally convivial. Almost anybody can learn to use them [printing presses], and for his own purpose. They use cheap materials. People can take them or leave them as they wish. They are not easily controlled by third parties. . . . The alphabet and the printing press have in principle deprofessionalized the recorded word."[89]

In the same way, the telephone "deprofessionalized" rapid long-distance communication. The telegraph (and radiotelegraph) required trained and licensed intermediaries—telegraphers—to interpret messages: telegraphers were the modern-day equivalent of the ancient Egyptian scribe. Two-way radio communication has long been restricted to those who can obtain the requisite government licenses. Only the telephone, among these technologies, is truly people-friendly. Not only is it cheap and simple to use, but it gives the user exactly what he or she wants of it, rather than what some third party wants the user to get. By this definition, the Internet would also qualify as convivial.

Put another way, the telephone permits unmediated conversation. Given the social importance of conversation or the sharing of person-to-person communication to the species from prehominid days onward, it is not difficult to see why such a useful extension of this fundamental means of social intercourse would be warmly welcomed.

At yet another level, people appreciate the telephone almost unreservedly because it is a tool that works for *them*, when so many other technologies require them to work for the machine. It is not necessary to alter ordinary behavior in any way in order to make use of the telephone; one simply uses it when it's needed. "Innovations" such as voice mail and telemarketing, which tend to reverse this equation, are disliked and reviled, with justification.

It should be acknowledged here that, when it was first introduced, the telephone did have its share of critics, especially in Britain, where it was feared that it would promote the breakdown of barriers between the classes. After all, if anybody could have a telephone, the lady or gentleman of the manor would have *no idea* who might be on the other end of the line when he or she answered the incessant ring, much less what the caller might say. Of course, the solution was to not have a telephone, or to keep it safely tucked away in the servant's quarters, and the spread of the telephone in Britain was significantly slower than in North America in part for this reason. In America, the class argument was discreetly framed in terms of unnecessary invasions of privacy, which proved a considerably weaker deterrent to adoption of the device.

All of this serves to highlight the fact that the telephone was to become (and remains) a powerful democratizing influence. It accomplished this, albeit unconsciously, by the stratagem of having all telephone users *subscribe* to the system. It was this fact of common subscription to a shared system that was the equalizing factor. The system treated everyone the same: nobody's voice, nobody's ring, was louder than anybody else's on the telephone network. The appliances themselves, thanks to the system's demands for technical uniformity, were pretty much alike, whether they were to be found in a palace or a tenement. British communications theorist Colin Cherry observed:

> [W]e are so accustomed to such ideas today that we may forget that such liberties of approach by one person to another, irrespective of rank, did not exist in such egalitarian forms in the nineteenth century. Letters of introduction might have been needed, for example. The telephone service, at one stroke, introduced this totally new concept—that persons of all ranks could be members of the same subscriber organization, with common rights, rights of "membership."[90]

In the early 1970s, the telephone industry attracted its share of criticism from the antiestablishment counterculture in North America. But it was the establishment and not the technology that came under fire. The technology, in fact, was a source of deep fascination for an emerging subculture of technological whiz-kids known as "phone hackers" or "phone phreaks." Homemade "blue boxes" which could manipulate the long-distance phone network by imitating its coded audio signals were distributed widely throughout North America. They allowed the user to make long-distance calls without paying for them. Phone phreaks took pleasure in manipulating the system for its own sake, in bouncing a call through the network, off satellites, across oceanic cables and perhaps back to the phone in the next room. It was the computer switching and the network system that intrigued them: few were in it for the money, although it is estimated by AT&T that $20 million in free calls were made using blue boxes in each year of the 1970s, until the company upgraded its security measures, ending the fad. Many phone phreaks found themselves deeply involved in the computer industry as it, and they, matured during the seventies and eighties. Two well-known blue-box manufacturers, remembered for the high quality of the products they sold around the San Francisco Bay area, were Steve Jobs and Steve Wozniak, who went on to found Apple Computer.

In terms of its impact on business, the telephone extended and amplified the impact of the telegraph in creating global markets. But Marshall McLuhan has argued that it also had an important impact on organizational structures. The telephone introduced

> "a seamless web" of interlaced patterns of management and decision-making. It is not feasible to exercise delegated authority by telephone. The pyramidal structure of job-division and description and delegated powers cannot withstand the speed of the phone to bypass all hierarchical arrangements, and to involve people in depth.[91]

One of course challenges McLuhan at one's peril; however, there would appear to be good evidence that the impact of the telephone on organizational structures was in many cases just the opposite. While it permitted geographic decentralization, it encouraged organizational pyramid building. Urban police forces, for instance, rebuilt their operations around the telephone call box. Police officers, who had previously patrolled streets more or less autonomously to prevent crime and occasionally to apprehend suspects, were by 1917–18 more likely to be assigned to squad cars stationed at call boxes, where they would wait for an order from the police dispatcher to respond to a crime in progress or already committed. The telephone centralized authority and control at the precinct office rather than on the street and moved police away from crime prevention to a more purely reactive position. Steven Lubar writes, in *Infoculture*:

> Many businesses used [the telephone] to centralize operations, a process already started by the telegraph and by other trends in American management. Salesmen were required to report in every day and get instructions. The telephone, *Telephony* magazine opined in 1906, "has curtailed the functions and responsibilities of a district manager as the cable has those of an ambassador." The phone made it easier to separate manufacturing plants from the offices where company sales and management staff worked, helping to create downtown office districts.[92]

Telephone historian Herbert Casson claims that, in the early days in which he was an observer, the phone also had a salutary effect on manners. Speaking of the telephone operator, invariably a woman, Casson says:

> To give the young lady her due, we must acknowledge that she has done more than any other person to introduce courtesy into the business world. She has done most to abolish the old-time roughness and vulgarity. . . . She has shown how to take the friction out of conversation, and taught us refinements of politeness which were rare even among the Beau Brummels of pre-telephonic days. . . . This propaganda of politeness has gone so far that today the man who is profane or abusive at the telephone, is cut off from the use of it. He is cast out as unfit for a telephone-using community.[93]

If there is anything we don't like about the telephone, it is its insistent nature. McLuhan points us to what must be one of the most bizarre demonstrations on record of this aspect of the phone, as described in the pages of the *New York Times*:

> On September 6, 1949, a psychotic veteran, Howard R. Unruh, in a mad rampage killed thirteen people, and then returned home. Emergency crews, bringing up machine guns, shotguns and tear gas bombs, opened fire. At this point an editor of the *Camden Evening Courier* looked up Unruh's name in the telephone directory and called him. Unruh stopped firing and answered, "Hello."

"This Howard?"

"Yes . . ."

"Why are you killing people?"

"I don't know. I can't answer that yet. I'll have to talk to you later. I'm too busy now."[94]

Had the incident occurred thirty years later, it is likely the person greeting Unruh on the other end of the phone would have been a telemarketer or pollster. Perhaps then Unruh would have better understood his own rage!

The abuse of the private and personal quality of the telephone—the quality that makes its ringing so irresistible—is a serious issue and one that is increasingly being addressed by governments. It is perhaps paradoxical that the intimacy of the telephone should be a characteristic of what is a public space. But society has defined the telephone system as a public utility and the network as public space from the earliest stages of its development. Governments have either taken over responsibility for telephones outright, as in Britain and much of Europe, or they have granted closely regulated private monopolies, as in Canada and the United States, on the understanding that the space created by the network is public and the monopoly is there to serve the public interest. The 1902 government of Sir Wilfred Laurier in Canada, for example, passed legislation which required Bell Canada to provide service to any person within its monopoly territory "with all reasonable dispatch," and forbade changes in rates without government approval.

As we've already noted, the telephone industry was slow to realize that the device's most valuable application would be in personal communication, in other words, that they were creating a new form of public meeting place or public space with their links of wire. To the extent that they did realize it, they resisted the trend. Telephone companies everywhere initially did all they could to discourage "frivolous" chatter on their equipment, insisting that the resources must be used for legitimate business purposes, much like the telegraph. Women, in particular, were discouraged from using the phone, it being assumed that their conversations would be of small substance! But ordinary users prevailed in their superior though "nonexpert" understanding of the telephone's real utility, and by the 1920s phone companies had seen the handwriting on the wall and were beginning to encourage social uses of the appliance. And women at home with children continued to use it, as they had from the beginning, for the very useful purpose of keeping in touch with family and friends in the community.

Surprisingly few studies have been done of how people actually make use of the telephone in their daily, domestic lives. Those that have been done indicate that it has become an integral part of the social structure, whose value is difficult to measure in monetary terms. With the Australian telecommunications authority considering allowing a request from local telephone companies to charge on a fee-for-time basis as opposed to the existing flat rate for local calls, sociologist Ann Moyal was recently prompted to gather some research data on how Australian women use the phone, in the belief that "important social data should be added to the equation." Moyal found that the cross section of women studied made from two to six calls a week for "instrumental" purposes such as setting up appointments for themselves and their families, volunteer activities or in pursuit of information. These calls normally lasted from one to three minutes. On the other hand, the women made between fourteen and forty-two personal or "intrinsic" telephone calls a week, many of them long distance. Respondents confirmed "that the prime importance of the telephone in their daily life related to 'sustaining family relationships' and to their contact with children, parents and to a less regular extent, with siblings, grandchildren and other members of the family." One particular family relationship stood out above the others in terms of telephone time consumed: "A singular proportion of these calls and time is devoted to communication between mothers and daughters, who establish telephone contact daily or regularly throughout the week, and maintain an intimate and caring telephone relationship across their lives." Communication was particularly intense when the daughter was in a childbearing or childcaring mode.

The value of this kind of communication for social stability and quality of life would be difficult to overestimate, and yet it is precisely the kind of telephone use that has been derided over the years, even when it has not been actively discouraged by the telephone companies. Moyal calls it "a pervasive feminine culture of the telephone in which kinkeeping, nurturing, community support, and the caring culture of women forms a key dynamic of our society."[95] Legislators and regulators who act in ways which will have the effect of restricting this kind of communication do so at their peril, Moyal suggests.

The virtual space created by a telephone network, the space in which conversations take place and through which data flows on its way to its destination, has been variously called the "telecosm" and the "telesphere." The law in democratic societies treats it as sacrosanct, in the same way that the law treats real public space as affording inviolable communal rights to all citizens. The law also recognizes that there are limits to those rights, and responsibilities that come with them. Defacing public property is an offense; citizens are expected to put their litter in trash bins. No one has the right to tell me how to deport myself in a public space—unless what I'm doing prevents others from enjoying their rights to the space. The same or analogous rights and responsibilities exist in the virtual public space of the telephone network. Some of the affronts to the conviviality of the telephone system, such as voice mail and automated call answering and recorded advertising messages, are merely the equivalent of boorish behavior in real space. But others, such as unsolicited marketing calls and repeated calls from collection agencies, should be illegal (as the latter are, in some jurisdictions), since they present a threat to the continued, healthy functioning of the network. Just as persistent solicitation or threatening behavior causes people to avoid some urban public spaces, abuses of the telephone network result in more and more levels of interference between users, in the form of call-screening devices of various kinds. Just as real public space infested with annoying and threatening people can fester and die and be lost to the public, so are virtual spaces at risk, and in need of policing.

Nowadays we think instantly of blue-uniformed, gun-toting officers when we think of policing, but professional police forces are of course a recent innovation. Before the bobby appeared on London streets in the days of Wellington and Napoleon, policing public spaces was handled largely by the citizenry which owned them and used them. A little "self-policing" in the public space of the telephone network, in the form of letters to offenders and other lobbying activities, can go a long way toward keeping it a safe and healthy environment. It would also be appropriate in an era in which we are discovering that institutional responses to public problems are not always the best idea: their well-meaning ministrations can be more dangerous than the condition they aim to cure. We are all citizens of the telesphere, and we need to take our individual responsibilities seriously.

CHAPTER TEN

The Invention of Radio

With the invention of radio, we come to the appearance of a communications technology that has two unique defining characteristics, both subjects of fascination, one innately occurring and the other imposed artificially for complex political and economic reasons. The first of these characteristics is its ability to communicate information at great distances without wires or other material connections between sender and receiver. The second is its adoption as a medium of mass communication, wherein messages originating at a central transmission point are broadcast to many "dumb" or passive receivers which may be located anywhere within the transmitter's range.

The quality of "wirelessness" is one of great scientific and technological significance. The adoption of a unilateral or one-way model for radio as a public appliance, on the other hand, raises issues of historic social and political significance that spill into our own era, and speak directly to our relationship to the new computer-mediated communications technologies. Understanding radio and its history is the key to a comprehension of the major social and technical issues surrounding both television and the Internet. The next several chapters will therefore be devoted to this engrossing and illuminating saga.

We will begin with Guglielmo Marconi, though that is not, strictly speaking, the beginning of the story. The world celebrates Marconi as the "father of radio" not because he invented it, but because he recognized its potential as a communications technology. The phenomenon of radio waves and their creation by electricity, as he freely acknowledged, had been discovered by others. They were a cosmopolitan group.

In constructing experiments to test mathematical predictions published by James Clerk Maxwell in 1873, Heinrich Hertz found that waves of electrical energy did indeed emanate from an electrical spark, as Maxwell's equations had predicted, and that they spread in concentric circles through space, traveling at the speed of light. He published his findings in 1888, but he died before he was able to pursue the practical possibilities held out by this phenomenon.

Few people, in fact, seem to have grasped the real import of Hertz's discovery. There was one amazingly prescient exception: British scientist Sir William Crookes, writing in a contemporary magazine, saw "a new and astonishing world" emerging from Hertz's work. "Rays of light will not pierce through a wall, nor, as we know only too well, through a London fog. But the electrical vibrations of a yard or more in wave length . . . will easily pierce such mediums, which to them will be transparent. . . . Here, then, is revealed the bewildering possibility of telegraph without wires."[96]

By the time Marconi began his experiments in wireless telegraphy in 1894, the key discoveries and inventions out of which the technology would be built had all been made. It remained for the young inventor to put them together in a practical package, and he could hardly believe his good fortune at having been given that opportunity. "I could scarcely conceive," he said years afterward, "that it was possible that their [radio waves'] application to useful purposes could have escaped the notice of eminent scientists." But invention requires more than technical knowledge and imagination; it demands, as well, an ability to correctly anticipate how the invention might best be used, or, in Marconi's words, its "application to useful purposes."[97] Marconi's great insight, his masterpiece of creative thinking, was in seeing radio's potential value as an instrument with which it would be possible to communicate messages *to distant recipients whose position was unknown*. No other means of communication permitted this seemingly impossible feat.

Marconi himself perfectly defined the secret of his success as a pioneer in a speech he gave late in his life to the Royal Institution of Great Britain: "Long experience has . . . taught me not always to believe in the limitations indicated by purely theoretical considerations or even by calculations. They are, as we well know, often based on insufficient knowledge of all the relevant factors. I believe, in spite of adverse forecasts, in trying out new lines of research however unpromising they may seem at first sight."[98]

He had early seen the most exciting possibilities for his apparatus in providing ship-to-shore communication, a need that could not be served by either telegraph or telephone, and with that in mind he chose to develop his ideas in the world's greatest maritime nation, Great Britain. With the backing of his mother's Anglo-Irish family (the Jamesons of Irish whiskey fame), Marconi patented his wireless telegraph in England. The year was 1896 and he was just twenty-two.

Within three years of securing his patent, he was operating a handful of small coastal radio stations and had equipped several ships with radio gear. It was by most standards a thriving business, though overextended in the manner of many a start-up firm: by 1898 his Wireless Telegraph and Signal Company was capitalized at $500,000. But competition was snapping at his heels, in Germany from the principles of what would eventually become the Telefunken company, and in the persons of Reginald Fessenden and Lee De Forest, both of whom were making significant advances in radio technologies in the United States. Marconi's marketing instincts and his determination to maintain leadership in the fast-developing field led him to announce his intention to link North America to Europe by radio, a feat deemed by conventional wisdom to be absurd.

Since radio waves were known to travel in straight lines, and since the curvature of the earth makes it impossible to draw a line of sight between two points on the surface more than about one hundred miles apart, one hundred miles was, in the unanimous scientific opinion of the day, the limiting distance for practical radio communication. Between North America and England there was a mountain of ocean more than 160 miles high! Marconi, however, suspected that radio waves would travel along the earth's surface, following the curvature according to well-understood laws of diffraction of light. Indeed, he had conducted experiments in which he had received signals at twice the distance line-of-sight calculations told him were possible. He went to his board of directors with a bold proposition: he wanted funding released to build two stations, one on either side of the Atlantic Ocean.

Neither Marconi nor his scientific detractors were correct in terms of what we now know about long-range radio propagation. Radio waves do indeed travel in straight lines as the scientists said, but they also bounce off the underside of the ionized (electrically charged) layer high in the atmosphere known as the ionosphere, and are thus reflected back to earth far beyond the horizon. In some conditions they will be reflected back up to the ionosphere by the earth's surface, leading to transmissions of several "hops" covering globe-girdling distances.

Marconi's initial choice of location for his North American station was Cape Cod, Massachusetts. There, he had a replica of his new Poldhu, Cornwall, station built, near the village South Wellfleet. Before the tests could take place, the Poldhu antenna array blew down in a storm. A smaller, substitute antenna was quickly lashed together. Then the antenna at Cape Cod was demolished in a September gale; a temporary array was set up, a fan of wires strung between two, 150-foot masts. A November northeaster brought that down before it could be put into service.

Haunted by visions of being beaten to the punch by competitors in America or perhaps in Germany or elsewhere, the inventor hastily packed a few wicker trunks with wire, some kites and balloons and a selection of receiving devices and set out for Newfoundland, geographically the closest spot in North America to his Cornwall station and therefore, presumably, the easiest transatlantic hop for radio waves. There, he was warmly welcomed by Governor Cavendish Boyle and Prime Minister Sir Robert Bond, who gave him use of an abandoned military hospital on Signal Hill, a windswept promontory six hundred feet above St. John's harbor, looking straight out to sea in the direction of England. He could scarcely have found a more appropriate location for his experiment. Signal Hill had been used in communication since the sixteenth century, when a signal canon was placed there. During the Napoleonic Wars, signal flags had been flown there, warning ships at sea about navigation hazards and weather. When Marconi arrived, work had just been completed on Cabot Tower, a monument commemorating the four hundredth anniversary of explorer John Cabot's landing on the beaches below.

By December 9, Marconi and his two assistants had finished assembling their equipment in a ground floor room of the barracks, and Marconi cabled Poldhu to begin transmitting the letter *s* continuously between the hours of 11:00 A.M. and 3:00 P.M. from December 11.

Gale-force winds played havoc with kite-borne antennas lofted by Marconi's men, but then, just after noon on December 12, 1901, Marconi heard, very faintly but nevertheless distinctly, the tap, tap, tap of the letter *s* being repeated over and over. It could only be the station at Poldhu, two thousand miles away. His assistant, George Kemp, clapped on the headphones and confirmed what he had heard. Faint signals were also heard on the following day. Marconi cabled word of the success to London and then informed the press on December 14. The news caused a sensation that rippled around the globe.

On December 15, the *New York Times* began its story with: "Guglielmo Marconi announced tonight the most wonderful scientific development in modern times." But it caused consternation in the London offices of the Anglo-American Telegraph Company, which owned the enormously expensive England-to-Newfoundland undersea cable over which Marconi had sent his telegrams. Marconi's success confirmed their worst nightmares: the upstart new radio technology would begin competing with them for overseas traffic, with the huge operating advantage of not having to lay cable. The company immediately wired Marconi, ordering him to cease and desist, or face a lawsuit for damages on grounds that Anglo-American had monopoly rights to telegraphy in Newfoundland. The damage was real: Anglo-American's stock on the London exchange had plummeted steadily in the days following Marconi's announcement. At stake was a business that, by 1900, involved twelve transatlantic cables carrying about twenty-five million words a year at twenty-five cents a word.

Marconi greeted the threat with a dismissive laugh, but the governor and council of Newfoundland were suitably affronted on his behalf.

On December 20, the council passed a resolution stating: "The Council are much gratified at Signor Marconi's success, marking as it does, the dawn of a new era in transoceanic telegraphy, and deplore the action of the Anglo-American Company." A celebratory banquet was held in St. John's and the council visited the wireless station en masse in a show of solidarity.

It was a telling episode,[99] and it served to underline a watershed between nineteenth- and twentieth-century technology, between the monumental and imposing, and the small and ephemeral. No one has described its significance so well as Marconi's daughter Degna Marconi, in her delightful biography, *My Father, Marconi*:

> In the eyes of the world what Marconi had done with two balloons and six kites was magic, an occult modern mystery. The Field achievement [of the first transatlantic undersea cable] was by comparison pedestrian and cumbersome. He himself had written when he was taken aboard the *Great Eastern* to see the ship's crew through their heavy labours, towards the end of the cable-laying voyage, that there had been on board "ten bullocks, one milk cow, 144 sheep, 20 pigs, 29 geese, 14 turkeys, 500 fowls as live stock and dead stock in larger numbers, including 28 bullocks and eighteen thousand eggs." Nor was this all. When they dragged the ocean floor to bring up the "slimy monster," they had to use twenty miles of rope twisted with wires of steel in order to bear the strain imposed by thirty tons of cable. It took two hours even to lower it to the bottom. As against this, Marconi had instantaneously caught his intangible train of waves with a kite that could be packed in a wicker hamper and weighed a few ounces.[100]

In some scientific quarters, the news caused plain disbelief. Why was no independent witness present, just Marconi himself and a co-worker? How was it possible for radio waves to be heard so far beyond the horizon? Thomas Edison, asked for comment by a New York newspaper, said: "I do not believe that Marconi has succeeded as of yet. If it were true that he had accomplished his object I believe he would announce it himself over his own signature." When asked to comment on Edison's skepticism, Marconi told a reporter the signals were absolutely genuine, and that Governor Boyle, at Marconi's request, had cabled the fact to King Edward.

As recently as the 1980s (the last time anybody checked) there were still scientists and engineers who believed it was impossible for Marconi's Poldhu station, transmitting in daylight on a frequency assumed to be about 820 kilohertz (366 meters), to have reached Signal Hill, and it does seem a technically problematic proposition in light of what is known today about absorption of such long-wave signals by the atmosphere during daylight hours. It is possible that Marconi was hearing one of the many harmonics of the main Poldhu signal, reaching him on much shorter wavelengths not so much affected by daylight and the ionizing effects of sunlight on the atmosphere.

In any case, the controversy became moot three months later, when Marconi sailed from Southampton to New York aboard the liner *Philadelphia*, which he had equipped with the latest in receiving equipment and an elaborate masthead antenna. On that trip, signals from Poldhu were heard loud and clear during daylight hours up to a distance of seven hundred miles, and more than two thousand miles at night. This time, there were plenty of witnesses.

When it came to finding a commercially and geographically salubrious site in North America for a permanent terminus for commercial transatlantic service, Marconi made a surprising choice. Geography indicated Newfoundland as the best location (assuming Anglo-American's objections could be overcome), but Marconi was now confident that his signals could reach much farther, into more commercially promising regions of the continent; commerce suggested Cape Cod, his original selection, nearer to the big American markets. But Cape Breton Island in Nova Scotia won the contest thanks to some aggressive entrepreneurial thinking on the part of the Nova Scotian and Canadian governments.

Word had spread quickly of Marconi's decision to leave Newfoundland and there were many offers of assistance: Alexander Graham Bell cabled him from his lab in Cape Breton to offer a place to work. Cape Breton Member of Parliament Alec Johnston knew that the inventor would disembark from his steamer at North Sydney on the island, where he would catch the train for New York. He met Marconi on the dock and immediately suggested that Cape Breton would make an ideal North American terminus for his continuing experiments. There were no telegraph monopolies in Canada, he explained, and the island's eastern coastal cliffs afforded a clear shot to Poldhu. Within hours, the Nova Scotian premier G. H. Murray was in North Sydney, twisting Marconi's arm as well. The train for New York left without him. Cornelius Shields, the manager of the Dominion Coal Company, which owned the local railway and much of the land in the area, was dragooned into service and Marconi was bundled aboard a special railway car, to be shown likely transmitter sites between Sydney and Louisbourg.

Between the villages of Bridgeport and Glace Bay, the rail line crossed a flat headland high above the ocean, and the train was stopped. While the engine idled, chuffing smoke and seeping steam, Marconi and his companions strode through the grass and scrub to the cliff's edge, the ocean stretching before them into a blue haze. When Shields saw Marconi's enthusiasm, he offered on the spot to give him the land, free of charge.

That evening, over dinner in the Sydney Hotel, enthusiasm waxed further. The Nova Scotians told Marconi they believed the federal government in Ottawa would be willing to finance construction of his station, and if they were not, money could be raised locally. The next day, Marconi was whisked by rail to Ottawa, where he received an immediate appointment with Prime Minister Sir Wilfred Laurier. By the following night, contracts had been drawn and the Government of Canada had pledged $75,000 to cover the complete construction costs of establishing a wireless station at Table Head, Nova Scotia. Marconi continued his interrupted journey to New York, no doubt suitably impressed with the diligence of Canadian officialdom, and pleased to have had substantial portions of his company's research and development costs underwritten.

By 1902, Table Head was in routine communication with Poldhu, the first regular transatlantic radio link.[101] Two years later, the station and its four tall wooden antenna masts were torn down and relocated farther inland, on land now known as Marconi Towers. It was paired with a new station at Clifden, Ireland, and they became the twin terminuses of the world's first commercial radiotelegraphic service.[102] To send a message cost seventeen cents a word, the price of a decent breakfast in many a restaurant—a not-insignificant sum. However, it compared well with the undersea cable prices of the Anglo-American Telegraph Company, which were twenty-four cents a word.

Marconi's considerable gifts as a businessman were not in managing corporate bureaucracy; that, he left to a growing rank of subordinates. But he was astute at choosing employees for key positions and was capable of engendering fierce loyalty thanks to an absence of what the British call "side," coupled with his willingness to get down in the trenches and operate the equipment or help out with repairs at the isolated radio stations he'd built. In his early life of meteoric success (including a Nobel prize in physics in 1909), he overcame an innate shyness to enjoy the attention he attracted, especially from women. He was a regular in the pages of *Vanity Fair*, an early twentieth-century celebrity likely to be mobbed wherever he went in the world. His first wife divorced him after tolerating many a romantic dalliance; both she and Marconi would remarry, he to a woman twenty years his junior. Soon after, he converted to Catholicism, and soon after that he installed a permanent radio link between the Vatican and the Pope's summer retreat at Castel Gondolfo, using the "impossibly" short wavelength of fifty centimeters. He would continue experimenting with short wavelengths which he felt were being unjustifiably neglected by the engineering community he had helped to establish.

In 1923, apparently after much soul-searching, Marconi concluded it was his patriotic duty to join Italy's Fascist party. Mussolini was understandably delighted to have the support of such a revered international figure, and throughout the subsequent years until Marconi's death in 1937, he made frequent use of the inventor as a roving diplomat and apologist for Fascist foreign policy.

CHAPTER ELEVEN

Radio Goes International

Marconi and his London backers had always had a global vision for the company, and they pursued it with astonishing diligence. By 1899, an American arm of the company had been established and an ambitious, worldwide shore-station construction program was under way. By 1903, the Marconi companies dominated wireless telegraphy to a degree that was beginning to cause international concern. Fiercely protective of its many patents and its commercial leadership, the company refused to sell its equipment: it was available only for lease (along with a Marconi-trained operator), an arrangement which allowed the user a certain number of words per month, beyond which surcharges kicked in. Its shore stations, scattered by now around the world, also refused to carry messages other than distress signals originating from non-Marconi equipment. A word was coined for all of this: "marconism." It meant that Marconi was not selling equipment so much as "communications," an effective and farsighted business strategy. From a modern perspective, it amounted to a means of controlling the technical standards by which communication took place. Closing the system architecture is an approach to market control that has been used throughout modern communications history, notably by Bell Telephone.

Given Marconi's suffocating dominance in strategically located shore stations and market command in shipboard equipment, it was exceedingly difficult for other manufacturers to break into the industry. This raised wider geopolitical issues: Britain already controlled a majority of the world's growing web of transoceanic telegraph cables; other nations, particularly Germany at this time, were loath to allow her to monopolize international communications by dominating wireless telegraphy as well.

Problems were cropping up on a more mundane level as well. The airwaves were becoming more and more chaotic in the complete absence of law, national or international, to regulate the use of transmitting equipment. The temptation to play with the equipment when it was not being used for official messages was too great to resist for most operators. In his *History of Radio*, Gleason Archer records:

> Operators, whether on Naval or merchant ships, attained importance out of all proportion to their actual daily duties. They were in a class by themselves. Soon they began to run wild. . . . Human nature in its most arrogant form soon bedevilled the new medium of communication. Gossip between friends could and did crowd out messages of life and death on the high seas. . . . The only way in which urgent messages could be gotten on the air when two gossiping cronies held the ether was to lay a book or other weight upon the transmitting key and thus blanket them with such a roar of interference that they were obliged to desist. This process, however, was time-consuming and provocative of feuds. . . .[103]

Accordingly, the International Radiotelegraphic Conference was organized in Berlin in 1903, and attended by the Great Powers and a host of smaller nations. Standards for certification of wireless operators were drawn up. It was agreed by the nations that CQD would be the international distress signal, and that signals preceded by it would be given immediate priority over all other traffic. The more difficult issue of opening up the architecture of the worldwide radiotelegraphic network to allow free exchange of messages among different carriers was debated hotly, with Britain and Italy obdurately supporting the Marconi position against Germany and other backers of open exchanges. The conference ended on a rancorous note when the deadlock could not be resolved, with the result that Marconi maintained its stranglehold on commercial radiotelegraphy. But at the second conference in 1906, with Germany, France, Russia, the United States, Spain, Hungary and Austria aligned behind the idea, the open exchange of messages was adopted. Coastal stations would now have to accept messages from ships at sea and transmit traffic to them without distinction as to the manufacturer of their equipment.[104]

The question of public access to the airwaves was debated throughout this period with some vigor. The issue was framed in the context of the rights of "amateur" radio and its growing ranks of hobbyists. Earlier in the century, there had been no clear distinction between amateurs and professional or commercial operations. Equipment being used by amateurs was often superior to that of commercial operators; as do-it-yourself hobbyists, they could adopt all the latest advances in tuners and other technology without having to worry much about patent rights. "How-to" plans were provided in the growing literature of magazines and books on radio. The first of the periodicals, *Modern Electrics,* had a circulation of nine thousand in its second year of operation (1909); by the end of 1910, it had leaped to thirty thousand. But by about 1908, commercial radiotelegraphy had become a substantial industry with bright profit prospects for the future, and friction grew inevitable.

With little in the way of domestic or international regulations in place, amateurs inevitably interfered with commercial stations. Clinton DeSoto describes the situation in his history of amateur radio in America, *Two Hundred Meters Down*:

> If a commercial station wanted to do any work, it was usually necessary to make a polite request for the local amateurs to stand by for a while. If the request was not polite, or if an amateur-commercial feud happened to exist, the amateurs did not stand by and the commercial did not work. Times without number a commercial would call an amateur station and tell him to shut up. Equally as often the reply would be, "Who the hell are you?" or "I've as much right to the air as you have." Selfish? Undoubtedly. And yet, the amateur did have equal right to the air with the commercial, from any legal or moral standpoint. He was seldom

> interrupting important traffic—contrary to accusations that have been made, there is no authoritative record that amateurs ever seriously interfered with any "SOS" or distress communication: on the contrary, there are instances when the constantly watchful amateurs heard distress calls which were not picked up by the regular receiving points. And he was even then doing a useful work developing new and better radio equipment through his patronage of the manufacturers of parts and apparatus.

Nevertheless, the fact that there was money to be made in commercial services would help to spell the end of the freedom of the so-called amateur—actually, the *citizen*—to roam the radio spectrum at will. (This topic is explored in greater depth in Chapter 13, "The State Muscles In.") Ironically, in the doomed fight against regulation that ensued, it was the Marconi company that proved to be the amateur's greatest ally: the company saw in the burgeoning numbers of amateurs both a pool of trained prospective employees and a market for their radio components. Marconi argued that the interference problem was not so much the fault of amateur operators as inferior (i.e., non-Marconi) equipment being used by the commercial operators who were doing most of the complaining. It was a self-serving argument that nonetheless was substantially correct.

Throughout the history of radio's technical development, there has been tension between two approaches to achieving elbowroom on the airwaves. If radio can be thought of as a new continent and Marconi as its Columbus, then the early attempts to reach lower and lower into the long wavelengths where Marconi and others believed signal reliability would be greatest can be likened to extending the frontier of settlement. Later, radio experimenters would find fertile new territory in ever-shorter wavelengths, higher up the spectrum in the direction of visible light. At the same time, refinements in both receiving and transmitting circuitry and components made it possible to squeeze more and more signals into adjacent frequencies at any point on the spectrum, without causing interference: "population density" could be increased throughout the territory.

Of course, there was no way of knowing of these technical developments in advance, and the public access controversy was carried on for many years on the faulty premise that usable radio spectrum was severely limited by the perceived need to use only long waves. As well, the lack of selectivity in receivers and the extremely greedy bandwidth characteristics of the early spark transmitters were taken as givens.[105]

While the territorial dispute simmered, whatever residual doubt may have lingered about the commercial future for wireless telegraphy evaporated early in 1909, in one of those events that shape the future of a medium by capturing the public's attention and making its potential plain. In January of that year, the White Star passenger liner SS *Republic* had been groping her way through dense winter fog off Nantucket Sound. Suddenly a gray mass loomed ahead and with a grinding crash the SS *Florida* drove her prow into the steel of the *Republic*'s hull, tearing a mortal gash.

Holed and sinking, the *Republic* was able to call for help thanks to an on-board spark-gap radio rig and a doughty operator named Jack Binns. Newspaper accounts breathlessly told the story of how Binns was able to crank up his spark-gap transmitter and began sending the signal "CQD," over and over. The cry for help was heard by a powerful Marconi-equipped United States Coast Guard station at Siasconset and retransmitted to ships in the area. These included another White Star vessel, the *Baltic*, some two hundred miles away. The *Baltic* raced to the scene of the accident, using bearings provided by Binns. Before the *Republic* sank, all her surviving passengers and crew were transferred to the *Baltic* and *Florida*. Only five lives were lost among the 460 passengers and crew aboard the *Republic*, and the world applauded. Marconi stock prices surged.

A colorful account in a contemporary children's encyclopedia concludes: "It is a thrilling thing to think of, these ships drifting in the fog, talking to each other over sixty miles of space, carrying along an invisible ocean new life for a thousand beating hearts." Radio operators like Binns were the heroes of juvenile fiction, the darlings of newspaper reporters, the astronauts of their day: to be one, it was necessary to have the right stuff.

Binns himself had a small subsequent career in vaudeville, demonstrating his signaling talents, and was contracted by a publisher to provide forewords for *The Radio Boys* book series. "The escapades of the boys in this book are extremely thrilling," he wrote, "but not particularly more so than is actually possible in everyday life."

Three years later, as the mighty *Titanic* lay dying under a starry sky in frigid April waters off Newfoundland, the SS *California*, close enough to affect a rescue, steamed on despite the sinking liner's increasingly frantic distress calls. *California*'s wireless operator was off duty, asleep in his bunk. One thousand five hundred and thirteen lives were lost. That clinched it for radio: international regulations soon prescribed twenty-four-hour manning of shipboard stations and stations for every vessel carrying more than fifty passengers. Marconi stock shot up from £55 to £225 on the London exchange even before survivors arrived in New York harbor, and American business went on a speculative binge of investment in the new technology that would not be repeated until the days of Apple and Intel, Yahoo! and Netscape.

CHAPTER TWELVE

How Radio Works

Hertzian Waves, Fleming's Valve and De Forest's Audion

All of this started, as we've briefly noted, with a remarkably prosaic piece of technology which was first demonstrated by Heinrich Hertz in 1886. Hertz, in seeking a practical demonstration of electrical theory as laid out in the mathematical equations of James Clerk Maxwell, devised a rudimentary transmitter of electrical waves which was nothing more than wire loop with a small gap in it. When electricity was applied to the circuit, it jumped the gap with a spark, generating electrical waves in the surrounding space. Hertz discovered that those invisible waves could be detected at a considerable distance (all the way to the other side of his laboratory!) if one constructed a receiving loop of the same dimensions as the transmitting loop, also with a small gap in it: the electrical waves traveling through space from the transmitting coil's spark would set up a current in the receiving coil, causing a tiny spark to jump across its gap. It was rather like the action of tuning forks upon one another, only it happened in inscrutable, mysterious silence, and at the speed of light. In electrical terms, it was as if a wire (the receiving coil) had been moved through an electromagnetic field (the radio waves moving past it): a current was induced to flow in the wire, just as Faraday would have predicted.

Historians, like Marconi himself, have sometimes marveled at the fact that Hertz did not seem to see the potential for his experimental device as an instrument of wireless communication. But Hertz was a scientist and not an engineer; he was out to fry even bigger fish. His confirmation of

Maxwell's equations proved the existence of electromagnetic fields, and thereby undermined the foundations of the seemingly impregnable three-hundred-year-old edifice of Newtonian physics. Electromagnetic field theory would lead to a hundred inventions of which radio may not even be the most important.

Science's success has been based on its ability to isolate areas of nature and study them intensively, ignoring for the purposes of simplification the linkages of the delineated study area with the rest of Creation. In this context, electromagnetic theory described in terms of billiard-ball electrons, solar system atomic structure and waves through "ether" worked well enough to allow practical engineers like Marconi and his successors to build instruments and appliances that performed useful functions. But the descriptions were very far from the reality as understood by science even as early as the first transatlantic radio message. From the perspective of today's physics, what happened when Hertz induced a spark to jump the gap in his receiving loop is, on the one hand, immensely more complicated than anyone could have suspected and, on the other, more intimately connected to our lives.

Einstein and the quantum physicists who followed him demolished the notion of electrons and other subatomic particles as bits of matter existing in the void of space, or in some insubstantial, indefinable substance called "ether." Both matter and empty space, we have come to understand, are illusions and ether is merely a convenient fiction. What, then, was going on in Hertz's radio apparatus? It certainly had nothing to do with orbits and billiard balls. Physicist Hermann Weyl wrote:

> According to [quantum mechanics' update of Faraday's field theory] a material particle such as an electron is merely a small domain of the electrical field within which the field strength assumes enormously high values, indicating that a comparatively huge field energy is concentrated in a very small space. Such an energy knot, which by no means is clearly delineated against the remaining field, propagates . . . like a water wave across the surface of a lake; there is no such thing as one and the same substance of which the electron consists at all times.[106]

According to quantum field theory, the propagation of radio waves through space is the distillation of patterns of energy out of the electromagnetic fields which are the fabric of space/time. There is a real sense in which Hertz's transmitter and receiver were connected despite the space between them—in which *every* transmitter and receiver everywhere are always connected—by the cosmic web of potentialities waiting to be realized. What happened in Hertz's receiving loop was that a potential created by the application of electricity to the transmitting loop was realized, the information for that realization having been carried across the intervening distance by wavelike disturbances in the fabric of space/time. Radio communication is not the sending out of something from a transmitter to be detected by a receiver, but the actualizing of an energy or information link that always exists as a potential. When we shine a flashlight in the dark, we do not think of the process as "transmitting photons"; we think of it as revealing what is there to be seen. In the same way, radio reveals, and in doing so taps directly into the fabric of existence in the universe to facilitate communication. Small wonder we find it so fascinating.

Marconi's early experimental work in radiotelegraphy was simply an elaboration of Hertz's basic experiment, inspired by an inventor's instinct for what would work in the real world outside the laboratory. He first attached a long vertical wire to one side of the transmitting gap in Hertz's loop (an antenna), and attached the other side to the earth. No doubt it was the knowledge that the telegraph used an earth return for its circuits that inspired the thought. By doing the same with the receiving loop, he was gradually able to increase the distance over which he could successfully transmit from three hundred yards to two miles, then across the English Channel and then, finally, across the Atlantic.

One of the early engineering problems facing Marconi was the fact that spark-gap signals were extremely broad-banded: each transmitter used a spectacularly wide chunk of the radio spectrum, and the result was that no two stations could operate in close proximity to one another without causing intolerable interference. In short, radio signals were not tuned to specific frequencies. Sir Oliver Lodge, whose early experiments with Hertzian waves had in part inspired the young Marconi, developed a method for tuning both transmitters and receivers to a resonant frequency, outside of which radio waves were greatly attenuated. Radio transmitters thus equipped used much less power and ate up much less spectrum. Marconi improved on the idea, developing the modern "tuned circuit" and patenting it in 1900.

As these early advances were made in the techniques of radio wave transmission, parallel improvements came to receiving. Early in his experiments, Marconi switched to a more sensitive radio wave detection device called a "coherer," which exploited the elementary fact that an electrical current passing through a wire creates a magnetic field. He sealed nickel and silver filings in a cigarette-sized glass tube with a silver plug at each end, and placed the tube in the gap in the receiving loop (which had by now evolved into a coil, or series of loops). When the presence of radio waves induced a current in the coil, the current would find its way into and through the filings in the glass tube, setting them up in an orderly alignment, or causing them to "cohere" or clump together, forming a good conductor of electricity.[107] A tiny, hammerlike tapper was used to unclump or "decohere" the particles so that they could detect the next incoming waves, supplied by a closing of the Morse key at the transmitter. Radio waves, in effect, caused the glass tube and its filings to act as an on-off switch. An electric current from an auxiliary supply, when attached to the coherer, would be switched on and off according to the presence or absence of radio waves in the receiving circuit, and this auxiliary current could be used to ring a bell. If the stroke of the bell was also made to tap the glass and decohere the metal filings, the presence of a radio signal would be indicated by a rapid coherence and decoherence and a ringing of the bell. A similar chain of events takes place today when the homeowner activates an automatic garage-door opener with a small radio transmitter in her car: the transmitted radio waves activate a small, sensitive relay (the coherer), which in turn switches on household electrical current to a motor, which winches up the heavy garage door.

The coherer worked, but only just. It was an electromechanical solution to an electronic problem. An electronic solution would have to wait for the vacuum tube.

An interim answer of great importance was provided by the discovery in 1903 of the crystal detector by the American physicist Greenleaf

Whittier Pickard. Pickard noticed that certain mineral crystals were good conductors of electric current in one direction, but poor conductors in the opposite direction, in other words, "semiconductors." Why this should be, he did not know, but he immediately saw that these crystals might be used in radio circuits, to change the high-frequency alternating current of radio waves into direct current. That conversion was a crucially important step in developing improved receivers, since direct current was many thousands of times easier to detect with electrical metering devices than alternating current, which typically changed its polarity or direction of flow many times a second and thus appeared to most early instruments as no current at all. Furthermore, a direct current would actuate electromagnets such as those in telephone earpieces and headphones and loudspeakers, whereas alternating current would not, due to the same problem of rapidly switching polarity. Thus, the direct current generated within the receiving device by radio waves, or "detected," in the lingo of the medium, would be audible in earphones or a loudspeaker as a sustained note as long as the current was present, rather than a simple "click" marking the onset of an alternating current. The crystal detector worked like this: when a crystal was presented with radio waves (an alternating current) via an antenna, it would begin passing on half of that current—say, the positively polarized half—and filtering out the other half (the negatively polarized current), creating a "direct" or unidirectional current. That current fluctuated in force or amplitude—it "pulsated," in the jargon—but only within the limits of its single polarity. (Imagine a perfectly symmetrical string of hills and valleys lined up side by side on a postcard. Now fill the valleys with water, halfway up the hillsides. The curved peaks visible above the flat waterline would be what a "pulsating" direct current of positive polarity looks like on a test instrument—an oscilloscope.)

If the direct current "detected" by the crystal was led along a wire to the electromagnet within a telephone-style earphone, it would cause the earphone diaphragm to vibrate in sympathy with the tiny variations in its amplitude, creating sound waves and thus making the radio waves audible. When no radio waves were present, of course, the crystal would pass no current, and the earphone would be silent. In effect, the crystal detector (and all other detectors) behaves like the flap valve in a hand pump on a water well: the valve lifts open in one direction only and thus allows water to move up the pipe when the pump handle is worked creating a suction (and lifting the valve flap), but closes itself against water trying to escape back down the pipe into the well when suction is released, that is, when the handle is moved up in preparation for the next downward stroke. The up-and-down motion of the pump handle is analogous to an alternating current; the unidirectional flow of the water is like a direct current (and pulsates in the same manner), and the flap valve plays the role of a semiconductor, converting one to the other.

A wide range of mineral crystals were found to operate this way, including galena, iron pyrites, carborundum and silicon. For many years the crystal detector, though delicate and finicky to adjust, was a mainstay of the market in inexpensive radio receivers and deserves recognition for its important role in popularizing the medium. It was, in fact, the precursor to and inspiration for that ubiquitous present-day semiconducting device, the transistor.

Shortly after the crystal detector came the vacuum tube revolution, which would at last let radio explore its full potential. It would also make long-distance telephony possible, and give birth to electronic data processing and ultimately the digital computer. It is a classic tale of how the scientifically informed tinkerings of engineers can lead to useful new devices.

Thomas Edison was in the early stages of development of electric light bulbs when he noticed the mysterious "Edison effect." Parts of the inside of the glass bulbs were somehow becoming coated with a thin black film, even though all the air had been evacuated from the bulbs to prevent the carbon filaments from burning out. What could be causing this peculiar phenomenon? Edison suspected it was caused by "molecular" bombardment of the glass from the filament, and he probed the bulb with a wire attached to a galvanometer. Sure enough, the galvanometer needle was deflected, betraying the presence of an electrical current flowing through the vacuum from the filament to the probe wire. This was the summer of 1880. A little more tinkering and he might well have noticed that the current would flow only in one direction within the tube, and thus have discovered a practical system of wireless telegraphy using vacuum tube detectors more than a generation ahead of time.

But Edison was focused on making his electric light bulbs commercially viable: the extra element he inserted into the tube didn't solve his blackening problem, so he turned to other matters of more immediate practical concern. He revisited the puzzling phenomenon in 1882 and again early in 1883, and learned that the current passing through the tube varied in amplitude with the voltage applied to the filament. On the basis of that information he patented what was essentially a laboratory toy, a device for measuring voltage using an incandescent lamp with a second element inserted. It was the first patent in electronics. He demonstrated it for the world to see at the 1884 Philadelphia Electrical Exhibition: still, no one understood the true implications of what he had observed. The puzzle of the minuscule electric current flowing within an evacuated light bulb was left to an English investigator to solve.

John Ambrose Fleming was a young engineering teacher at Cambridge, a former student of the great James Clerk Maxwell, when he was asked by the newly organized British Edison Electric Light Company in 1882 to become a scientific consultant—in effect, a consulting engineer. He began almost immediately to examine the Edison effect. Meanwhile, the German scientists Julius Elster and Hans Geitel conducted a number of experiments on electrical conduction in an evacuated bulb, and reported that current flowed in one direction only. Again, no practical application of this phenomenon was suggested. Fleming, always, it seems, overloaded with a combination of teaching, textbook writing and consulting work, eventually picked up the threads of the mystery again in 1889. He had a number of bulbs built by glassblowers, using a variety of electrode arrangements. His experiments confirmed the one-way flow of electricity in a vacuum containing a heated electrode and a second element. He had come within a hairbreadth of inventing the electronic tube, but once again he was sidetracked by other matters for several years.

In 1895, Fleming had another batch of assorted lamps built, conducted another set of experiments and published another long paper: this time he noted that a vacuum tube—or "valve" as he called it—could, because of the one-way conduction phenomenon, be used to "rectify" an alternating current, or, in other words, convert it to direct current. This was because electrons would flow from the filament to the second element or "plate" within the tube only when the two elements had opposite electrical charges, i.e., when the filament was negative and the plate was positive. When both the plate and filament were negative, no current would flow (electrons would not be attracted to the plate). Thus, only half of the positive-negative waveform of alternating current would pass through the tube and that half was direct current, i.e. current of a single, rather than alternating, polarity or charge. (It would be two more years until physicist J. J. Thomson proved the existence of the electron: Fleming attributed the current flows he'd monitored to "negatively charged molecules.")

Fleming's findings received little public notice, overshadowed as they were by the astonishing discovery of X-rays and Marconi's exciting demonstrations of wireless telegraphy. The scientist's strange inability to grasp the practical applications of his discoveries have been a source of intense, vicarious frustration for historians of his life and work. George Shiers, writing in *Scientific American*, is typical:

> With Edison's sharp intuition, or the deep perception of Thomson, or perhaps more freedom from pressing duties, Fleming could have leaped into the twentieth century simply by combining the elements of his lamps. He could have suggested the possibilities of using his glow lamp as a rectifier for small currents (such as radio waves). . . . Even more important, he could have placed a zigzag wire (which he had put into one of his models) between the other electrodes (just to find out what would happen, as Edison would probably have done) and thereby have anticipated Lee De Forest in inventing the three-element vacuum tube.[108]

Shiers seems to sigh, before continuing:

> None of these leaps however, would have been characteristic of a man as methodical and cautious as Fleming.

For the third time in his sixteen-year investigation, Fleming put the bulbs away in favor of other matters.

By 1899 wireless telegraphy was well on the road to full-scale commercialization, and Fleming was contracted as a consultant to the Marconi Wireless Telegraph Company and put to work designing the power supply for the experimental wireless station at Poldhu, Cornwall. It was during this work that he had, at long last, "a sudden and very happy thought" that his test lamps were "the very implement required to rectify high-frequency oscillations"—in plain language, to detect radio waves. Finally, he was close to making his historic invention. He quickly wired together a simple spark-gap transmitter and a receiving circuit, using a two-element tube almost identical to the one Edison had constructed ten years earlier. The experiment worked immediately: Fleming had discovered a new means of detecting radio waves. He called it, without much apparent thought to marketing, the "thermionic valve."

The radio waves collected by an antenna were led to the "plate" or metal cylinder surrounding the filament inside the "valve." Whenever the radio waves were present and in the positive half of their alternating cycle, a current would flow (electrons would be attracted) from the heated filament (which was negatively charged) to the plate. When radio waves were in the negative phase of their alternating cycle, no current would flow. And, of course, no current would flow when no radio waves were present. The presence of direct current could easily be monitored with a meter such as a galvanometer, or it could be heard in a telephone headset. Fleming's valve was both sensitive and durable, and it gradually replaced earlier detector types in Marconi equipment beginning in 1907.

The American inventor Lee De Forest took the Fleming valve a giant step further by experimenting with a third element within the tube, which he called a "grid." The resulting device has been called one of the most important inventions in history. And for poor Fleming, it was a source of frustration and unhappiness for the remainder of his life. He could not accept that De Forest, a brash, egomaniacal and none-too-scrupulous young American, had done anything more than make a minor modification to his thermionic valve. Given the record of Fleming's meticulous experiments over so many years, there's some justification for this position. Nevertheless, De Forest patented the three-element tube (in 1907) and Fleming did not.

De Forest had been working on the problem of an improved method of detecting radio waves for several years, even experimenting with the rectifying properties of gas flames. In 1905, after hearing of Fleming's work, it occurred to him that if a grid—a wire screen—were to be placed between the filament and the plate in a Fleming valve, it ought to be possible to alter the strength of the electron flow in the tube by applying different charges to the grid, which would act as a kind of electric gateway between filament and plate. De Forest's revolutionary insight was that the grid would make it possible to greatly amplify a weak, variable current, such as a radio signal. It was no problem to get a very strong electron flow streaming from the filament to the plate: the weak current (e.g., radio waves) applied to the grid could now be used to regulate that flow, by changing the grid's polarity (opening and closing the gate) as the weak current fluctuated through its positive and negative cycles. The strong primary current would in this way be made to duplicate the characteristics of the weak grid current, only at much greater strength or amplitude. The incoming radio signal would be both detected *and amplified*. De Forest hurriedly drew up plans for the device and took them to a glass maker.

In his autobiography, he recalled:

> I hurried to my laboratory carrying in my pocket the world's only supply of three-electrode vacuum tubes. They were just two in number, and I was eager to test them to determine whether my latest invention, a tiny bent wire in the form of a grid inserted between the carbon filament and the platinum plate . . . could bring in wireless signals over substantially greater distances than other receiving devices then in use. . . . To my excited delight I found that it did; the faint impulses which my short antenna brought to that new grid electrode sounded many times louder in the headphones than any wireless signal ever heard before. I was happy indeed.[109]

This simple device, which had taken so long to arrive at and which

De Forest called the "audion," was the foundation of all modern electronics, including the electronic computer, until it was superseded by the transistor.

Lee De Forest, who did not count modesty among his slim repertoire of social virtues, was apt to call himself "the father of radio" at the drop of a hat. Many others have awarded that title to Marconi. But if it's radio as we currently understand the word, the radio we hear when we switch on a Walkman or a bedroom clock radio, Reginald Fessenden probably has as much right to the title as anyone.

While Marconi and De Forest were still preoccupied with solving the problems of wireless telegraphy, Fessenden was experimenting with voice transmissions by radio. By 1901, using equipment of his own invention, he had been able to hear his assistant's speech superimposed on the crash and thunder of a spark-gap signal. He then went to work on the transmitting side of the equation to get rid of the characteristic din of spark transmission while preserving the voice, and by 1906 was ready to demonstrate a practical "radiotelephone." By inserting a microphone circuit in a radio transmitter's output, he was able to "modulate" the radio waves, or superimpose on their shape the subtle variations in sound of a human voice or a musical instrument. At the receiver, these variations would be captured in the detection circuit and passed along to the earphone, where they would be reproduced.

Fessenden was born in Quebec and educated at Trinity College School in Port Hope, Ontario, and Bishops' College in Lennoxville, Quebec. He is credited with more than five hundred inventions in the fields of radio, sound and optics. He had the early good fortune to find the backing of two Pittsburgh bankers who supported him during his period of greatest productivity in radio, eventually sinking more than $2 million into his experiments. In 1902, the three of them formed the National Electric Signaling Company, with Fessenden's patents as the nugget of value. Fessenden was to give his unfortunate partners many a sleepless night as he burned through huge sums of capital in obsessive pursuit of his goal of refining radiotelegraphy and perfecting the transmission of voice by radio.

An early assistant, Roy Weagant, recalled:

> He could be very nice at times, but only at times. His voice boomed like a bull. . . . He had no regard for cost; he once sent for a 100-horsepower boiler and had it shipped by express. The more anything cost the better he liked it. He was sometimes unsound technically, but he more than made up for it by his brilliant imagination. He was the greatest inventive genius of all time in the realm of wireless.[110]

Another assistant, Charles J. Pannill, worked with Fessenden from 1902 to 1910:

> He had a great character, a splendid physique, but what a temper! Many of us were fired on more than one occasion, and usually on some slight provocation. When he had cooled off, he was sorry and hired us back, always at higher pay. . . . Fessenden could think best when flat on his back smoking Pittsburgh stogies; that's when he got his real ideas.

In 1905 Fessenden's company built a state-of-the art experimental station at Brant Rock on the Massachusetts coast. It was equipped with Fessenden's latest invention, a high-speed alternator-type transmitter that would prove to be the climax of spark technology, soon to be superseded by the high-power vacuum tube. On Christmas Eve 1906, with no fanfare, Fessenden made the first real radio broadcast. An early historian of the medium reports:

> Early on that evening, wireless operators on ships within a radius of several hundred miles sprang to attention when they heard the call "CQ,CQ" in Morse code. Was it a ship in distress? They listened eagerly, and to their amazement, heard a human voice coming from their instruments—someone speaking! Then a woman's voice rose in song. It was uncanny! Many of them called to officers to come and listen; soon the wireless rooms were crowded. Next someone was heard reading a poem. Then there was a violin solo; then a man made a speech, and they could catch

> most of the words. Finally, everyone who had heard the program was asked to write to R. A. Fessenden at Brant Rock, Massachusetts—and many of the operators did. Thus was the first radio broadcast in history put on.[111]

Fessenden went on to perform a series of tests that greatly improved both the range and quality of his radio broadcasts. Wireless technology had been lifted to a new level, an advance comparable to the leap from the telegraph to the telephone. However, the fact scarcely seemed to register on the public consciousness and it was not until 1920, fourteen years after the first broadcast, that radio really caught fire as a consumer product.

Until voice transmission became widespread, radio was to remain in the realm of the laboratory curiosity for most people. For those who understood how it worked, it held endless fascination, and amateur radio operators abounded. But the fact that communicating was done in Morse code and the equipment, even when it was not homemade, required a degree of expertise to operate, meant that for another decade and a half radio was well-buried in the consciousness of the majority of people. We might say in today's idiom that the interface between sender and receiver was just too complicated.

Nevertheless, radio magazines abounded and there were articles in all the popular science periodicals with instructions on how to construct and operate transmitters and receivers and build antennas. The author of a popular how-to series of books was soon earning royalties in six figures. Parts manufacturers were reaping a bonanza and, in the process, patent rights were often scandalously abused. Lawsuits proliferated.

Carneal has an evocative description of the era that will find resonance in the experience of many a late-twentieth-century parent of computer kids:

> At just about this time [1910] the word "radio," heretofore a mysterious "force" that only the scientist understood, sprang into being. It was ushered in by a motley group of people, mostly boys and young men, working all alone on crude homemade apparatus

in the isolation of their own homes. These young people, scattered over the country, were engaged in the strange pastime of "fishing" things out of the invisible space around them, at first dots and dashes and later words and music that come from nowhere. . . . Parents, at first indulgent of the strange and not inexpensive devices their youngsters brought into the home, began to gape in wonder as they donned their headsets and actually listened to music coming to them out of the air. They did not understand in the least how it happened. It seemed mysterious beyond words. Their amusement died out as they found themselves listening to an explanation of the vast intricacies of "radio."[112]

CHAPTER THIRTEEN

The State Muscles In

The transformation of radio from a freely accessible, two-way technology to a primarily one-way, heavily regulated broadcast medium took place in two main stages. First, governments everywhere took radio out of the hands of ordinary citizens. Then, in the United States, they turned it over on an all-but-exclusive basis to commercial, profit-making enterprises; in Britain, it was put in the hands of a government agency. By the end of the transformation, radio had become a "manipulative" rather than "convivial" technology in Illich's terms. The transition merits detailed examination, since it is a story replete with cautionary tales and the lessons of history, which are directly applicable to today's media environment. The saga hinges on the trauma of World War I.

When "the war to end all wars" began, amateur radio operators in Europe and North America were required to dismantle their stations for the duration, to eliminate the possibility of accidental or deliberate disruption of military radio traffic. All radio communication, including private commercial radiotelegraphy carried on by Marconi and a handful of other companies, was placed under the control of the military. To expedite the production of the great quantities of wireless equipment required by the armed forces, all patent disputes were frozen and manufacturers were permitted to use any and all technical innovations, whatever their source.

During the war, wireless was of limited use in ground engagements due mainly to the weight, bulk and unreliability of field sets and the fact that the enemy could listen in on transmissions. Its potential for military applications was noneless made clear before the fighting had ended. Radio's most significant contribution was in providing immediate feedback from artillery spotters flying over enemy lines in aircraft equipped with relatively compact, low-power transmitters. They could see where a shell had landed and radio back instructions for correcting the aim. At sea, where weight and bulk were unimportant, and where constant movement over great distances was a prime factor in warfare, wireless revolutionized naval strategy and tactics by permitting worldwide coordination of battle fleets. Ships no longer had to maintain visual contact with one another to coordinate operations, and they were able to benefit almost immediately from intelligence gleaned by their command establishments on shore.

When the war ended, governments everywhere showed a reluctance to relinquish to their citizenry control over radio. It had become a technology of strategic significance for the Great Powers; leadership in the field of radio communications was an important military asset and would likely become even more important as the science evolved. The transoceanic telegraph cables which until now had formed the backbone of the world's communication system had shown themselves vulnerable to warfare: when fighting began in 1914, Britain herself had cut or otherwise disabled most of the global undersea telegraph network to keep it from falling into enemy hands. Overland cables were likewise vulnerable, particularly in Europe, where a line of any length inevitably crossed several frontiers. In a world that had quickly grown to depend on rapid international communication, radio was now seen as a crucial backstop to the telegraph system.

From the point of view of the British and American postwar administrations, putting the medium back in the hands of what was perceived to be an undisciplined rabble of amateur operators simply would not do. Moreover, the tangle of conflicting patent claims and lawsuits would have to be sorted out, and quickly, so that technical development could proceed. And finally, it was now seen as imperative, for reasons of national security, that foreign nationals be prevented from owning any significant portion of the radio industry.

The United States and Britain found diametrically different solutions to these postwar problems. In Britain, the government of the day placed radio in the hands of the Post Office, which had long regulated the telegraph and telephone businesses. Private commercial broadcasting was outlawed, and British Marconi resumed its accustomed armlock on the international radiotelegraph business.

A similar plan for public control of radio was proposed in the United States and had the backing of the navy and significant support in Congress. Congressional hearings were held in December 1918. Navy secretary Josephus Daniels testified: "We strongly believe that having demonstrated during the war the excellent service and the necessity of unified ownership, we should not lose the advantage of it in peace." He argued that the navy "is so well prepared to undertake this work and to carry it on that we would lose very much by dissipating it and opening the use of radio communications again to rival companies."

Leading the opposition to the bill in committee testimony were the head of the American Radio Relay League, representing American radio amateurs, and the president of American Marconi, which was essentially a branch plant of British Marconi and the country's principal commercial radiotelegraph operator prior to the war. Informed opinion on Capitol Hill at first predicted adoption of the bill, but opposition was vocal and well organized. Before the end of the month, the committee had voted unanimously to shelve the measure; the idea of state monopoly in a thriving industry was too much for American legislators to stomach.

This, of course, left the strategic military issue unresolved. Sooner or later, President Woodrow Wilson would have to return commercial assets to their owners (presumably, as soon as the armistice under negotiation in Paris had been signed), and foremost among the owners was American Marconi. Wilson's position was described by one of his wartime advisors in a 1937 interview:

> [The] President had reached the conclusion, as a result of his experience [at the armistice talks] in Paris, that there were three dominating factors in international relations—international transportation, international communication, and petroleum—and

> that the influence which a country exercised in international affairs would be largely dependent upon their position of dominance in these three activities; that the British obviously had the lead and the experience in international transportation—it would be difficult if not impossible to equal her position in that field; in international communication she had acquired the practical domination of the cable systems of the world; but that there was an apparent opportunity for the United States to challenge her in international communications through the use of radio; of course as to petroleum we already held a position of dominance. The result of American dominance in radio would have been a fairly equal stand-off between the U.S. and Great Britain—the U.S. having the edge in petroleum, Britain in shipping, with communications divided—cables to Britain and wireless to the U.S.[113]

The issue became all the more urgent when General Electric reported that British Marconi had placed an order for several of the advanced, high-power transmitters GE had built on the model of Fessenden's Brant Rock station. It was an enormous order, worth more than $5 million. The alternators, massive pieces of machinery that resembled the electrical generators found at a small hydroelectric station, were the state-of-the-art in radiotelegraphy: these more powerful models could be heard around the world, and they were far superior to the spark-gap equipment still being used by Marconi.[114]

Alarmed at the potential strategic consequences of the deal, the U.S. Navy, in the person of Admiral William Bullard, director of naval communications, stepped in behind the scenes at the behest of President Wilson to abort the sale in the name of national security. In response to General Electric's plaintive cry that it would have to cease production of the alternators and lay off workers without a sale to *someone,* and that, in any case, nothing short of a state monopoly could succeed in usurping Marconi's dominance, Admiral Bullard proposed setting up a government-sponsored, private American corporation with the resources necessary to compete internationally.[115]

The proposal was to establish a government-sanctioned cartel (although, of course, it would not be identified as such in public) under the name of the Radio Corporation of America, or RCA for short, which would be owned by General Electric and other participating companies. The cartel would be the customer GE needed to keep its war-bloated manufacturing arm afloat. Company officers jumped at the suggestion.[116]

As a prelude to establishing RCA, General Electric, with the active encouragement of the U.S. government, negotiated the purchase of all the assets of American Marconi. "Negotiation" may be a trifle euphemistic for what occurred in the paneled boardrooms of British Marconi when GE honchos arrived for talks: it was made clear to the British company that the government of the United States would under no circumstances allow Marconi to continue to dominate wireless telegraphy in America, and that the offer being made by General Electric was the best they could expect. It was well known that there were powerful interests within the American government pushing for outright state control of radio, whereby Marconi's American interests would have been nationalized in any case. Faced with what was essentially an ultimatum, the company agreed to divest itself of its American assets.

The strategic political question nagging the Americans had thus been addressed. But it did not stop there. A year later British Marconi, France and Germany's Telefunken all found themselves on the receiving end of intense diplomatic pressure from the United States and its new radio conglomerate, over plans to develop a powerful wireless station in Argentina. Telefunken had already begun construction on a station there; the British wanted to beat them to the punch if at all possible, as did the French. France and Britain were outraged, then, when the Americans presented an ultimatum demanding that any station in Argentina or anywhere else in the Western Hemisphere be operated by a board of directors on which the United States would be given veto power. Over several weeks of negotiations in Paris in 1921, the Americans prevailed. This was an explicit embodiment of the Monroe Doctrine, under which the United States had since the 1820s claimed all of Central and South America as its exclusive economic and political sphere of influence, the theater of its "manifest destiny."

The patent mess was also on its way to a resolution. When RCA was formally chartered in October 1919, GE granted it reciprocal rights to all of its radio patents and inventions, and an immediate merger with American Marconi was undertaken. Within weeks, RCA had signed cross-licensing agreements with AT&T and its subsidiary, Western Electric, which allowed the companies use of each other's patents in the area of manufacture and sale of electrical appliances of all kinds. (About a thousand patents were involved.) RCA was made exclusive wholesale-selling agent for the radio receivers manufactured by all three companies. Similar cross-licensing agreements were signed with other holders of important patents. The budding American radio industry had to all intents and purposes become an exclusive club, to be operated for the benefit of its members and in its directors' view of "the national interest."

In 1924, the niggling problem of what to do about two-way radio in the hands of citizens—amateur radio—was dealt with by an international convention on the allocation of radio spectrum. Amateurs were henceforth to be denied access to radio frequencies below two hundred meters, which meant, in terms of contemporary engineering knowledge, they no longer had access to the only part of the spectrum usable for long-distance radio communication.[117]

However, as radio amateurs began despondently tinkering with the "useless" chunk of spectrum to which they had been relegated, they soon perked up. It turned out that the higher they went in frequency, the farther they could transmit and the less power they needed to establish contact with distant stations. As they crept up the spectrum, cautiously at first, then with rising excitement, through the present-day AM radio bands and on to eighty- and fifteen-meter wavelengths which are now occupied by international shortwave broadcasters, they were soon setting distance records using power outputs that seemed infinitesimally small by commercial radio standards. Moreover, they discovered that at higher frequencies, static interference was no longer the bane of existence it had been on the low frequencies. And antennas, really efficient antennas, were much simpler to build since they were substantially smaller due to the shorter radio wavelengths involved.

The discoveries made by the amateurs in this new frontier for radio revolutionized the industry. Both RCA and British Marconi backed away from well-advanced plans to build worldwide radio networks based on massive, low-frequency alternators, and returned to the drawing board to design a high-frequency system based on vacuum tubes and directional "beam" antennas that proved to be much cheaper and more reliable. Radio equipment became smaller, lighter and a great deal more effective as the general upward movement of radio services through the spectrum proceeded through the late 1920s and 1930s. Today, the low frequencies so popular in the days of Marconi and Fessenden are all but vacant—a static-ridden, abandoned landscape where amateurs, ironically, are once again beginning, in very small numbers, to tinker and explore.[118]

But it is radio as a broadcast medium that mainly concerns us here, and we return to that subject to discover how it is that the radio spectrum, clearly a public resource, has in most places in the world come to be occupied principally by commercial operators, whose primary goal is to serve the private business interests of commercial sponsors.

CHAPTER FOURTEEN

How the First Mass Medium Was Born

Left like a jilted lover on the outside of the cozy ménage involving RCA, GE and AT&T was the Westinghouse Electric and Manufacturing Company.[119] It was a snub that would change history, for it then fell to Westinghouse, by force of necessity, to introduce to the world the most powerful medium of mass communication it had yet seen.

Westinghouse naturally viewed the patent-sharing cartel in radio with alarm. It had entered World War I as a relatively minor player in radio, but emerged after two years as a manufacturing power to be reckoned with. Now it faced being frozen out of the postwar radio business. The company set out in hopeful pursuit of international radiotelegraphy agreements in Europe, only to find itself beaten to the punch wherever it turned by exclusive rights deals negotiated by the fast-moving RCA. Westinghouse briefly tried to turn a profit with its ship-to-shore stations by providing ships at sea with regular newscasts, but there were too few subscribers. It soon capitulated to RCA, giving up its ambitions in international radiotelegraphy in return for an agreement from RCA to purchase 40 percent of its radio components from Westinghouse. Westinghouse reluctantly became a major shareholder in RCA alongside AT&T and General Electric, and the long-standing, debilitating patent mess was finally and definitively resolved.[120]

But the deal left Westinghouse with only the domestic market in which to offer radiotelegraphy services. It was a dismal prospect, since the country was already well provided with land-line telegraphy and telephone networks: radiotelegraphy would be a very small niche indeed where wired communications systems were so advanced. The only other way to make money at that stage of the industry's development was in the manufacture of radio components, an aspect of the business the RCA cartel had already thoroughly organized. It was in these circumstances that Westinghouse, almost by accident, first turned its attention to an entirely novel use for the medium—domestic radio broadcasting as a commercial venture.

One of Westinghouse's senior engineers, Dr. Frank Conrad, had for some time been experimenting by broadcasting music from the amateur station he operated from his garage in Wilkinsburg, Pennsylvania. He was deluged with mail from listeners all over the continent, many of them asking to hear particular tunes at particular times. It soon became an overwhelming job to fill the many requests and he was forced to put his transmissions on a regular schedule which ran from 7:30 to 9:30 on Wednesday and Saturday evenings. He called them "broadcasts," borrowing the term from agriculture: the spreading of seed over a wide area. The recordings were loaned to Conrad by a local music shop, which in due time requested an on-air mention. Soon it was evident that recordings that were played on the programs were outselling others at the music store. Sales of radio receivers shot up in the area, and a local department store placed a newspaper advertisement that encouraged readers to buy one of their crystal sets in order to listen to Conrad's broadcasts.

It was the department store ad, placed by the Horne Company in the Pittsburgh *Sun*'s September 29, 1920, edition that finally made the light go on in the mind of Westinghouse vice-president Harry P. Young. He recalled the moment of inspiration in an address to the 1928 graduate business administration class at Harvard University:

> We were watching [Conrad's broadcasting experiments] very closely. In the early part of the following year [1920] the thought came which led to the institution of a regular broadcast service. An advertisement of a local department store in a Pittsburgh newspaper, calling attention to a stock of radio receivers which could be used

> to receive the program sent out by Dr. Conrad, caused the thought to come to me that *the efforts that were then being made to develop radiotelephony as a private means of communication were wrong, and instead its field was really one of wide publicity,* in fact the only means of instantaneous mass communication ever devised [emphasis added].[121]

A colleague of Conrad's, Samuel Kintner, provided a slightly different perspective on the moment in a 1932 memoir:

> Their advertisement in the Pittsburgh *Sun* . . . caught the eye of [Harry P. Young] vice-president of Westinghouse. The next day he called together his little "radio cabinet," consisting of Dr. Frank Conrad, L. W. Chubb, O. S. Schairer and your speaker. He told us of reading the Horne advertisement and made the suggestion that the Westinghouse Company erect a station at East Pittsburgh and operate it every night on an advertised program, so that people would acquire the habit of listening to it, just as they do of reading a newspaper. He said: "If there is sufficient interest to justify a department store in advertising radio sets for sale on an uncertain plan of permanence, I believe there would be a sufficient interest to justify the expense of rendering a regular service—looking to the sale of sets and the advertising of the Westinghouse Company for our returns."[122]

A little more than two weeks after the appearance of the Horne advertisement in October 1920, Westinghouse had applied for a radio station license and had been assigned the call letters KDKA by the U.S. Department of Commerce. Two weeks later, the station was on the air in time to carry the election returns in the presidential contest between Warren G. Harding and Charles Evans Hughes.[123] It was estimated that between five and ten thousand people were able to hear its signal.

At the time, the idea of commercial sponsorship in all of its manifold possibilities had not clearly emerged. However, what was becoming increasingly clear to Westinghouse was that the real future of radio as a profit maker lay not in one-on-one communication, in which content was provided by individual operators seeking to communicate with each other, but in point-to-multipoint communication, or broadcasting to a mass audience. In broadcasting, Westinghouse would own the content, and could therefore sell it (to advertisers) at a profit. It would own not only the hardware of the system, but the software too.

Before broadcast radio could become universally popular, however, something would have to be done about the receivers of the day. The problem was that a typical radio set required two large batteries to operate, one a dry cell and the other a lead-acid battery which needed to be periodically recharged and topped up with water, an inevitably messy process. (The first AC-powered sets did not appear until 1928.) There were two, sometimes three, variable capacitors involved in the tuning process, each of which had to be adjusted by a big knob on the front panel. Each of the vacuum tubes might also have a separate control. Changing wavelengths might involve removing one set of coils from inside the cabinet and replacing it with another. Some sort of outdoor antenna was usually required for all but nearby stations. Radio enthusiasts might enjoy fiddling with all those knobs, wires and adjustments, but the ordinary listener found it all far too complicated.

What was needed, then, was a user-friendly receiver, a simplified interface between the provider and end user of the broadcast information. And Westinghouse had just such a stripped-down, simple-to-operate set ready for market, as recalled by an early radio historian:

> No "forest of knobs" here; no complicated table of settings; merely one circuit and one handle to vary it. I can recall the personal scorn with which this single circuit receiver was viewed by "old-style" radio engineers, i.e., myself, for it was held that this was going back to the days of 1900. . . . Actual use showed that for handling by people who knew nothing of radio's technicalities the single circuit was just what had been needed.[124]

Westinghouse had stumbled on a new business plan for its radio operations, one that seemed to have some real promise. Within a year, three other Westinghouse stations were broadcasting, and national demand for the simplified receivers had outstripped the supply. Now other broadcasters began flocking to the airwaves.

Since the U.S. Radio Act of 1912 placed no restrictions on ownership of a license beyond American citizenship, the new broadcasters were a colorful group. Several distinct categories emerged. First, there were the big manufacturing interests like Westinghouse, GE and RCA. Then there were the department stores like Gimbels and Wanamakers which operated stations for self-promotion. Some hotels had stations. There were stations in laundries, chicken farms and a stockyard. In 1922 eleven American newspapers held broadcasting licenses, mainly, one suspects, out of self-defense, just as newspapers today have been quick to populate the World Wide Web. Churches and universities operated stations. And there were many so-called ego stations operated by wealthy individuals in the spirit of noblesse oblige, or just for the hell of it. A common experience shared by all of these groups and individuals was shock at discovering how expensive it could be, especially once performers' unions began raising copyright issues and charging for appearances.

RCA, the mother ship, finally entered the broadcast arena on December 1921, with station WDY in Roselle Park, New Jersey, and an ambitious schedule of recorded music and live lectures, variety and drama. While RCA had to some extent inherited the slow-moving bureaucratic propensities of its gargantuan parent, General Electric, there was at least one employee of the new company who had been deeply chagrined by the fact that Westinghouse had stolen a march in putting a broadcasting station on the air. David Sarnoff had, with the shotgun marriage, moved from being commercial manager at American Marconi to taking on the same position at RCA. He was soon promoted to general manager of RCA. One of the more remarkable figures in the annals of American industry, Sarnoff, the son of penniless Jewish immigrants from Russia, had by dint of brains and determination and a knack for self-promotion risen from newsboy to telegraph messenger to Marconi radiotelegraph operator, and finally was taken under the wing of the great Marconi himself, to become a trusted adviser and protégé.

Sarnoff's greatest personal asset was his comprehensive understanding of the technology of radio and its practical applications. He had worked in every corner of the business. As early as 1916 he had seen such advances as Fessenden's Christmas Eve broadcast and better receiver designs converging to make an entirely new industry possible. In that year he had written for his boss at American Marconi a memorandum on the subject that laid out the future development of broadcasting before anyone had even used the word.

"I have in mind," Sarnoff wrote,

> a plan of development which would make radio a "household utility" in the same sense as the piano or phonograph. The idea is to bring music into the house by wireless. . . . The receiver can be designed in the form of a simple "Radio Music Box" and arranged for several different wavelengths, which should be changeable with the throwing of a single switch or the pressing of a single button. . . . The same principle can be extended to numerous other fields as, for example, receiving lectures at home which can be made perfectly audible; also events of national importance can be simultaneously announced and received. Baseball scores can be transmitted. . . . The manufacture of the "Radio Music Box" including antenna in large quantities would make possible their sale at a moderate figure of perhaps $75.00 per outfit. . . . It is not possible to estimate the total amount of business obtainable with this plan until it has been developed and actually tried out but there are about 15,000,000 families in the United States alone and if one million or 7% of the total families thought well of the idea it would, at the figure mentioned, mean a gross business of $75,000,000 which should yield considerable revenue. . . .[125]

With the world mired in the horror of Flanders, Marconi operations under government control and the company focused on producing hardware for the military market in Europe, it was not a propitious time for Sarnoff's plan, and no action was taken. With the formation of RCA, he submitted it to his new boss, who, with a wary eye on Westinghouse, had given him the go-ahead.[126]

Broadcasting's growth in that year (1921) was nothing short of phenomenal, though it may seem less astonishing today from the perspective of readers who have experienced the similarly breathtaking growth of the Internet: the number of new stations receiving licenses in the United States grew from two in August to twenty-three in December, to ninety-nine in May of 1922. The number of radio receivers in the United States grew from 50,000 in March 1921 to 750,000 just fourteen months later.

Radio Broadcast magazine, in its debut issue in May 1922, reported:

> The rate of increase in the number of people who spend at least a part of their evening in listening in is almost incomprehensible. . . . The movement is probably not even yet at its height. It is still growing in some kind of geometrical progression. . . . It seems quite likely that before the movement has reached its height, before the market for receiving apparatus becomes approximately saturated, there will be at least five million receiving sets in the country.

In fact, the 5 million mark was passed less than four years later, and by 1927 there were 6.5 million sets in use in the United States, one for every 17 persons. In 1940 there would be 50 million sets, one for every 2.6 persons. RCA, the major distributor of radio receivers, recorded awe-inspiring sales: 1921: $1,468,919.95; 1922: $11,286,489.41; 1923: $22,465,090.71; 1924: $50,747,202.24.

Another article in the same issue gave this picture of the booming industry in 1922, a portrait which is eerily evocative of the early development of the modern computer industry:

> The manufacturers of radio receivers and accessories are in much the situation that munition makers were [in] when the war broke. They are suddenly confronted with a tremendous and imperative demand for apparatus. It is a matter of several months at best to arrange for the quantity production of radio receiving apparatus if the type to be manufactured were settled, but the types are no

> more settled than were the types of airplanes in the war. The manufacturing companies are, therefore, confronted with carrying on their experimental work, devising new types and at the same time producing the best they can in such quantity as they can, and they must do all of this while building their organizations, working out their policies and keeping an eye on the Government so that they can keep in accord with regulations.[127]

Despite, or perhaps because of the rapid growth, the industry had yet to resolve the critical issue of how to pay for broadcasting. Westinghouse, RCA and other equipment manufacturers had an obvious economic interest in spurring the numbers of listeners: there was plenty of indirect revenue to be derived from receiver and component sales. Department stores employed their stations as advertising vehicles for their own wares. Newspapers opened stations with a view to cross-promotion, counting on increased paper subscriptions to pay the radio bills. But there were also a large number of new stations run by individuals who had no obvious source of financial support for their broadcasting efforts. The great debate of the day in the pages of radio periodicals and the more mainstream media was: How should radio broadcasting be financed?

Commercial sponsorship, although it was by no means unheard-of, was not the most popular approach to the problem; few thought it a likely source of substantial sums of revenue. How much could you realistically charge for an advertisement that might be heard by a few hundred people, or at best a few thousand? RCA's David Sarnoff hated the idea on principle, believing it would destroy radio's potential for education and quality entertainment. He felt radio should be financed by the big manufacturing companies like his own, and Westinghouse, as a public service. They would, after all, profit handsomely from appliance sales. (This sentiment may not have been altogether altruistic. The astute Sarnoff must have realized that introducing direct advertising revenue to broadcasting would make the field accessible to others who did not have the luxury of owning radio factories from which to derive indirect revenues.)

Sarnoff's public views on advertising were shared by (then) Secretary of Commerce Herbert Hoover. In a speech in 1922 Hoover declaimed:

"The ether is a public medium and must be used for the public good. It is quite inconceivable that it should be used for advertising." *Radio Broadcast* magazine, fast becoming the unofficial bible of the industry, editorially supported public financing through an endowment model similar to that which had successfully built a system of libraries throughout North America, and many fine universities in the United States. Direct government financing was also proposed in the magazine, though somewhat gingerly as befitted a mass-circulation outlet in an era of rampant anti-Bolshevist sentiment in America:

> A weird scheme this [government financing] will undoubtedly appear to many, but upon analysis it will be found not so strange, even to those who have no socialistic tendencies. In New York City, for example, large sums of money are spent annually in maintaining free public lectures, given on various topics of interest; the attendance at one of these lectures may average two or three hundred people. The same lecture delivered from a broadcasting station would be heard by several thousand people. Because of the diverse interests of such a large city as New York, it would probably be necessary to operate two or three stations, from each of which different forms of amusement or educational lectures would be sent out. The cost of such a project would probably be less than that for the scheme at present used and the number of people who would benefit might be immeasurably greater.[128]

Alas, commercial-free radio in America was an idea whose time had come and gone years earlier when the first Westinghouse station, KDKA, drilled its modest Pittsburgh test well into the reservoir for advertising demand and came up with a gusher.

CHAPTER FIFTEEN

Broadcasting's Pot of Gold

The ultimate solution to the financing conundrum was to come from what seems at first blush an unlikely source: telephone industry giant AT&T, Alexander Graham Bell's alma mater. The company had for years been interested in vacuum tube technology, and its engineers had steadily refined vacuum tube amplifiers to make transcontinental long-distance telephony possible. In the spring of 1922, AT&T announced it would open an experimental radio broadcasting station in New York, using the most advanced technology then available.

It is a curious fact, and one that illustrates how little the future of broadcasting was generally understood, that the patent cross-licensing agreements between RCA and AT&T in 1920 and 1921 gave AT&T the exclusive right to employ the pooled patents for radiotelephony. (RCA retained exclusive rights for radio*telegraphy*.) Voice communication by radio was considered a natural extension to AT&T's existing telephone interests; commercial radio clearly was still being thought of primarily in terms of point-to-point communication as late as 1921. However, when new prospects for revenue were scented, the telephone company was quick to extend its interpretation of the covenants to include all of commercial broadcasting. It was AT&T's position for many years—one hotly contested within the industry—that any station engaged in broadcasting using any of the hundreds of patents tied up within the cross-licensing webs of RCA and its allies (which included Westinghouse) owed a royalty to AT&T.

Captured in this category was virtually every broadcast station in North America. And it was in large measure to protect its rights in this area that the company entered broadcasting when it did.

AT&T's broadcasting business plan was an interesting one. It was based on the economics of a long-distance telephone service, in which customers are charged by the minute for use of telco facilities. Extended to radio broadcasting, this meant that AT&T would be in the business of selling time on its transmitters; advertisers would be invited to purchase a block of time and provide their own programming within that block. (This is similar to the commercial arrangements for infomercials on today's television stations.) Air time that had not been sold to advertisers would be filled by AT&T's own programming resources to maintain audience numbers.

It is evident from the company's announcement of the new service in February 1922 that it had also hit upon the momentous idea of networking among stations to accumulate audience numbers large enough to be of interest to major sponsors. It deserves recognition as one of the most brilliantly imaginative business ideas of the twentieth century. The news release read:

> This is a new undertaking . . . and if there appears a real field for such a service . . . it will be followed as circumstances warrant by similar stations erected at important centers throughout the United States by the AT&T Company. As these additional stations are erected, they can be connected by the toll and long-distance wires of the Bell System so that from any central point, the same news, music and other program can be sent out simultaneously through all these stations by wire and wireless with the greatest plausible economy and without interference.[129]

The first of the AT&T stations would be WBAY, which was soon folded into a second, technically superior station, WEAF. (General Electric weighed in at about the same time with its first station, WGY. It was a heavyweight that could be heard on the West Coast, throughout Canada and in Alaska.) By 1925 the company was regularly assembling networks of up to twenty-five stations, linking them with its long-distance telephone lines.

Twelve days after it went on the air, AT&T WEAF sold its first block of time to a Long Island real estate developer called Queensborough Corporation. A ten-minute segment was purchased by the company in each of the five days between August 28 and September 1, 1922, to sell its suburban condominiums. In keeping with the contemporary notion that content was king on radio and that listeners would not stand for a straight sales pitch, the advertisement was couched, though none too subtly, in a tribute to the American author Nathaniel Hawthorne, for whom part of the housing development (Hawthorne Court) had been named.[130]

On September 21, 1922, WEAF carried three commercials: Queensborough Corporation, Tidewater Oil and American Express.

And in the September 1922 issue of *Radio Broadcast* appeared what may be the first published rant against radio commercials, under the headline: "Should Radio Be Used for Advertising?":

> Anyone who doubts the reality, the imminence of the problem, has only to listen about him for plenty of evidence. Driblets of advertising, most of it indirect so far, but still unmistakable, are floating through the ether every day. Concerts are seasoned here and there with a dash of advertising paprika. You can't miss it; every little classic number has a slogan all its own, if it is only the mere mention of the name—*and* the street address, *and* the phone number—of the music house which arranged the program. More of this sort of thing may be expected. And once the avalanche gets a good start, nothing short of an Act of Congress or a repetition of Noah's excitement will suffice to stop it.

Later, when broadcast advertising had achieved the status of a fundamental engine of growth in the consumer society and the raison d'être for broadcasting itself, the idea of eliminating it would come to be seen as quaintly fantastic at best and downright unpatriotic at worst. In retrospect, though, it is difficult to find a point at which the idea of direct advertising on radio gained acceptance. Even in industry circles it was widely viewed with distaste. As late as 1925, a blue-ribbon committee of industry experts set up to mediate ongoing disputes between RCA, GE and Westinghouse, on the one hand, and AT&T, on the other, had this to say about advertising, in a report prepared for the companies involved:

> There is a natural conflict of interest between serving the broadcast listener [the purchaser of radio receivers] and serving the advertiser using broadcast facilities. The listener desires a program of the highest quality as free as possible from all extraneous or irrelevant material, particularly such as may be psychologically distracting from an artistic performance because of its commercial tinge. . . . The tendency of advertising programs is toward direct advertising . . . the tendency of stations devoted entirely to pleasing the listener and without toll payment for advertising programs is to minimize to the utmost any inartistic or irrelevant text.

The committee recommended that RCA avoid direct sponsorship of programming by advertisers on grounds that "the Radio Corporation of America is the largest radio sales organization in the United States. . . . [It] requires the good will of the broadcast listener and this might be jeopardized by the responsibility for the commercial success of toll advertising programs."

It also presented a concrete proposal for an idea that had been entertained in various forms since the advent of radio: a program-providing foundation, paid for by radio manufacturers and related industries on the basis of a percentage of their sales. The foundation would also accept advertisers' money, but would only permit indirect sponsorship of programming (i.e., the sponsorship format currently used by the U.S. Public Broadcasting System as opposed to the direct advertising appearing on other radio and television outlets).

The plan became a victim of the frenetic pace of evolution in the industry, overlooked rather than discarded, as issues of more pressing priority continued to demand attention. Nevertheless, as late as 1926, when Sarnoff and RCA launched NBC, the prevailing attitude on the boards of directors of both RCA and the new broadcasting network was quite clearly one of broadcasting as a public service, supported, but not driven, by commercial sponsorship. Unsponsored programming of special artistic merit, including broadcasts of symphony concerts, theater and opera, were frequent and education programming held a prominent place in the network schedules. It was, perhaps, a point of view that could only be nurtured at the breast of a quasi-monopoly like RCA.

Historian Gleason Archer notes that in its initial months of operation, NBC recorded mounting deficits, and goes on to observe: "A deficit, however, was an expected development. The directors . . . were inclined to view the matter as a semi-philanthropic activity which the leading corporations of the radio industry were joining hands in supporting. It was regarded, and characterized by [Chairman] Owen D. Young as 'an investment in the youth of America.'"[131]

The advisory board selected to guide programming on the new network reflected that point of view: it was decidedly heavy on university presidents, symphony conductors, clerics and philanthropists, and light on lawyers and businessmen.

However, the arrival on the scene two years later, in 1928, of the upstart Columbia Broadcasting System with its ambitious young owner, William Paley, forced NBC to take a more aggressive stance. Paley was able to put CBS well into the black in its first year of operation by focusing on commercial sales, and by capitalizing on the networking experience won by NBC and its predecessor, the AT&T network, over such a long time and at such great expense. Before long CBS profits were matching those of NBC, on a small fraction of NBC's total revenues. CBS had demonstrated that there existed a greater public tolerance for commercials than NBC and its parent RCA had believed possible. Paley very quickly became wealthy enough to support an extravagant Manhattan lifestyle, including a sumptuous townhouse and a private chef who had once worked for Enrico Caruso.

There was, as well, a shift in the nature of radio programming that arrived with the runaway North American hit show "Amos 'n' Andy." In it, two white comedians, Freeman Gosden and Charles Correll, played black buffoons with unblushingly stereotyped parody. Even as an independent syndication distributed to a handful of stations on shellac-recording disks, the show drew huge audiences which high-toned programmers at NBC could not ignore. Advertisers were dazzled by the numbers. Finally, NBC purchased the show in 1929 for $100,000 a year. It was the network's first foray into pop programming, and within months "Amos 'n' Andy" was drawing forty million listeners, or half the radio audience.

Over at CBS, the always-alert Bill Paley went on a crash program to shift his network's focus to popular programming as quickly as he could sign the artists. Paul Whiteman and his orchestra were hired for "The Old Gold Program" at $35,000 a week. Bing Crosby, a Hollywood crooner with a drinking problem, was rounded up and dried out for fifteen minutes a week on the network. The "True Story" drama series, based on a lurid pulp magazine, went into production. Soap operas dominated the afternoons. Comedians abounded.

With this kind of programming beginning to dominate the two early networks, advertising seemed less out of place than it once had in the context of more substantial fare. Network programmers and advertisers had at last apparently located the pulse of mainstream America's cultural sensibilities, at least as they existed in the 1930s. Alternatively, one could say that they had discovered a formula in which commercial messages seemed less intrusive, more suitable and appropriate. For, while some of the "populist" programming was extremely popular, much of it was not, and yet virtually nothing of the original middle-brow format of plays, lectures, serious music and informed discussion remained. Before long, the well-worn apologia for the draining of nourishment from mass programming would begin to be heard—that is, that commercial radio gives audiences what they ask for, what they want. It was, and remains, a specious argument based on a tautology, to wit: only programs that the public wants survive; if the program survives, the public must want it. Often, in reality, authentic public tastes had less to do with programming "success" or longevity than the skill of publicists and marketing specialists, network economic considerations, executive whims and, above all, the needs of advertisers.

The model of the sponsored time-block or program had placed radio content in the hands of advertising agencies, which acted as marriage brokers between the networks and potential sponsors. The agency naturally saw its first obligation as being to serve the interests of its client, the advertiser. Serving the public was a secondary, peripheral consideration. While its early proponents had seen radio as a potentially powerful tool of public enlightenment and cultural enrichment, to the ad agency it was, rather, "a latchkey to nearly every home in the United States," a sort of psychological Santa Claus (or burglar) slipping through the radio into the home, leaving behind commercial messages. To quote a leading marketing specialist of the day:

> For years the national advertiser and his agency have been dreaming of the time to come when there would be evolved some great family medium which should reach the home and the adult members of the family in their moments of relaxation, bringing to them the editorial and advertising message. Then came radio broadcasting, utilizing the very air we breathe, and with electricity as its vehicle entering the homes of the nation through doors and windows, no matter how tightly barred, and delivering its message audibly through the loudspeaker wherever placed. . . . In the midst of the family circle, in moments of relaxation, the voice of radio brings to the audience its program of entertainment or its message of advertising.[132]

And finally, the Great Depression would appear to have softened public attitudes to advertising on radio. At a time when radio provided the only professional entertainment many North Americans could afford, there was a natural tolerance of, even gratitude for, the corporate spending on advertising which made so many programs possible. With the end of the Depression came all-consuming war, and by the time it had ended the fact that radio had once had a genuine opportunity to be something other than a strictly commercially driven medium was but a fading memory of a hazy dream.

Why does it matter? Why is it significant that radio came to be sponsored by advertisers rather than by public funds or some other endowment system? It matters because combining commercial sponsorship with broadcast technology creates an irreversible slide toward the lowest common denominator in programming. It is inevitable, since broadcasting, by definition, makes no provision for addressing minority or niche interests. To be successful, broadcasters had to satisfy as many people of as many diverse interests and backgrounds as possible, thereby increasing the number of listeners. But they had just one vehicle with which to do that—their station or network and its programming lineup. Therefore, broadcasters had no choice but to tune that vehicle to the satisfaction of the greatest number of listeners throughout the broadcast schedule. In other words, the broadcaster had no choice but to aim for the lowest common denominator, wherein lay profit and satisfied shareholders.

It is also of no small significance that commercially sponsored programming exists primarily and fundamentally as a means of selling the products or services of the sponsor. This fact inevitably shapes the content of the sponsored program: certainly the show will in no way be allowed to contain material which the sponsor might find objectionable for any reason, and, where possible, it will also be expected to further the commercial interests of the sponsor in some direct or indirect way. Thus, in ways both blatant and subtle, commercialization of the medium shaped the program content it delivered.[133] This applied even to news programming. Throughout the political tinderbox of the 1930s, NBC frequently canceled programs that might "undermine the public confidence and faith"; Cincinnati powerhouse station WLW had an explicit policy: "No reference to strikes is to be made on any news bulletin broadcast over our station." This was censorship of real significance, since by 1939, 70 percent of the American populace reported that radio was their main source of news.

In a broader sense the commercialization of radio had effectively taken what had been a shared public resource and made of it a tool for the furtherance of the interests of one segment of society, i.e., business, and one social paradigm, i.e., the "consumer society." Whether or not that was in the interests of society at large remains a matter for debate: it is a debate, however, that one is unlikely to hear in a commercially sponsored broadcast.

In Britain, radio's development had taken a much different course following World War I. As in the United States, there had been the pressing problem of large radio manufacturing facilities built up during the conflict, with no peacetime market to serve. And, as in the United States, the manufacturers were anxious to develop broadcasting as a means of supporting equipment sales.

However, from an overall strategic point of view, there were important differences between the two countries. Chief among these was the fact that while the United States was trying to develop a globally competitive industry, Britain was in the position of a world leader trying to defend the status quo. There were strong vested interests in both countries, but in Britain the more conservative among them would have the upper hand. The Royal Navy shared the view of American admirals that private radio would interfere with military communication. The Post Office had long had control over telephones, telegraphy and radiotelegraphy and felt it only natural that its bureaucratic prerogatives should extend to the control of radiotelephony, or broadcasting, as well.

The British upper classes were generally disdainful of radio broadcasting. In G. E. C. Wedlake's *SOS: The Story of Radio-Communication,* the story is told of how Sir Harold Nicolson came to hear Prime Minister Neville Chamberlain's declaration of war against Germany over the radio, on September 3, 1939. Nicolson was visiting friends in West End London: "His host had no radio set, but the housemaid had one, and just after eleven she brought it into the room where they were sitting. It seems strange that as late as 1939 only the servant in a well-to-do household had a radio receiver. It showed, in fact, that in certain circles radio has still not been accepted. It was something for the masses; intelligent people did not listen to it."[134]

A handful of experimental broadcast stations appeared in Britain following the war, run by Marconi, Western Electric, Metropolitan-Vickers and others. Broadcasts from Europe, particularly the Sunday concerts from The Hague, were listened to avidly. British broadcasters were handicapped by a ludicrous Post Office requirement that they go off the air for three minutes out of every ten so that an operator could check to see whether interference was being caused with any commercial radiotelegraph station. Despite this, the public response was only slightly less enthusiastic there than in North America, and it was clear that the government was going to have to allow broadcasting sooner or later. And so in April 1922 a gathering of interested parties was organized by the Post Office. The manufacturers were there, of course, as well as the armed services, the Foreign Office, the Colonial Office and the Board of Trade. The Post Office claimed jurisdiction over broadcasting based on the authority granted it in the Telegraph Act of 1869 and the Wireless Telegraph Act of 1904. There was grumbling over the fact that neither act mentioned anything about radiotelephony and the issue was raised in the Commons, but it was never tested in court.

As it had been in the United States, the ownership of patents was a crucial issue. Marconi, which controlled most of them, made it plain that if other companies were to be given government permission to operate broadcast stations, they would not be given access to the patents they would need to build either transmitters or the receivers they hoped to sell to the public. The Post Office, for its part, was opposed to giving Marconi or any other single company a monopoly on broadcasting. It was decided that broadcasting in Britain would be carried on by a single company operated by a board of directors drawn from the Post Office and the country's six biggest radio manufacturers. Given the prevailing attitude among the ruling classes, there was never any doubt that it would be noncommercial radio; private companies would have to make their money by selling receivers and other equipment. The broadcaster itself would be financed by a license fee of fifty pence charged with every receiver sold. Half that money would go to the Post Office and half to the broadcaster. When the company's charter expired in 1926, it was reorganized as a Crown corporation and given additional financing from the government treasury. Its staff by then had grown from four to nearly six hundred.

The first BBC station opened on the roof of Marconi House in The Strand, London, on November 14, 1922, with a one-hour musical concert. Other stations were quickly opened in Birmingham, Manchester, Cardiff, Bournemouth, Newcastle, Glasgow and Aberdeen. The restrictions under which the BBC had to operate give some idea of the latent hostility to the medium among those who governed it. For a start, it could broadcast only between 5:00 P.M. and midnight. Thanks to effective lobbying by the nation's newspapers—which were major customers of the

Post Office telegraph department—the BBC would not be permitted to broadcast news before it had appeared in print, which in practice meant before 7:00 P.M. The BBC could broadcast live speeches, but could not comment on them; it could broadcast the atmosphere and activities surrounding major sporting events, but not the events themselves. It was, above all, forbidden to broadcast anything "controversial."

Hemmed in by all these rules, it is a wonder the BBC could build enough credibility with its audience to survive, and perhaps it wouldn't have if there had been competition from the private sector. But survive it did, to become the world's most respected broadcast organization. The competitive, innovative environment in American radio ensured that the United States would dominate world markets in radio manufacturing, but the quality of programming in Britain was unsurpassed. World War II was the corporation's turning point. Its wartime broadcasting gained it the respect and admiration of Britons of every class, and of people around the world. It never looked back. Today, when Japan and other Asian countries dominate the manufacturing sector in broadcasting, and the United States remains saddled with the abysmal level of quality in programming that is dictated by commercial considerations, it might well be argued that Britain made the wiser choice in the long run.

Canadian broadcasting, destined to become in many ways the most technically sophisticated in the world thanks to the demands placed on engineers by geography and language (Canada is officially bilingual), has had a schizophrenic relationship with commercial sponsorship. The first national radio network in the country was constructed by the Crown-owned railway, the Canadian National, as a commercial enterprise. It was nationalized in 1932 to become the foundation for a new public broadcaster. Private broadcasting has since that time been permitted to thrive alongside the public broadcaster (the Canadian Broadcasting Corporation, or CBC) and at times both CBC radio and television themselves have been partly commercially sponsored as well. Competition from privately owned national networks has been prohibited in radio; in television, one private network (CTV) has been licensed. Currently, CBC TV is supported partly from the public purse and partly from advertising revenue, while CBC radio is commercial-free.

By the late 1920s, the die had been cast where "public" use of the airwaves was concerned. On the one hand were islands of state control like Britain; on the other, what had been a vast public resource was now devoted almost exclusively to the furtherance of commercial economic interests through direct, on-air advertising. But commercial influence did not stop there; it spilled over into every aspect of the medium, as it would in radio's stepdaughter, television.

CHAPTER SIXTEEN

Television

Television is the problem child of communications technologies. Anticipated as a natural and inevitable outgrowth of radio and motion picture technologies from the earliest days of radio experimentation, it was to incorporate radio's principal drawbacks as a one-way medium, and add several serious problems all its own. At the same time, it was seldom, if ever, able to attain the high standards of artistic merit reached in the cinema, thanks to its small screen and to the adoption of radio's advertising-supported model. As a very "cool" medium, in McLuhan's terms, it demanded complete attention and created a vegetative subspecies of humanity aptly named the "couch potato." Far from bringing the best of Western culture to its audience, as it had the potential to do, it systematically debased public tastes as it sought out the lowest common denominator in order to maximize audience size. Its executives argued that it gave people what they wanted. However, TV is not squalid because people are vulgar; it is squalid because people share the same primitive impulses and desires, though they are widely and wonderfully diverse in their higher interests and aspirations. Television exposed generations of children en masse to puerile, commercial-ridden pap and shocking violence, and then insisted that it was all harmless, good fun. There have been high points as well, most of them connected with TV's ability to bring important news into the living room, but even these have been ambiguous triumphs. The images carried on television news shortened the Vietnam War, but television conspired with the military to mask the horror of the Gulf War behind the telegenic myth of "surgical" smart bombs and civilian-friendly cruise missiles. TV shocked the United States out of its complacency toward Southern school segregation, but at the same time perpetuated racial stereotypes in its dramas and sitcoms. TV helped speed the collapse of the Soviet Empire in Eastern Europe, but then promoted a consumer ethic emerging democracies could ill afford.

On balance, it is possible to argue—certainly in North America—that through most of its existence television has been more bane than benefit to society, a fact that makes it unique among the technologies of communication. It is only in recent years, as television has shattered into hundreds of specialty channels, weakening the grip of the network structure and providing viewers with a semblance of real choice in programming, that the medium has begun to show some net benefit to society. The social and political impact of both radio and television will be discussed in some detail in Chapter 18; the focus here will be on the technology behind television, a subject which holds the keys to many of the medium's shortcomings.

In 1873, Willoughby Smith, a senior technician employed by the Telegraph Construction and Maintenance Company of Britain, the firm that had laid the first successful transatlantic cable in 1865, was assigned the task of finding a way to continuously test undersea cables while they were under construction. To work out an idea, he needed a material that had a high resistance to electrical current, but was not a complete insulator, which could be used as a stand-in for many miles of copper cable. He decided to try bars of crystalline selenium. In his crowded lab, the selenium was placed on a windowsill. Before long, Willoughby was surprised to notice that the current flowing through his test circuit seemed to be changing with fluctuations in light coming through his window; the more light, the more voltage his meters registered. Evidently, the resistance of selenium dropped markedly when it was exposed to light.

It was an exciting discovery. Alexander Graham Bell was among the inventors who went to work on applications for it, and during a lecture in 1878 he told his audience that it ought to be possible to "hear a shadow fall" on a piece of selenium wired into a telephone circuit. He would develop that idea into what he came to consider his best invention, the "photophone," a wireless telephone. Bell mounted a mirror on a tightly drawn diaphragm of canvas, and shone a strong light on it. Words spoken into a tube placed close to the mirror caused it to vibrate, in effect modulating the reflected light with the voice's vibrations. On the receiving end, a parabolic mirror collected the light reflected from the mirror and focused it on a selenium cell. When the received light varied, so did the current passing through the selenium from a battery attached to it. That variable current, when applied to a telephone receiver, perfectly reproduced the words spoken at the transmitter. The only drawback, and it proved to be fatal, was that the photophone worked only over short, line-of-sight distances, in perfect atmospheric conditions.

Bell's work on the photophone, however, was not wasted. An English scientist named Shelford-Bidwell, in tinkering with a replica of the photophone, had the inspiration that it ought to be able to transmit images; that is, changes in the intensity of light captured in a black-and-white photograph could be converted into a fluctuating electrical current by selenium, and that current could be sent down a telephone wire. At the receiving end, the photograph would be reconstructed by reversing the process and changing the fluctuating current back into varying intensities of light. Shelford-Bidwell had invented the fax machine. In describing it to the Royal Society in London in 1881, he said with commendable modesty: "I cannot but think that it is capable of indefinite development, and should there ever be a demand for telephotography, it may turn out to be a useful member of society."

The problem Shelford-Bidwell had solved was a daunting one; normal telephony by comparison was simple. All a telephone had to do was carry a continuous stream of data as it emerged from the speaker's mouth in discrete syllables or bits. But what if what the speaker was saying had been presented all at once, as a page of dense text instead of a linear stream of bits? Clearly, the entire page could not be transmitted en bloc; it would have to be sent down the line letter-by-letter and word-by-word so that the page could be reconstructed at the other end. That was essentially the difficulty facing the transmission of pictures, and the solution lay in finding a way to break them up into bits appropriate for transmission in a linear stream of electrical impulses and then reassemble into a two-dimensional image at the other end. Shelford-Bidwell's contribution was the idea of *scanning*.

Shelford-Bidwell's contraption can be considered in direct line of development with television, in that it provided a method of converting photographic information into electrical energy for transmission. It worked like this: a photograph was developed onto a glass plate. A strong light was shone through the plate onto a rotating cylinder about the size of a soup can, in the center of which was a selenium cell. The cylinder was pierced by a pinhole, and as it rotated, it moved slowly from left to right, scanning the photograph onto the surface of the selenium cell. With each rotation, the pinhole traversed or scanned a narrow band of the photographic image, directly adjacent to the line scanned on the previous rotation. The light falling on the selenium changed according to the shading of the photograph, and the selenium cell produced fluctuations in electrical current that were sent down wires to the receiving apparatus. The receiver used a stylus that mimicked the transmitter's scanning of the photograph, moving back and forth across a sheet of paper that had been chemically treated to make it sensitive to electricity, darkening when exposed to a current. Since the stylus was wired into the circuit carrying the fluctuating current from the transmitter, as long as it could be synchronized to move across the paper in exact concert with the motion of the scanning cylinder at the transmitter, the photograph would be accurately reproduced. Shelford-Bidwell fudged a bit here: his desk-top demonstration model had transmitter and receiver hooked up to the same motor, ensuring that they would be in perfect synchronization. This, of course, would have been impractical in a commercial model.

By 1905, Dr. Arthur Korn of Germany had perfected the idea. An ordinary photograph was wrapped around a cylinder and placed in the transmitter. As the cylinder rotated, a bright point of light moved along it, and the reflection of that light was captured by a selenium cell for transmission as fluctuating current down telephone lines. At the receiver, a sheet of photographic paper was wrapped onto a cylinder and loaded into the machine. The incoming current regulated the intensity of a small point of light scanning the spinning photographic paper, which was then developed in the usual way. The quality proved good enough for reproduction in newspapers and the wire photo was born, a major boon to that image-hungry industry.

It was only a small step to adapt the photo fax to radio transmission, and in 1924 the British Marconi Company and RCA staged a transatlantic swap of photos of President Calvin Coolidge and the Prince of Wales. It is a little-remembered curiosity that for a time the idea of the home facsimile machine for receiving broadcast "newspapers" in overnight transmissions seemed poised to become the Next Big Thing in communications technology. By 1940 there were forty commercial photo fax radio stations in the United States, and more than ten thousand receivers. The industry died with the spread of television, a superior broadcast medium in that it could provide moving as well as still pictures (though they could not be stored or printed).

The problem of transmitting moving pictures, though it sounds daunting, was really only a little more difficult than still pictures. The earliest working televisions, in fact, had a lot in common with the facsimile machine. In 1884, Paul Nipkow of Germany patented a system that used a perforated disk spinning in front of a selenium cell, to change a moving image projected onto the disk into electrical information for transmission to a receiver. The perforations on the disk were arranged in a spiral, so that the entire image would be scanned from top to bottom with each rotation. The perforation at the outer rim of the disk scanned the top few millimeters of the image as it passed over it, then the next perforation scanned a few millimeters lower and so on. On the next rotation, the process was repeated. At the receiver, the fluctuating current from the selenium cell was reproduced as fluctuating light, by an electric lamp wired into the circuit. A second spinning disk, identical to the first and revolving in perfect synchronization in front of the lamp, reproduced an image of the transmitted, moving picture. If the disks could be made to rotate fast enough, the eye would see the image as being in continuous motion rather than a separate series of pictures, thanks to the feature of human vision known as "persistence"—the eye retains an image for about 0.1 seconds after the light fades from it. Thus, if the image is refreshed more than ten times a second, the eye will perceive continuous motion. (In practice, to avoid flicker the refresh rate needs to be twenty-five to thirty times per second.)

It was Nipkow's system, refined by J. L. Baird, that formed the basis for the BBC's earliest experiments in television between 1929 and 1935. By then it had become clear that adequate picture definition would require a scanning rate much higher than thirty lines a second; three hundred was closer to the magic number, and that speed of operation proved impractical for the mechanical system. The problem awaited an electronic solution.

That solution is commonly credited to a television research group at EMI (Electrical and Musical Industries) in Britain, and in the United States to Russian-American émigré named Vladimir Zworykin. Zworykin had fled the Russian Revolution, taking with him an idea for an all-electronic television system, and had landed a research job with General Electric, where he worked on his scheme. But he was not satisfied with the level of support he was getting from GE, and when fellow émigré David Sarnoff of RCA offered him a position, he jumped ship. Sarnoff was convinced that television would eventually make radio obsolete: the "supplantive theory" of technology, he called it. He was wrong, of course, but his enthusiasm for the new medium was enough to secure Zworykin's loyalty. That, and a promise of $100,000 in development money. In 1932 Zworykin was able to demonstrate a complete television system at the RCA labs, using 120 scanning lines per second, which was quickly upgraded to 343 lines. (In 1936, the BBC adopted the EMI system which used 405 lines per second.[135]) At the New York World's Fair in 1939, Sarnoff personally introduced the age of television with live broadcasts from the elaborate RCA pavilion, including a cameo appearance by U.S. President Franklin Roosevelt.

Many, if not most, modern references give credit to Zworykin, Sarnoff and RCA for the invention of electronic television, and not without reason. Sarnoff was a consummate mythmaker and had a powerful radio and television network at his disposal to assist him in promoting RCA's achievements, real and fanciful. His personal mythology was an example of his talent: it included the "fact" that he had almost singlehandedly taken care of radio traffic between the doomed *Titanic* and, later, the rescue ship *Carpathia* while a junior operator for the Marconi Company in 1912. His biography as prepared by the RCA public relations department contained this passage:

> On April 14, 1912, he was sitting at his instrument in the Wanamaker Store in New York. Leaning forward suddenly, he pressed the earphones more closely to his head. Through the sputtering and static . . . he was hearing a message: "S.S. *Titanic* ran into iceberg. Sinking fast." For the next seventy-two hours Sarnoff sat at his post, straining to catch every signal that might come through the air. That demanded a good operator in those days of undeveloped radio. By order of the President of the United States every other wireless station in the country was closed to stop interference. . . . Not until he had given the world the name of the last survivor, three days and three nights after that first message did Sarnoff call his job done.[136]

The paragraph contains several errors of fact that could only have been deliberate. April 14 was a Sunday and the department store was closed when the collision occurred at 10:25 P.M. Sarnoff put his headphones on the Monday morning following the sinking. Had Sarnoff remained at his post as described, he would have been there continuously for a hundred, not seventy-two hours, an unlikely feat. He would likely have been unable to hear *Carpathia* and would have acted as a relay for stronger ships' stations and shore operations. In fact, the main role played by Sarnoff (and two other Marconi operators at the Wanamaker rooftop station) was to compile transcriptions of radio traffic for the Hearst newspaper *American*. Five days after the sinking and before *Carpathia* had reached New York, radio interference had grown so intense that Marconi shut down all but four of its own stations, and the Wanamaker store was among those closed. It is another fact of the historical record that a complete list of survivors and victims never was received on shore until the *Carpathia* docked, despite the fact that several well-heeled *Titanic* survivors were able to use the *Carpathia*'s radio to book rooms at New York hotels.

The true story of the invention of electronic television is that all of the important patents to the system eventually marketed by RCA were held by an independent mathematician and inventor from the unlikely state of Utah, who rejoiced in the name of Philo T. Farnsworth. Farnsworth conceived of the electronic image scanner and the picture tube as a youngster of sixteen, and had patented these and many other associated circuits before he reached twenty. His genius put him on a direct collision course with the industry colossus, RCA, and its hard-driving president.

Thanks to the rather lenient terms of the antitrust "consent decree" of 1933, Sarnoff's RCA and AT&T continued to share virtually all of the important patents in the field of radio, and were thus able to decide who would and who would not be allowed to license rights to those patents. It was an important means of controlling competition, and Sarnoff, according to legend, would often assert, "RCA doesn't pay royalties; we collect them!" He had every intention of extending this arrangement to television.

Sarnoff sent Zworykin to pay a visit to Farnsworth's West Coast lab to find out what he could about the television system the young man was developing. Zworykin left impressed by what he'd been shown, but confident he could engineer a way around Farnsworth's patents by building on his own ideas for the "iconoscope" picture tube. Farnsworth had in any case refused an offer to sell his invention to RCA, as he would continue to do, preferring to maintain control of its development himself. He was not the first (though he may have been the last) independent inventor to snub his nose at the mighty RCA, and Sarnoff thought he knew how to deal with him.

The techniques were by now well-established: flood the media with stories of RCA's leadership in the field to starve the competitor of capital investment and shake his morale; spend whatever it takes to engineer alternative solutions to the technical problems at issue; attack the upstart's patents in court, using the full weight of RCA's awesome legal resources; continue offering the carrot of a patent buyout. Invariably, sooner or later the lone inventor would crack or run out of money or both, and RCA would buy up his intellectual property.[137] It was a strategy RCA had used against Edwin Armstrong, perhaps the most brilliant and certainly one of the most prolific inventors in the field of radio, responsible, among many other breakthroughs, for FM (frequency modulation). Armstrong, following years of bitter, ruinously expensive litigation to protect his patents from RCA (and others), killed himself in 1954.

However, Farnsworth's patents were to prove legally impregnable, and Zworykin's attempts to engineer alternatives were unsuccessful with the iconoscope, which proved too crude to be practical for commercial applications. When the RCA labs did come up with a successful camera tube in the "image orthicon" in 1937, Sarnoff felt confident enough to begin planning television's official inaugural for the 1939 World's Fair. AT&T had by then patented another crucial element of the TV network, coaxial cable capable of carrying television signals over long distances, and RCA could look forward with confidence to a patent-sharing in television on the model of the successful radio arrangement.

But AT&T, under increasing antitrust pressure from Washington, stunned Sarnoff and the rest of the electronics industry by offering a cross-licensing patent agreement to Farnsworth, in a dramatic public gesture during a hearing of the Federal Communications Commission. Farnsworth immediately accepted. Then came more bad news for RCA: lawyers preparing patent papers for the image orthicon discovered patents held by Farnsworth for the past four years, covering all of the tube's important functions. These were of course challenged, but with the distressingly familiar result of Farnsworth's priority being upheld in court.

Finally, RCA capitulated and agreed to negotiate a licensing arrangement with Farnsworth. The deal was signed without fanfare in a Radio City boardroom in New York in 1939. It was the only instance in which RCA had found it necessary to license rather than purchase intellectual property, and it was a bitter pill for Sarnoff to swallow. But it allowed television to finally reach the consumer market.

Early electronic television was relatively simple compared to today's sophisticated color technologies, but the operating principles remain much the same. The image to be transmitted was projected onto a small screen inside a vacuum tube. The screen was coated with millions of tiny, light-sensitive metallic particles. Where the Nipkow system had scanned the image using the clever mechanical expedient of a spinning, perforated disk, electronic systems employed an electron gun. Similar in principle to the filament of a vacuum tube, the electron gun is an emitter of a focused beam of electrons. It was built right into the tube containing the photosensitive screen, and together they comprised the principal element of television cameras for more than forty years. (Modern cameras use solid-state electronics.) The camera tube's electron gun was able to scan the image projected onto the photosensitive screen by means of electromagnetic collars which deflected the electron beam back and forth across the screen from top to bottom several hundred times a second. As each individual segment of the screen was bombarded with electrons, it emitted electrons according to the brightness or darkness of the picture element projected upon it. These electrons were collected at the back of the screen as electric current. It is this current which was used to modulate the signal transmitted to the television receiver.

The receiver's picture tube (kinescope or cathode ray tube) was similar to the camera's. An electron beam was made to sweep across the luminescent material coating the viewing surface of the tube, in exact synchronization with the electron beam in the camera, replicating more or less exactly the brightness of each camera-scanned picture element. It sped back and forth from top to bottom of the screen, many times per second. (The synchronization was handled by a "synch code" transmitted along with the picture information.) Today's color television works in an analogous way, only with separate electron guns for different primary colors.

World War II shut down television broadcasting in Britain and severely restricted the schedule of NBC in the United States. The BBC had been broadcasting two hours a day, six days a week since November 1936 (noncommercially), and NBC had formally begun its TV operations three years later in April 1939, broadcasting from the Empire State Building. NBC programs were sponsored by advertisers from the outset. In 1940, there were about 10,000 TV sets in the London area, and about 3,000 in New York. Following the war, interest in television skyrocketed in the United States: in 1949 there were a million receivers in use; by 1951 10 million and by 1959, fifty million, twice as many as were in use in all other countries of the world combined.

It is perplexing how television, a medium with so much potential, became such a letdown in practice. We've already touched on the answer in observing that television shares two serious technological handicaps with radio: as a broadcast medium (and especially as a commercial broadcast medium), television must accumulate mass audience. Because it was chained to point-to-multipoint technology in an environment of limited numbers of stations, it could do this only though "mass appeal" programming. The other shared handicap was that, as an analogue medium anchored in linear time, television broadcasting began its day at sign-on and ground inexorably through to the national anthem at sign-off (or to an all-night orgy of infomercials). The audience had no choice but to watch by appointment only, tailoring their lives to the demands of the schedule.[138]

In a way, television was also the first entirely passive form of mass entertainment. It requires no involvement from its audience—certainly no serious intellectual involvement—and in fact demands all but complete immobility and mental disengagement in front of the glowing screen. Only the thumb moves, on the zapper: "When I was your age, we had to *walk* to the television to change channels," the wisecrack goes. Radio, theater, books and cinema all require the participation of the audience at one level or another, if only in applause or the exercise of imagination; other media involve a conversation between the user and the source. We bring to a book, for example, our literacy, our imagination, our interpretive skills. The cinema, at its best, engages the same talents in its audience, as well as stimulating their visual aesthetic sensibilities. TV asks only that the viewer show up and it will do the rest. Even laughter is supplied. As audience stupefaction sets in, as it must, television must create tricks and techniques to keep its viewers from drifting off—techniques like the video jolt, the false climax, the flash of nubile flesh, the shock of violence. In so-called talk programming, civility and erudition are not effective: aggressive confrontation, bald assertion and the rhetorical expletive are what's demanded.

There is another sense, however, in which television is not a "passive" medium at all. It engages us by the millions, and brings us back for more. It obviously is having some sort of impact on us as we watch. An important clue to understanding this seeming paradox lies in McLuhan's cryptic dictum to the effect that we don't watch TV; TV watches us. Or as New York ad executive and TV critic Tony Schwartz put it: "TV is not a window on the world, it's a window on the consumer."[139] In other words, television should be seen as a mental programming or conditioning agent, which is actively invading us whenever we subject ourselves to its siren appeal.

This is seen most explicitly in its advertising, of which Neil Postman writes:

> The television commercial is not at all about the character of products to be consumed. It is about the character of the consumers of products. Images of movie stars and famous athletes, of serene lakes and macho fishing trips, of elegant dinners and romantic interludes, of happy families packing their station wagons for a picnic in the country—these tell nothing about the products being sold. But they tell everything about the fears, fancies and dreams of those who might buy them. What the advertiser needs to know is not what is right about the product but what is wrong about the buyer. . . . The consumer is a patient assured by psycho-dramas.[140]

It could be argued that this is true of all information media, that they all, to a degree, invade us when we open ourselves up to them. But McLuhan and current media theorists like Derrick De Kerckhove argue that television is fundamentally different in the way in which it communicates. Whereas a book, for example, engages us through our intellect, television acts directly on our neuromuscular system. Each rapid-fire edit, each "jolt" provided by TV, sets up in our bodies what is known in clinical psychophysiology as an "orientation response." This subliminal reaction, which can be monitored with the appropriate equipment, prepares us to either examine the object or event or withdraw from it. Clinical observation has determined that it takes, on average, about half a second for an individual to absorb the nature of the "occurrence" and decide how to react. In making that response decision, the tension of the "orientation response" is resolved. Television, De Kerckhove and others have argued, is designed to deny its audience that half-second response time and the subsequent resolution, and thus to maintain a high level of tension in which rational thought is suppressed.

With the viewer's higher mental functions effectively short-circuited, television is free to convey its message by direct communication with the body. Anyone who doubts the impact of the body, i.e., the neuromuscular system, on brain organization and learning need only recall the process of learning to ride a bicycle. The first attempt is always a failure, but as the patterns of synaptic responses in the brain are established and reinforced, subsequent attempts become more successful. Eventually, the brain has been effectively "rewired" and riding becomes second nature. Conditioned neuromuscular responses caused by television viewing also affect brain organization and learning. Of course, messages conveyed through the body cannot be as subtle or sophisticated as messages directed to the intellect, which may explain the fact, for instance, that Americans who watched television coverage of the Gulf War were found to have concrete knowledge of the conflict in inverse proportion to the amount of coverage they watched. One suspects that the degree of support for Allied involvement in the war would have been highest among those who watched a great deal of the TV coverage. The message of television coverage, after all, was that the war was being fought between the forces of good and evil. That this simplistic message was very carefully contrived for television (by the Pentagon, but also by advertising and public relations firms in New York and elsewhere acting on behalf of participating governments) is well documented. It is a message that could not have been conveyed to such effect in any medium that encourages rational reflection. They don't call it the "idiot box" for nothing.

If it is the speed of editing and the frequency of jolts on television that makes it different, how did it get that way? The answer is, advertising. Producers of television advertisements discovered all of the essential tricks of speaking directly to the body, bypassing the mind, early on in the game. It is a skill that has been honed to a fare-thee-well over five decades of trial and error and ruthless, Darwinian competition among Madison Avenue practitioners. It remains more art than science, despite such techniques as electronic monitoring of focus groups, and those who are good at it can charge astronomical fees. Just as the advent of commercial sponsorship on radio led to a change in radio programming to forms of expression which more closely matched the commercials (see

Chapter 16), so, too, television commercials have had a profound impact on the production techniques used in all forms of television programming, including news. It is not necessary to propound a conspiracy to account for this change: the symbiotic relationship between programming and advertising in the broadcast media is explanation enough. In television, the first priority is to have effective commercials. If that means quick cuts and many jolts per minute, then that is what will be done. If entertainment or news content then looks somehow dated or if the commercial jars or offends in context with slower-paced, more thoughtful programming, then the programming, and not the commercial, will have to be changed. The melding of the production values of advertisements and nonadvertising content is largely an unconscious process, but it is inevitable and irreversible. Because it has completely transformed American television, and because American television dominates the world market, the impact has been transferred to some degree to television producers everywhere, even those who produce programming for noncommercial outlets.

With all its problems, it could be argued that television has formed a kind of prison of the mind for several generations of viewers. But it has not been escape-proof. Television's blandishments have been strenuously resisted since the medium's birth, and in recent times, technology has come to the aid of those who object to the total subservience the medium demands. Firstly, the problem of linearity of schedules and the "appointment viewing" it demands (viewers must make themselves available at the time the program is broadcast) has been addressed in two ways, neither of them entirely satisfactory. The first is the invention of the videocassette recorder or VCR; the second is the development of the specialty channel on cable (or satellite) television.

With the VCR it is no longer necessary to be on hand in front of the TV at a specific time in order to catch a favorite show: it is possible (in theory, at least) to program your VCR to record it and play it back later on, a process the industry refers to as "time-shifting." This is far from a fully adequate solution; programming functions of VCRs became synonymous with bad interface design and were found to be so annoyingly complex by most owners that they were seldom used for recording. As well, so much promotion goes into network TV schedules that a taped program seems somehow dated, out of context and unappealing; the sad fact is that few programs have enough intrinsic value to warrant the trouble and expense of putting them on tape. Most of them have value only as a diversion of last resort.

A majority of owners use their VCRs primarily for watching prerecorded rental movies. In most households, this means that television viewing per se declines, since the VCR is typically connected to the family's only (or main) set. This is indeed an unpleasant development for television programmers and advertisers. Advertisers are further disadvantaged by VCRs because they allow their users to either delete commercials in the recording process or to "fast forward" through them during time-shifted viewing.

Even more significant has been the advent of cable television. What we now call simply "cable" got its start as CATV (Community Antenna Television) immediately after World War II, as a means of providing television signals in locations shielded from on-air transmissions by mountains or other geographic barriers. CATV coaxial cables linking homes to the cable operator's remote, very tall antenna system had bandwidth aplenty and cable operators soon got the idea of adding their own channels to those picked up off the air. From this, the specialty channel developed, able to serve niche audiences because of its low operating costs (relative to broadcast TV) and dual revenue sources—advertising charges and subscriber fees. The all-news channel was born in the guise of Ted Turner's Cable News Network, or CNN, to be followed by the all-music channel and channels devoted to travel, cooking, shopping, the courts, Parliament and Congress, science fiction, comedy, animation, sex, movies old and new, NASA, women's issues, history, the performing arts, every variety of sports, foreign-language programming, rock videos, television listings, the weather, fitness and health, children's programming and so on. Before long, the field had exploded from twelve or fifteen selections on an average cable system to more than one hundred, with five hundred channels touted as being just around the corner. This greatly expanded both program choice (though a great deal of the specialty content was simply recycled network programming) and scheduling opportunities: most cable channels operated on a rotating schedule in which programs were repeated several times in a week.

Program choice offered the viewer saw further expansion in 1995, when direct-to-home (DTH) satellite television came of age. Satellite TV had been available in North America for a decade, but market penetration had been severely limited by the problem of the eight-foot parabolic dish antenna required to receive signals from relatively low-powered satellites. Its customer base had been confined mainly to rural households where cable was not available, and suburban areas where lots were large and zoning bylaws lenient. Owning such a dish was an outward and visible sign of the importance of television in a family's life, and was, in the eyes of some, the reverse of a status symbol: in cartoons in the *New Yorker* the satellite dish was an essential artifact in any hillbilly landscape. But a new generation of powerful satellites using digital technology required only a discreet, eighteen-inch receiving dish, and DTH television quickly became a direct competitor to cable, offering essentially the same services only in greater abundance and with superior (digital) technical quality. In Britain and Europe, where small-dish DTH technology had been in place for some time, North American-style cable access was introduced at about the same time, becoming a strong competitive factor and offering in some jurisdictions both cable TV and telephone service on the same line.

Another problem of linear, analogue broadcast technology (a problem, at least, for the viewer) is that advertising on television has to be intrusive, has to interrupt the flow of programming. In a linear environment, it cannot be otherwise. Viewers are forced to sit through it in order to get to the next scene or program. A deceptively innocuous gadget called the wireless remote control—the "zapper"—has addressed that issue. Once again, however, it is far from an adequate or satisfying solution to the basic problem of analogue linearity. Nevertheless, it can be argued that the zapper is a tool of enormous significance, because it does much more than make it possible to change channels or mute sound without having to rise from one's easy chair. The zapper put a crude editing tool in the hands of the viewer, giving him or her, after decades of enforced passivity, some rudimentary control over the content on the screen. Like the video games children of the eighties and nineties have grown up with, the zapper may be said to have helped prepare audiences for the advent of truly interactive media—the digital media supported by the Internet, CD-ROMs and so-called full service digital delivery systems being tested by cable companies and telephone systems. To this extent, the humble zapper may be seen as the foot in the door to a wholesale conversion of television to interactive, digital media—its undoubted fate.

For all of the plethora of choice, and for all of the Band-Aid patches to its problems, from the zapper to the specialty channel, television programming has remained stubbornly and disappointingly prurient, banal and trivial. In the new world of audience fragmentation across an increasingly broad base of program alternatives, the "vast wasteland" only seems to expand, as deserts will when there is overgrazing and deforestation along their rims. It becomes inescapably clear that the key to improved content on TV is not simply in having more channels, and that the medium, in its commercial manifestation, is irredeemably flawed by its technology. In a sense, the problem boils down to simply one of time: trapped in linear time, broadcast television has to fill all those long spaces between commercials for a full broadcast day, 365 days a year. Multiply that by the number of networks and add to the equation the technologically driven pressure to serve only the most broadly based tastes, and the result is predictable—and viewable every day.

The different experiences of the two television systems, the American commercial broadcasting networks and the public broadcasters like the BBC, can be illustrated by the fact that the BBC unabashedly, though by no means exclusively, catered to what North Americans might consider an "elite" audience both in the arts and in news and information programming. This focus would not, of course, have been practical for a commercial broadcaster. It was also possible for the BBC, as a public broadcaster, to mandate several different strata of programming, from highbrow to pop, on different television channels (and radio frequencies). This created a steady market for programming across the full range of tastes. Commercial television, permitted after 1952 (commercial radio had to wait until 1972), was limited for decades to a single channel while the BBC had two, and the result was that in a predominantly non-commercial environment programmers at the advertising-supported outlets had to strive to meet the relatively refined tastes firmly established among viewers by the BBC. The effect was that Thames Television and Channel Four, to name two of the commercial outlets, often matched the BBC in program quality.[141]

The Canadian compromise of a CBC (Canadian Broadcasting Corporation) television network supported partly by commercial advertising revenue and partly by government subsidy provides ample proof, if any was needed beyond common sense, that public and private broadcasting cannot coexist within a single organization. The compromise was arrived at after a Royal Commission study had recommended Canada establish a noncommercial, BBC-style public broadcasting system as a bulwark against the overwhelming cultural pressure exerted by the United States, just to the south and with ten times the population and an entertainment industry to match. Canadian broadcasters, who had been getting rich rebroadcasting American programs, violently objected and were listened to by a series of sympathetic Conservative governments in Ottawa. The upshot was a CBC with only a handful of owned-and-operated television stations, the rest of the transcontinental network being served by privately owned stations that had agreed to carry a certain amount of CBC programming each day.

CBC television's legislated commitment to the altruistic principles of public broadcasting prevent it from wholeheartedly embracing the kind of content that would maximize commercial revenue. But the pressing need for substantial income beyond its government subsidy drive it to corrupt its cultural and public service programming, diluting it to standards acceptable to a very broad audience. The result is a program schedule that draws flak from both ends of the critical spectrum: some deride it as being fusty and elitist, others for being crassly commercial. Increasingly, it is being seen by tax-weary Canadians as simply irrelevant and therefore expendable. The same is not true of CBC radio, which divides its widely diverse but always intelligent programming between two networks in each of the official languages, and which carries no commercials. Its large audience is ferociously loyal. As the CBC is subjected to increasing financial pressures from a federal government intent on minimizing expenditures, the corporation reacts, quite naturally, by cutting back, not in revenue-generating areas such as professional sports coverage, but in precisely the noncommercial areas that give it what relevance as a public institution remains to it. Radio, despite its success, is cut as severely as the much less popular television service. It is a recipe for frustration, schizophrenia and, ultimately, failure—an ignoble experiment entered into by a series of Canadian federal governments unwilling to resist self-serving pressures from commercial broadcast interests.

As it was with radio, commercial support proved to be a Faustian bargain for television. A commercial broadcast medium in a market economy, it is worth repeating, is inevitably little more than a vehicle for advertising in which all other content survives only to the extent that it furthers the primary goal of bringing an audience to the ads. Broadcasters like to say that advertising supports programming, but that is incorrect. Programming supports advertising, and if it does not, it is cancelled. There was a time when this might have been an arguable proposition. Even Philo Farnsworth, who hated much of what he saw on television and often wondered aloud if his efforts had been worthwhile, was impressed when he joined millions of other viewers in July 1969 to watch Neil Armstrong take mankind's first steps on the moon. "This has made it all worthwhile," he is reported to have told his wife. But such unambiguously redeeming episodes can be counted on the fingers of one hand. Television in North America, like radio, was effectively hijacked by commercial interests early on in the game in the name of more and better sensation and spectacle, which it delivered, and audiences have been paying the Devil his due ever since.

The end of television's agonies, however, is in sight. As the century closed and the Internet loomed ever larger and inter-media convergence became a tangible reality (about which, more later), even television executives were waking up to the fact that their medium was not dying, but dead. In the United States, the venerable NBC was the first of the historic networks to realize that the future of all home information and entertainment media was digital and that delivery would be by cable modem or on the Internet. Moving first into cable and then into cable-Internet hybrid content in partnership with Microsoft, NBC was followed closely by ABC and its parent Disney corporation.[142] The networks were preparing themselves for what now seemed a sobering inevitability, thanks to the continuing evolution of digital delivery technologies—a world in which every World Wide Web site has the potential to become a television station. By 1999, when U.S. household Internet penetration ranged from 60 to 75 percent in most urban areas, more than two-thirds of respondents to surveys affirmed they would rather give up their television sets than lose their Internet access.[143] Very early in the new millennium, television's vaunted five hundred-channel universe will seem charmingly quaint, as viewers contemplate viewing options in the thousands or even tens of thousands, delivered to their combination TV-Internet appliances by high-speed cable and telephone lines, or by satellite. There is no need to speculate about the range and depth of content that will be available: a few hours spent on the Internet even today will demonstrate how truly extraordinary it is going to be.

CHAPTER SEVENTEEN

The Politics of Broadcasting

Woodrow Wilson, who had grasped radio's power as a new factor in world affairs, harbored serious reservations about it. In a speech in Des Moines, Iowa, in the autumn of 1919, he coined an insightful metaphor in describing the new medium's impact: "Do you not know," he asked his audience, "the world is all now one single whispering gallery?" He added,

> Those antennae of the wireless telegraph are the symbols of our age. . . . All the impulses of mankind are thrown out upon the air and reach to the ends of the earth; quietly upon steamships, silently under the cover of the Postal Service, with the tongue of the wireless and the tongue of the telegraph, all the suggestions of disorder are spread throughout the world.

The revolutionary tumult then engulfing Russia vivid in his mind, Wilson warned of the spreading "poison of revolt, the poison of chaos."

Wilson was acknowledging the enormous power of radio broadcasting as an instrument of propaganda, a power the world had scarcely begun to exploit at the time of his speech, and which few others had recognized. Today, among the democracies of the world, communications technologies tend to be accepted as tools of liberation, weapons against oppression, guarantors of freedom. But not all communications technologies have the same impact on democratic institutions, and the broadcast technologies like radio can serve any master, democratic or authoritarian.

In actively promoting the growth of radio, governments the world over were motivated to a greater or lesser degree by the fact that the medium allows them to communicate directly with their citizens and so to influence them in their favor. Radio's importance as an instrument of social and political control is attested to in the fact that the first target of insurrectionists everywhere since about 1925 has been the country's radio (and now television) stations. The problem, on the other hand, is that radio is not confined by political boundaries, and broadcasts of foreign origin can sometimes have as much impact as those from domestic authorities. In Germany prior to World War II, the penalty for passing on news heard on a foreign radio station was five years in jail, and the government reduced the odds of this happening by manufacturing a *Volks* radio that was capable of receiving only those frequencies occupied by German stations. As well, every district in the country was assigned an official called a *Funküberwachthund*, whose job it was to wire up public squares and other meeting spaces with loudspeakers whenever radio programming deemed important by the government was aired. The Soviet Union severely restricted private ownership of radios and instead wired much of the country (and the rest of Eastern Europe) with loudspeakers attached to central party-controlled receivers. The Soviet household radio appliance had the virtue of being simple to operate: it had only a volume control.

Television, like radio, has never been content with servicing exclusively local or domestic markets. Its power as an agent of change has made it impossible for either government or business to ignore its potential for opening new markets and promoting friendly attitudes worldwide. Until recently, however, the fact that it was confined to the VHF region of the radio spectrum, where signals travel by line of sight and are thus confined to distances within the antenna's horizon, made television an almost exclusively domestic medium. The global era of television was foreshadowed by AT&T's Telstar, which successfully relayed pictures from the United States to Britain and France in brief experiments in 1962. It got underway in earnest in the mid-1960s with Syncom and Early Bird, the first geosynchronous communications satellites, stationed, respectively, over the Pacific and Atlantic. Despite the enormous cost involved, television networks on both sides of the Atlantic began using "the Bird" regularly, driven by the same imperative that forced widespread business use of the telegraph in the 1850s—if they didn't, their competitors surely would. Today, there are more than 160 commercial communications satellites in orbit, many of them carrying television signals. The impact of satellite technology was felt most immediately in the area of live news telecasting. By 1989, a planet-wide force of undeniable power and influence, satellite television with its new ability to ignore political boundaries and flout local censorship was in a position to claim a major role in accelerating the collapse of the Soviet Empire in Eastern Europe simply by facilitating access to news. Premier Gorbachev's science adviser reported that the Gorbachev administration in the final days of the Soviet Union had a deliberate policy of providing wide public access to foreign satellite television as a means of thwarting expected coup attempts by hard-line Communists. With foreign television news freely available, Gorbachev believed, no palace revolt could achieve the backing of the populace. He was proved correct on more than one occasion.[144]

Nevertheless, even in the democracies, it is misguided to call radio as it evolved following the 1920s, or television, which adopted an identical regulatory and administrative model, "democratic" media. In principle, they are quite the opposite. The most cursory look into what is involved in owning a broadcasting station, the sheer cost and the onerous regulatory requirements, is sufficient to demonstrate that while everyone is free to listen to radio or watch TV, only a select few are able to broadcast. And in no way can those who are granted that privilege be said to be representative of those to whom they transmit their programs. The de facto economic and regulatory screening processes see to that. The public at large has never had more than nominal direct input into the process by which license holders are selected, that input usually consisting of the opportunity to file an objection once every few years when a license is up for renewal by the regulatory authority. Nor does the public have any formal means of influencing the nature of broadcasters' programming in any but the coarsest, least discriminating ways—through boycotts of advertising and letter-writing campaigns. Positive programming suggestions must be offered in the deferential spirit of Oliver Twist and are anyway likely to be politely ignored.

The issue of access to the technology is one that defines the history of broadcast communication and has returned to haunt us in the era of the Internet. It is illuminating to look back at how it has been handled, and how it came to be that access to radio, which began as a two-way or bilateral medium available to the general public, was organized so that the public retained the privilege of receiving programming but lost the right to originate broadcasts.

From the beginning, access was a concern among legislators wherever radio was introduced. As we saw earlier, Britain gave the government Post Office a monopoly over broadcasting. In the United States, on the other hand, the first act to regulate radio, passed by Congress in 1912, prohibited the secretary of commerce from withholding a license from any U.S. citizen who applied. Radio was to be open to all. By 1923, so many stations were crowding the two frequencies then allotted for broadcasting that they often had to double up, sharing air time in their coverage areas. Disputes were inevitable and frequent. Something had to be done, though no one quite knew what. Commerce Secretary Herbert Hoover reacted by making more bandwidth available, that is, by increasing the number of frequencies on which stations would be allowed to broadcast.[145]

This eased the pressure, but it soon built up again since a feature of the receivers of the era was poor selectivity among adjacent signals. However, that was an engineering problem that had already been solved. One of the great inventive geniuses of radio, Edwin Armstrong, had capped a long series of technical triumphs with the design for a "superhetrodyne" receiver which was vastly superior to anything then on the market in terms of both sensitivity, the ability to detect weak stations, and selectivity, the ability to tune out interference from nearby signals.[146]

So while the technical issues affecting access had in part been solved by 1924, it was also true that the medium's commercial possibilities and political power were better understood. That inevitably changed attitudes to access, as well. By that year, fourteen hundred broadcast licenses had been issued in the United States. Secretary Hoover, acting with the insistent advice of industry lobbyists, announced that all available broadcast frequencies were now filled and that no further licenses would be issued. Interference among stations, "chaos" on the airwaves, was given as the reason for this fundamental shift in policy, which tilted the balance away from public access and toward commercial control. Not only were new licenses no longer available, existing licenses immediately acquired a substantial cash value, and hundreds of small, public service, noncommercial radio outlets owned by universities and altruistic individuals and organizations were bought up by the big commercial broadcasters in the succeeding months. Radio would henceforth be first and foremost a medium of advertising, in which nonadvertising content played the role of attracting audiences for commercial messages. Access to broadcasting facilities was proscribed in the way we have already described.

Nevertheless, in the glow of optimism that so often accompanies new technologies, it was widely seen as a triumph for democracy that, in 1925, President Calvin Coolidge was able to deliver his inaugural speech to an assembled radio audience estimated at more than five million. There were high hopes for the impact of radio on political discourse in general, for reasons outlined in an August 1932 *Saturday Evening Post* report on the Democratic National Convention of that year:

> The Democratic Convention was held in New York, but all America attended it. . . . [Radio] gives events of national importance a national audience. Incidentally, it also uncovered another benefit radio seems destined to bestow upon us, the debunking of present-day oratory and the setting up of higher standards in public speaking. . . . Orators up to the present have been getting by on purely adventitious aids: a good personality, a musical voice, a power of dramatic gesture have served to cover up baldness of thought and limping phraseology. . . . The radio is even more merciless than the printed report as a conveyor of oratory. . . . It is uncompromising and literal transmission. The listeners follow the speech with one sense only. There is nothing to distract their attention. They do not share the excitement and movement of the meeting, nor does the

> personality of the speaker register with them. . . . Silver-tongued orators whose fame has been won before sympathetic audiences are going to scale down to their real stature when the verdict comes from radio audiences.

There is scant need to point out the irony in this argument, which paradoxically seems logically bulletproof. Of course radio must force on politicians a more reasoned, substantive style of discourse! Then why didn't it? The answer is in the phrase, "The listeners follow *the speech* . . ." In early experimental radio, speeches often were broadcast in their entirety. The commercialization of radio and the resultant assigning of monetary value to air time soon put an end to that and before long there was no "speech" for listeners to follow. Commercials placed a value on air time that hadn't been there before. Time became money: air time became a commodity, and as such it could most efficiently and conveniently be sold in regular blocks. Commercials forced a new approach to programming, in which content had to fit into a rigidly adhered-to schedule so that advertisements could appear in the time slots for which they had been sold. Content had to fit the format, rather than vice versa. News programming was no exception, and that meant that everything within a newscast had to be weighed for its information value in relation to everything else that was newsworthy at that moment. Excerpts from speeches became more and more truncated, as did excerpts from interviews. Only in the most extraordinary of circumstances was the politician given unmediated, uninterrupted access to the public on commercial radio. The glib phrase, the oft-misleading metaphor, the dramatic expostulation—the sound bite—was commercial radio's replacement offering for oratory.

Broadcasting's proponents have consistently argued that the unilateral or point-to-multipoint media serve the important function of bringing society closer together through shared experience provided by broadcast programming. The vision of the American or Canadian or Brazilian or Australian family settled around the living-room television set, sharing with millions of other families the same network programs, learning the same things, experiencing the same emotions, is a portrait painted by this notion. It is a cozy picture until one stops to consider what it is they are learning, and from whom, and to what end. The short answer is that broadcast outlets must, by their nature, further the goals of those who run them. Those goals may be entirely laudatory and wholly constructive, as they can sometimes be in a developing nation in which society lacks the necessary cohesion to form an effective nation-state. Under these circumstances, the power of broadcasting can be used very effectively in advancing socially useful goals.

But the cohesion provided by uniformity of experience is not the only, or even the preferred, form of socially useful solidarity. The great sociologist Emile Durkheim identified two kinds of solidarity in society.[147] The first was "mechanical solidarity," by which he meant the kind of cohesion imposed by criminal law and paternal authority of various kinds—the type of solidarity that leads to conformity, regimentation, hierarchies and strict discipline. The second was "organic solidarity," by which he meant the cohesion that is enforced by civil law, custom and consensus and which grows out of continuing conversation among individuals of different training, beliefs and understandings. Either can hold a nation together, and both are present in any society. Most of us would agree that we'd like to see rather more organic and less mechanical cohesion in our countries. Durkheim also made the obvious point that broadcast technologies like radio, TV, newspapers and movies tend to promote "mechanical" solidarity, while bilateral media like the telephone and mail (and now the Internet) encourage the organic variety. Compare the role of the (bilateral) fax and the Internet in the collapse of the Soviet Empire or the civil liberties movement leading to Tiananmen Square in China, with the role of (unilateral) broadcast radio in Nazi Germany and in Rwanda during the recent genocide.

The social solidarity, the so-called democratizing impact claimed as a benefit of broadcast technologies, clearly falls into the mechanical category, while bilateral technologies such as the telephone and the Internet, just as obviously promote organic solidarity, as facilitators of conversation and self-organization. Even Plato would have recognized the distinction: he pointed to the antagonism (especially in the *Gorgias*) between communication in the form of dialogue, which is the speech adequate to philosophical truth, and "rhetoric," the one-way speech by which the demagogue seeks to convince the masses. Hannah Arendt saw a massive shift having taken place in media-saturated Western culture, away from truth and in the direction of opinion, which she believed implied a shift from "man in the singular to men in the plural."

> In the world we live in, the last traces of this ancient antagonism between the philosopher's truth and the opinions of the market place have disappeared. Neither the truth of revealed religion, which the political thinkers of the seventeenth century still treated as a major nuisance, nor the truth of the philosopher, disclosed to man in solitude, interferes any longer with the affairs of the world.[148]

It is thus possible to argue that broadcast technologies have a more important role to play in immature or developing societies than in mature ones; that politically mature societies such as the developed Western nations are on balance better served by bilateral media such as computer and telephone networks than by broadcasting, whether by satellite and cable or over the air. Certainly it seems sensible to reexamine the roles of broadcast technologies and the resources dedicated to them in the technically advanced democracies, now that bilateral media are widely available and comprehensively efficient.

It should not be supposed that the individuals in charge of broadcast media have typically been cavalier about their obligations as stewards of an essentially authoritarian medium. The opposite would be closer to the truth, although it is true that they have responded in widely differing ways. The early history of radio broadcasting in North America was marked by a clear sense of social responsibility, exemplified in high-quality entertainment and a commitment to news programming and documentaries. This carried on into the early years of television, though under the corrosive influence of commercialism. By the mid-1950s, however, television entertainment programming had come to be characterized by a sharp slide into mediocrity, a trend which has been so frequently documented as to need no further description here. This was the dawning of Newton Minnow's[149] era of the "vast wasteland" of American television. Entertainment programmers responded to their social responsibilities simply by withdrawing from the field of controversial or in other ways remarkable content; in this way they were able to avoid accusations of abuse of privilege. Unfortunately, they simultaneously opened themselves to the justifiable criticism of neglect of the responsibility to provide quality programming.

Television network news programming managed to hang on to its integrity and to strive for excellence for another twenty years and more, reaching a zenith of performance in the closing years of the Vietnam War and in the era of Watergate. The inevitable decline established itself in the mid-1980s when new corporate owners at all three American networks decreed that news would no longer be treated as a loss leader, but would henceforth have to become a profit center capable of supporting its own operations. The idea of news programming as a socially responsible service provided by broadcasters in return for the privilege of using the public airwaves for their own profit, was dead. With it died such notions as CBS News' legendary pride in being "first with the best" whatever the cost. Television news organizations throughout North America are shriveled, despondent shadows of their former selves, and there is no reason to believe that they will ever recover.

Sad though the systematic dismemberment of television news organizations has been, it might be argued that it has been the inevitable consequence of the maturing of society. In this view, we no longer have great broadcast news organizations because we have shifted away from mechanical solidarity toward the more organic variety, in which predigested broadcast news is increasingly irrelevant. Certainly there is no question that television news throughout its history provided programming that could not escape the description of "authoritarian," in the sense that it actively promoted national political agendas based on views held by its practitioners and owners and, in times of crisis especially, permitted itself to be used as a mouthpiece for government. Usually, that authority was exercised benevolently, in socially responsible ways. But this was not always the case and abuses have been more widespread than is generally recognized. For instance, television news too often simply ignored the voices from society's social, economic, academic and political fringes. Vocal dissatisfaction with the news remained confined largely to a small minority of academics and former practitioners, thanks mainly to the fact that the sins tended to be those of omission and thus were familiar only to the initiated. As well, as a potent propaganda medium, and as the major employer of the very people who were in a position to know of its misdeeds, television was eminently capable of dealing with its critics.

Specialty news networks such as CNN and CNBC, in usurping the role of network news divisions as leading news providers, have been able to lessen the inevitable impact of their own intrinsic authoritarianism by vastly increasing the range of stories they cover, an ability they owe to the greatly expanded air time they devote to news. And because they are broadcasting news continuously, they are also able to devote large blocks of time to live coverage of events as they happen. The role of interpretation (and the potential for distortion) in this programming environment is considerably diluted: people do not require a great deal of interpretation of events they can see unfolding for themselves. Though they cannot avoid the fact that they are owned and controlled unilaterally, cable news operations can and do mitigate this with a more relaxed and inclusive process of story selection than is possible in the twenty-minute or half-hour network news format.

In the developing world, the issues raised by media of all kinds are naturally quite different from those which worry us in the so-called developed world. We noted earlier that television and other broadcast technologies can be especially useful as accessories to social cohesion in developing nations, where Durkheim's mechanical solidarity plays a dominant role over the more subtle organic solidarity. While this is true in principle, real-life experience with broadcasting in the era of satellite TV has been problematic for many Third World nations. The trouble is that television programming is expensive, and the cheapest, most abundant product on the market by far is Western, and principally American. With fear of American TV imperialism rife throughout the *developed* world, it is little wonder that in the Third World, where resources are generally not available to mount much indigenous programming of competitive technical quality, the concern is palpable. It is well understood that television is a potent agent of conformity; the concern is that the conformity induced will be to the often debased Western values inherent in

American programs. (It is not just television that is of concern. Other broadcast media—movies, radio, books, especially textbooks, news services and periodicals—all have the potential to erode local cultural values if they are not locally produced.)

To the extent that the problems of developing nations can be said to be more organizational than technological (technology is available, albeit at a price, but the skills and organizational infrastructure to make it work are in short supply), then it is important not to underestimate the need in these nations for bilateral communications services. For it is bilateral communication which supports and fosters self-organization in society. The irony is that bilateral communication by telephone, post, fax and computer network is viewed as potentially subversive by the authoritarian governments that rule in so many developing nations.

China is proving to be an interesting test case. Initially hostile to the Internet, the government changed its position and began actively to encourage Internet use in 1998. The country's largest telecommunications supplier embarked on a huge project to upgrade its Internet backbone using American technology, and a research and technical cooperation deal was signed with Japan. Industry reports were soon predicting that by 2005 China would have the world's second-largest Internet population, after the United States, with as many as thirty-seven million Chinese on line. The turnabout seems to have followed the realization that, with the cooperation of Internet Service Providers (ISPs), the network gateways, it is possible to track any user's movements on the Net—which sites are visited, for how long, etc.—by monitoring server logs. Cooperation is, of course, forthcoming since the government owns all the ISPs in China, and the law forbids the use of Internet access by anyone other than the registered ISP subscriber. (The law is frequently ignored.) It is also possible for ISPs to block user access to "problem" sites: that power was reportedly used in 1999 to black out CNN and BBC Web sites. As well, China has authorized its police to access personal e-mail. At the same time, the government began officially to encourage the development of Chinese-language Web domains with officially sanctioned content. In principle, each of these approaches to suppressing information can be easily subverted on the Net; as we'll see, it is the nature of the technology to resist attempts to block the flow of information. It will be fascinating to see to what extent traditional authoritarian approaches to censorship can be effective on such a scale in the world of the Internet.

With broadcasting a potential source of cultural subversion and bilateral media like the Internet providing the ubiquitous communications channels of political dissent, the lot of the authoritarian ruler in the Age of Information, like Gilbert and Sullivan's policeman, "is not an happy one."

PART TWO

The Digital Era

CHAPTER EIGHTEEN

"Reasoning Is But Reckoning"

The Philosophic Underpinnings of the Electronic Computer

We have reached the point in the story of the evolution of communications media when a great watershed, a continental divide, can be identified. It is marked by what may be the most important development in the entire history of communications technologies, though it is always risky to accord such significance to an event so near to us in time. We have alluded to it throughout the preceding chapters; now is the time to look into the nature of the change in detail.

The watershed is the widespsread adoption within communications media of the common symbolic language of digital systems. The change began slowly, with the construction of the first electronic digital computers during and immediately following World War II, and has continuously gathered momentum since then. Recently, we have begun referring to the phenomenon as "convergence," meaning the intersection at a single point of the technical development trajectories of TV and other broadcast media, the telephone and telecommunications, text-based publishing, graphics, still pictures and interactive media. That focal point is characterized by the shared use of the simplest and therefore most universally applicable language imaginable, the binary language of 1s and 0s as it is employed by digital computers.

Just as Swahili, the lingua franca of Africa, allows people from many linguistic backgrounds to communicate with one another all over that continent, binary code allows for communication between otherwise separate and distinct information media. A major portion of the world's vast and growing computing power is devoted to the business of translation of print, video, audio and graphic illustration from their original analogue forms into digital, binary code for editing and other forms of "value-adding" manipulation, for transmission and distribution of the enhanced product, and then for translation of that product back to the analogue or wave-based structures our human senses are equipped to respond to. In this context we can think of the digital computer as a sort of universal translating appliance of the kind that turns up from time to time in science fiction.

The story of where this remarkable machine came from and how and why it works is woven like a compelling subplot through the history of Western civilization. The digital computer, far from being an alien and bloodless tool of impenetrable complexity, is more than any other artifact of human invention an extension of our most impressive capabilities and a reflection of our most profound insights. It is impossible to comprehend the trajectory of modern communications technologies and the new economy they have spawned without understanding what makes the computer, above all, a very human instrument.

We should begin by pointing out that not all computers are digital: some are analogue. The difference between the two is significant: digital computers operate by counting; analogue computers operate by measuring, that is, by *analogy*. An old-fashioned pocket watch *measures* time by mechanically translating the distance covered by spinning wheels and cogs in its works into the changing position of the hour and minute hands in relation to the numbers around the watch face. A modern digital watch *counts* vibrations of a crystal which is excited by a tiny battery, and that is converted into a numeric readout by a microchip. In practical terms, the distinction means that digital computers can be much more precise than analogue computers, since measurement can never be 100 percent accurate, whereas counting can. What is measured, apparently accurately, with a yardstick may be significantly too long or short when measured with a micrometer. No two bushels of wheat or gallons of gas are precisely the same in weight or volume. The tiniest of inaccuracies at one scale are magnified into significance at another. Scale always matters in measurement, whereas it does not in counting. Digital computers are therefore excellent preservers of the integrity of the information they deal with: there is, in principle, no loss of information no matter how many manipulations or iterations it may go through.

Of course, counting only works if there are separate and distinct entities to be counted (one cannot very well "count" the water coming out of a hose!), and the trick that makes digital systems perfectly accurate is a requirement that such units be well defined in the system's "operating rules." A digital water meter counts the number of revolutions made by a wheel placed in the flow of water; it can be perfectly accurate in its own terms because it has in effect divided up the flow into discrete volumes required to cause a revolution of the wheel. It has, in Nicholas Negroponte's image, turned *atoms* of water into *bits* of information. The alphabet is a digital system in which the letter *A* can be represented a number of different ways—in different fonts, in different media and in different handwriting styles—but a convention of the alphabetic system is that all such variations, so long as they are legible, are accepted as *A*'s. In arithmetic, another digital system, a digit can "legally" be represented by symbols as diverse as characters written on paper, or an assemblage of marbles. In this way, wide variations in the content of individual units can be tolerated while preserving the accuracy of the digital system. The text in a book is created within a digital system; a painting is not. This helps to explain why it is possible to reproduce a four-hundred-year-old Shakespearian sonnet with complete accuracy in a book published in our time, while it is not possible to reproduce a Rembrandt portrait of the same era in the same book (or in any other medium) with anything approaching complete accuracy. To fully understand the Rembrandt, you must see the original.

How then do digital technologies reproduce sounds and images that are intrinsically analogue in nature? By measuring values of sound and color and light, and assigning numbers to those values. Those numbers can then be translated into binary form, and transmitted or manipulated or reproduced by the machinery of digital electronic systems. A digital recording of a solo cello suite is made by sampling the sound waves produced by the instrument many times a second, and assigning binary numbers to the pitch and amplitude of the sound at each sampling. A digital representation of a painting is made by dividing the image field into many small sectors and assigning numbers to the various qualities of the light in each sector. Digital technologies become more accurate in their translations of analogue media in direct proportion to their sampling rates, that is, the frequency with which they sample sound, for example, or the number of discrete segments into which they divide a picture. Thus, a digital reproduction of an analogue sound or image can be extremely accurate, but never an *exact* replica, since that would involve an infinte mumber of samples.

Digital systems, however, are much more interesting than their utilitarian role as analogue surrogates would suggest. They have the distinction of being a fundamental metaphysical category, which proposes that meaning can be captured in symbols (such as numbers or combinations of letters), which are then capable of manipulation according to prescribed rules.

The combination of a digital system and its formal rules of operation—the complete package—is usually referred to as a "formal system." Formal systems, because they are digital, are independent of the medium in which they are embodied. It is the *form* of the system and not the physical nature of the tokens or symbols employed that matters. The game of checkers, like chess, is a formal system. It could be played with hay bales and bed sheets in a cow pasture just as well as on a conventional checkerboard with plastic disks. Baseball, although also a game, is not a formal system. To try to play baseball in a swimming pool would radically alter the game and how it proceeds. For the same reasons, as the Shakespeare/Rembrandt example above indicates, we can reproduce Shakespeare on the Zipper in Times Square as accurately as on parchment, but not Rembrandt.[150]

The hidden power of formal systems reveals itself when the tokens or symbols being used are assigned meanings. Then the operations of the system become subject to interpretation, and if a formal equivalence between the operations of the symbols and the things they represent in the real world is maintained, those interpretations will have *meaning* in the real world. A formal system is therefore a way of abstracting real-world problems, resolving them at the abstract level and interpreting the results back into real-world terms. If this sounds complicated, it is not: arithmetic is such a system, in which five oranges contained in one basket can be added to three oranges contained in another basket using the symbolic, digital language of numbers (which are not oranges themselves but are capable of representing oranges). Without ever having to physically count all of the oranges at once, we know there are eight of them in total, because the result of our calculations using the formal system called arithmetic (i.e., $5 + 3 = 8$) tells us so.

The insight that language, with its rules of grammar and syntax, can in some respects be seen as a formal system led to speculation as early as the sixteenth century that thought itself, inasmuch as it is based on language, may follow some of the same rules as other digital systems such as arithmetic. The English philosopher Thomas Hobbes (1588–1679) made the case in 1651 with characteristic bluntness: "Reasoning is but reckoning," he stated. By this he meant two things: first, that thinking is "mental discourse." The only difference between thinking and talking out loud or working out arithmetic problems with pencil and paper is that thinking is conducted internally, thoughts being expressed not in written symbols or spoken words but in special brain tokens Hobbes called "phantasms" or thought "parcels." The second thing he meant was that reasoning at its clearest follows rules which lead to correct outcomes just as accounting does. In other words, reasoning is a mechanical procedure akin to operating a mental calculating machine: it is a digital process.

Hobbes expressed his notion of the formal nature of reasoning in *Leviathan*:

> When a man reasoneth, he does nothing else but conceive a sum total, from addition of parcels; or conceive a remainder, from subtraction of one sum from another. . . . These operations are not incident to numbers only, but to all manner of things that can be

> added together, and taken out of another. For as arithmeticians teach to add and subtract in numbers; so the geometricians teach the same in lines, figures . . . angles, proportions, times, degrees of swiftness, force, power and the like; the logicians teach the same in consequences of words; adding together two names to make an affirmation, and two affirmations to make a syllogism; and many syllogisms to make a demonstration.[151]

It is a rather mechanistic notion that is unlikely to have occurred to earlier thinkers produced by a less technologically oriented time, before Galileo had reinvented the world along rational scientific determinist lines. But it was perfectly in keeping with the contemporaneous views being expressed by René Descartes (1596–1650), who is generally credited with inventing both modern mathematics and modern philosophy.

No philosophical trifler, Descartes set out to rebuild the edifice of knowledge from first principles. In order to establish those first principles on a firm foundation, he attacked the enterprise by systematically doubting everything that can be doubted. Can I doubt that I am sitting in front of the fireplace in a dressing gown? he asked. Yes, I can, because I have dreamed that I was here when in fact I was asleep in bed. I also know that people may have hallucinations which seem remarkably real. If the material world cannot be trusted, Descartes then asked himself, how about the more abstract world of arithmetic and geometry? Surely it should be more reliable. But what if there were an evil demon bent on misleading me in my computations, causing me to make a mistake whenever I add 2 + 2 or calculate the area of a circle? If there were such a demon, it could be that everything I see is only an illusion concocted as a trap for my credulity. How can I know anything for sure?

In a more modern clothing, the question might be the subject of an *X Files* episode in which the beauteous Agent Scully's brain is removed while she sleeps and is placed in a vat of chemicals to keep it alive so that alien scientists can prod it with electric probes to cause sensations, which Scully takes to be reality. How can she know what's real and what isn't?

As it turns out, there is one thing we can all know for a certainty, Descartes said. I may not have a body—it may be an illusion like everything else in the physical world. But thought is different: "While I wanted to think everything false," he reflected in *Meditations*, "it must necessarily be that I who thought was something; and remarking that this truth, I think therefore I am, was so solid and so certain that all the most extravagant suppositions of the skeptics were incapable of upsetting it, I judged that I could receive it without scruple as the first principle of the philosophy that I sought."[152]

Here is the same argument in another form: I can doubt that my body, or my brain, exist. I cannot doubt that I see and hear and feel and think. Therefore, I who see and hear and feel and think cannot be identical to my body or my brain; otherwise, in doubting their existence, I would doubt the existence of myself.

This approach to first principles had the effect of placing mind above matter in the hierarchy of things, since one could only know about the latter through the workings of the former. Descartes expanded on this theme to propose two distinct substances in the universe: mind and matter. This dualist philosophy had the attraction of being more intuitively believable than either a strictly materialist point of view in which everything, mind included, is just a form of matter, or an idealist point of view in which all is merely mind. But it stirred up one of philosophy's longest-running and most vehemently contested debates. The question it raises is: How do mind and body interact, if they are indeed different "substances"? How can an event with no physical mass, no charge, no location or anything else physical, make a physical difference in the brain, or anywhere else? It is a debate that was to have obvious interest for the computer scientists who would arrive on the scene three hundred years later, eager to develop artificial intelligence based on silicon chips and electrons. Could such a creation ever be authentically intelligent in human terms if Descartes was correct about mind and matter being fundamentally different stuff? How do you build a mind? What is it made of? How is it connected with the machine?[153] Where Descartes and his contemporaries had started with mind as a given, artificial intelligence as a discipline arose out of the idea that mind could be seen as an emergent product of sufficiently complex systems of logic processors. That is, mind *grew out of* the operations of certain kinds of physical systems.

Descartes himself had a number of answers to the question of how mind interacts with the body, none of them very convincing. The one that had the most currency over the years was that the mind and body operated like two perfectly synchronized clocks. What appeared to be cause and effect in their interactions was really just synchronicity. In a time when the mechanical clock was the chief technological marvel, the argument had a certain authoritative ring to it.

Descartes' major contribution to mathematics was the invention of analytic geometry, which he conceived as a means of converting geometric problems of lines and volumes into algebraic notation. In other words, he found a way of solving geometric problems using algebraic methods and formulae. In a sense, he had applied digital (arithmetical) techniques to a field that had grown out of analogue measurement. Perhaps even more important than the invention itself was the way Descartes abstracted the notion behind it. What analytic geometry did was concentrate on the *relationships* between various elements of a problem, rather than on the elements themselves. The same approach, he realized, could be used in solving problems in physics as well—in fact, in any rule-based system.

The fact that real-world problems could be examined and solved using the abstract symbols of algebra led Descartes to further examine the relationship between symbols and the things they symbolize. In particular, he concluded that thoughts were symbols just like mathematical notation in that they were representations of things that existed in the real world, an idea which further reinforced the essential dissimilarity between mind and matter. Just as mathematical notation need have no direct connection with the objects it described in order to be of practical value, so, too, with thoughts. Thought was in fact a digital system.

The question remained: If thought was a digital system, what were the rules that governed the operations of that system? The German philosopher and mathematician Gottfried Wilhelm von Leibniz (1646–1716) is best remembered as the inventor of the mathematical system called "differential calculus," but he also put a lot of energy into answering that very question. He believed it should be possible to design a "universal calculus" by which all human reason could be abstracted, codified and reduced to digital notation. In pursuing this goal he invented symbolic logic, which, while regarded as a curiosity in his day and for many succeeding generations, would turn out to be a mathematical insight of incalculable importance.

A contemporary and rival of Newton (each had discovered differential calculus independently), Leibniz agreed with the great English scientist that the only reliable route to unimpeachable knowledge was through mathematics. But he saw application for this view in fields far removed from science. It seemed to Leibniz that if one could identify the fundamental elements of experience, and the bottom-line or irreducible forms of relations among them, one could by successive combination of symbols representing these experiences and relationships describe all possible knowledge with mathematical accuracy. The laws of reasoning were to be converted to mathematical formulae and all logical deduction would be reduced to algebra. A new universal language of great precision and power would be at work; it would be the pinnacle of achievement in the determinist philosophy that saw the world as a clockwork mechanism engineered to perfection by God.

To some, the idea of reducing human thought to algebra seemed faintly ridiculous if not blatantly outrageous, and Voltaire viciously lampooned Leibniz as Dr. Pangloss ("Everything happens for the best in this best of all possible worlds.") in *Candide*. Nevertheless, as a person of prodigious mental energy and stamina, Leibniz was confident that his universal calculus could be put together in short order: "I believe that a number of chosen men can complete the task within five years: within two years they will exhibit the common doctrines of life, that is metaphysics and morals, in an irrefutable calculus." Leibniz foresaw his system being widely adopted as a means of resolving disputes of all kinds, his assumption being that once the correct "answer" had been produced by his logical formulae, the disputants would cheerfully accept the result. He proposed that it be used initially to test the body of Christian doctrine and establish a core of knowledge that all Christian sects could agree upon.

There is a famous quotation which nicely captures the spirit of Leibniz's enthusiasm for the idea. "All inquiries which depend on reasoning would be performed by the transposition of characters and by a kind of calculus," he said, "And if someone would doubt my results, I should say to him: 'Let us calculate, Sir,' and thus by taking pen and ink, we should soon settle the question."

Though obviously misguided in his belief that constructing a universal logic—*lingua characteristica*—would be a simple chore, Leibniz sensed the power of digital, formal systems to amplify information:

> Once the characteristic numbers of many ideas have been established the human race will have a new *Organon* [Aristotle's rules of logical reasoning], which will increase the power of the mind much more than the optic glass has aided the eyes, and will be as much superior to microscopes and telescopes as reason is superior to vision.[154]

Not surprisingly, perhaps, these particular ideas of Leibniz were not paid much attention for the next two hundred years. But then George Boole, the self-taught nineteenth-century English mathematician (1815–1864), set out to construct "a mathematics of the human intellect" in the spirit of Leibniz, in which logic is expressed not in words but in precise mathematical symbols. Boolean logic proposed that all logical argument could be translated into a series of "yes" or "no" responses which in turn could be represented in binary terms. The numbers 1 and 0 were used by Boole to indicate existence or nonexistence when referring to classes, and truth or falsehood when applied to propositions. Boole showed that symbols of this kind obey the same laws of combination as symbols in algebra, which meant that they could be added, subtracted, multiplied and divided in almost the same way as numbers, to give a result that was either true (1) or false (0). Boole's logic in fact obeyed all of the ordinary laws of algebra with just one exception, that $x^2 = x$ (that is, in algebra it is not true that every x is equal to its square, whereas in Boolean logic, it is true). In ordinary English the equation means simply that the class of all things common to a class x and to itself is simply the class x.

Without straying too far into the thickets of the subject, a couple of examples may serve to illustrate the general thrust of Boole's scheme. If the symbol x represents the class of "all white objects" and if the symbol y represents the class of "all round objects," Boole used the compound symbol xy to represent the class of objects that are both white and round. He saw that since the class of objects that are white and round is exactly the same as the class of objects that are round and white, it is possible to write: $xy = yx$.

If x and y are mutually exclusive classes (in other words, no member of one class can at the same time be a member of the other), the symbols $x + y$ can be used to represent all objects which belong to class x or to class y. So that if x represents all men and y represents all women, xy represents all people. It must therefore be true that $x + y = y + x$. Furthermore, if z represents the class of all Canadians, then $z(x + y) = zx + zy$; in other words, the class of Canadian men and women is exactly the same as the class of Canadian men and Canadian women.

It may seem that these examples are merely a complicated way of stating the obvious. But there is a power in symbolic logic that allows it to provide genuine insights. A very simple example can be used to demonstrate this:

Everyone aboard the boat was lost.
Merrihew was aboard the boat.

Although we have never been told that Merrihew is dead, an analysis of the statements will demonstrate that he is. Increasing the level of complexity somewhat, one can see how formal logic can be helpful in reaching conclusions that are not immediately apparent when a problem is stated in ordinary English:

If the sentry was not paying attention at the time, the car was not noticed when it came in to the compound.
If the witness's account is accurate, the sentry was not paying attention at the time.
Either the car was noticed, or Merrihew is hiding something.
Merrihew is definitely not hiding anything.
Is the witness's account accurate?

The correct conclusion is, the witness's account is not accurate, but few among us will be able to derive that fact from a single reading of the premises. In Boolean notation, however, the problem is reduced to foolproof algebra. It is clear that even at this trivial level of complexity, a formal calculus can be helpful in solving problems with consistent accuracy. In fact, insurance companies faced with elaborate policies that needed interpreting in the face of complex claims have used Boolean logic to determine when they were liable and when they were not.[155]

There are few more abstruse fields of intellectual endeavor than formal symbolic logic in its modern manifestations. And so it may seem all the more astonishing that this most abstract of disciplines has turned out to have an enormous impact on the day-to-day lives of virtually everyone on the planet. The explanation is in the fact that Boole and Leibniz had made of logic a digital system operating by formal rules; in other words, a "formal system."[156] In today's terminology, they made logical problems accessible to solution by algorithms, an algorithm being simply a set of steps through which a mathematical solution can be reached. Boole's system, moreover, was a binary system that needed only the digits 1 and 0 to function, no matter how elaborate the question asked of it. As an algorithm for reason, when the appropriate time arrived, it would be ripe for adoption by computer scientists.

CHAPTER NINETEEN

The Amazingly Precocious Charles Babbage

It was the not-entirely-happy fate of the English inventor Charles Babbage (1792–1871) to become the first person to seriously attempt to employ the ideas of symbolic logic in constructing a true computer.[157] He was one of those prodigies of intellect that populated nineteenth-century Europe, and Britain in particular. He is the father of operational analysis or time-and-motion studies, producing the first text on the subject in 1832, nearly eighty years prior to the work of the better-known Frederick W. Taylor. He was a leading mathematician of his time. He investigated the British Post Office and made recommendations which led to the establishment of the penny post. He invented the railway cow-catcher and occulting beacon lights for marine navigation. As one fascinated by complex mechanisms, he was an accomplished lock-picker and an expert on ciphers. Not surprisingly, he knew most of the principal scientists of his era, and counted among his literary friends John Stuart Mill, Charles Dickens, Robert Browning, Thomas Carlyle, Lord Tennyson and Captain Fredrick Marryat. He explored live volcanoes and wrote authoritatively on dating of archeological finds, and published a paper on surface features of the moon, where a crater is now named for him.

The scion of a wealthy banking family, Babbage was able to make of himself a gentleman amateur of natural philosophy and technology in an age when all such knowledge was accessible to any intelligent man or woman, regardless of academic background. Babbage, as it happened, was schooled at Cambridge, but at a time when English universities studiously eschewed instruction in applied sciences (engineering, as we would say today) as being beneath them. Oxford and Cambridge played negligible roles in the Industrial Revolution. In Babbage's youth, compartmentalized studies or "disciplines" had yet to be imposed. "Physics, chemistry, science and engineering, literature and philosophy, art and industrial design, theory and practice—all constituted a continuum of knowledge and skill, within which men roamed freely," says historian Paul Johnson.[158]

As Babbage's century closed, academic turf would be divided and subdivided within universities, as knowledge and information became commodities—hard-won possessions to be jealously guarded within the individual disciplines. It was a rare economist who would know anything of chemistry; few chemists would have more than a passing knowledge of astronomy. Specialization led to rapid advances in many fields of science and technology thanks to the focused effort it produced, but it also led to development of arcane dialects and impenetrable terminology, which served to deny knowledge to nonspecialists and protect the specialist's territory from outsiders. It crippled interdisciplinary communication for the next 150 years. It is only very recently in our own era that progress has been made in reversing this trend, thanks in large measure to the communications technologies surrounding the Internet, which arrived coincidentally with a convergence of interest imposed upon the sciences forced upon them in part by the approaching outer limits of experimental knowledge.

It was also a feature of Babbage's world that the state of the contest between medicine and disease was such that enormous tragedy was a commonplace in the lives of even the privileged. In 1827, when Babbage was hard at work on his famous "Difference Engine," the most advanced calculating machine the world had ever seen, he had to deal in a single year with the deaths of two of his children, his father and his wife, Georgiana. Some biographers speculate that the stress unhinged him, with the result that he was from then on constitutionally unable to finish his machines, always finding a way to improve them that delayed the work of fabrication until new drawings had been prepared.

It is not surprising that he should have been attracted to a much-discussed challenge that affected all of the sciences, impeding their progress—that of swift and accurate computation of large numbers. The problem of error-riddled mathematical tables used by astronomers, navigators and engineers to simplify their calculations was a real and lively issue as the Industrial Revolution gathered momentum in England, and Babbage showed an early interest in the possibility of mechanizing their production to ensure accuracy. In 1818 he traveled to France to have a firsthand look at recent advances in mathematics there, and it was during this visit that he encountered the remarkable work of Gaspard de Prony. De Prony was a leading civil engineer in Napoleonic France, and a prominent advocate of the country's conversion to the metric system, as part of the Republic's program of rationalization of all things. De Prony was assigned the unprecedented task of producing for publication logarithms of whole numbers from 1 to 200,000 and trigonometric values for 3,600 divisions of a circle—a job that could normally be expected to take teams of skilled mathematicians many years to complete. As a first step, the mathematicians would have to calculate an algebraic formula called a polynomial that would approximate each value required. Then, long and tedious arithmetic calculations would have to be done, based on the polynomials, to derive each number included in the tables.

De Prony's solution allowed him to complete the job in a fraction of the time that would otherwise have been needed. It was to divide up the labor, creating an assembly line of mathematicians of varying levels of skills, a notion he'd picked up quite by accident, according to Babbage's account,[159] while browsing through a copy of Adam Smith's economic treatise, *The Wealth of Nations*. In a famous passage, Smith had illustrated the idea of division of labor as a key to enhanced productivity, by describing a pin factory in which each step of the process was handled by a different worker, rather than having the workers produce complete pins on their own. Smith had visited such a factory, and had concluded that dividing up the labor greatly increased overall output. In a flash of inspiration, de Prony determined to employ a similar attack on his own problem of the tables. In de Prony's calculating factory, a handful of highly skilled mathematicians created the polynomials. Arithmetical calculations arising out of them were divided into two tiers of difficulty and assigned to teams of calculators of appropriate skill levels. According to Babbage, most of the hordes of calculators on the lowest tier knew nothing of mathematics beyond simple, mechanical addition and subtraction.

De Prony had created a human analogy of the yet-to-be invented computing machine. Babbage, well primed in the theories of mechanical computation, recognized the connection and set about converting the human analogue into mechanical form. It would be one of the epic journeys in the annals of science, replete with Victorian melodrama and genuine tragedy.

The basic principle used by Babbage's Difference Engine is a simple one, adapted from an error-checking system conceived by de Prony to validate results produced by his army of human calculators. The table below has been set up for the calculation of squares (a number multiplied by itself) but similar, more elaborate tables can be made for any computation that progresses by regular increments.

In de Prony's case, if the "differences" followed their expected sequences, it was assumed the numbers that created them were also accurate. Babbage used the system in reverse. To get his Difference

number	square	difference	2nd difference
1	1		
		3	
2	4		2
		5	
3	9		2
		7	
4	16		2
		9	
5	25		

Engine to calculate and print a table of squares like the one represented by columns A and B, above, he would set the wheel on the "number" shaft of his machine to 1, the wheel on the "first difference" shaft to 3 and the wheel on the "second difference" shaft to 2. The first turn of the crank would print a "1" (the square of 1). The next crank would add 3 to the 1, printing "4" (the square of 2), while at the same time adding 2 to 3 (the first and second differences) to produce a 5, which was stored in the middle shaft. The next crank would add 4 plus the stored 5 and print the result of "9" (the square of 3), while at the same time adding 2 to 5 (first and second differences) to produce and store the number 7. The next turn of the crank would add 7 to 9 for "16" (the square of 4) while adding and storing 2 plus 7, and so on. The numbers would be transferred from shaft to shaft by means of carefully calibrated gear teeth.

Naturally, Babbage had to work with the technology and materials available to him in the early to mid-1800s, and so his amazing computation machine was designed to be constructed of thousands of carefully machined brass gears and cogs arranged in incredibly complex patterns and powered by a steam engine. Work proceeded for a decade beginning in 1822, financed by an increasingly restive British finance minister. The government had initially shown interest in the project because more accurate mathematical tables would have significant economic and military benefits, not to mention the fact that errors in marine navigational tables were frequently responsible for the loss of lives at sea. However a characteristic quip from Prime Minister Sir Robert Peel, one of several politicians from whom Babbage sought support in his requests for government backing, neatly sums up the difficulty faced by governments then, as now, in funding scientific research: "I should like a little previous consideration," he remarked to a friend, "before I move in a thin House of country gentlemen a large vote for the creation of a wooden man to calculate tables from the formula $x^2 + x + 41$. I fancy Lethbridge's face on being called to contribute."[160]

Babbage was in some ways his own worst enemy, a victim of the French proverb that says the best is often an enemy of the good; he was continually revising and simplifying his plans with the result that nothing got finished. Had he chosen to freeze development work at any time during the years he worked on the Difference Engine, it is entirely likely that he could have produced a working model within a reasonable time at a figure acceptable to the government. (And we would undoubtedly be living in a very different world today!)

Government funding was eventually halted, never to be renewed. Meanwhile, Babbage's fertile mind had come up with radical new designs for an "Analytical Engine," a machine which would not simply automate arithmetic processes, but also the logical control of the processes. He had found a way to mechanize the idea: *if/then*. Data would be fed into the machine by means of punch cards similar to those being used in the newly invented Jacquard loom; from there it would be processed by a "mill" or calculating and decision-making unit, operating under instructions from a "store" or memory, another bank of finely machined gears and interconnected shafts. Thanks to the adjustable nature of the store function, the machine could be programmed to handle all sorts of computational chores, and those processes could be automated with decisions based on its own calculations. Thus, the machine was capable of doing simple, autonomous deductive reasoning. For example: *If* the answer is x, *then* add y and proceed to operation z. An answer could be fed back into the machine as data for finding another answer in a complex problem. It was, in every important respect, a modern digital computer, except that it was designed to be constructed of brass and mahogany. If completed, it would have occupied most of an aircraft hangar (had such a thing existed) and required six steam engines to power it. The Analytical Engine was so far out of its time that it takes on a faintly preposterous, seductively humorous mien; one expects a Disney adaptation to reach the cinemas at any time.

As Babbage aged, and as his frustrations with funding and bureaucracy and the limits of machine technology mounted, he became a well-known curmudgeon, notorious for his energetic campaigns against organ grinders and other "street nuisances" in London. In return, they pitched stones through his windows and jeered at him in the streets. Despite his grumpiness, he had the good luck to attract the benevolent attention of Lord Byron's only legitimate daughter, Lady Ada Augusta, Countess of Lovelace, who was not only a brilliant mathematician, but beauteous and well heeled to boot.[161] She wrote a number of routines for the Analytical Engine, becoming the world's first computer programmer in the process. (The modern programming language, Ada, is named in her honor.)

The relationship between Babbage and Lady Lovelace is both poignant and deliciously shrouded in intrigue. She was the age his only daughter would have been had she not been taken from him by illness; he was several years younger than her absent father. In a world in which his inventions were ridiculed even by eminent scientists (the Astronomer Royal called the Difference Engine "humbug" and the Analytical Engine "worthless"), she understood perfectly how and why they worked and what they were capable of, and produced the single most authoritative book on the subject. Indeed, Babbage might have escaped history's appreciation altogether were it not for her careful documentation of his work.

Their increasingly intimate friendship took a bizarre twist some time in 1848: the otherwise rich and well-preserved correspondence between them contains a large lacuna in this period with only tantalizing hints of the reasons why so many letters should have been destroyed, presumably by Babbage himself. As the story has been pieced together, it appears that Lady Lovelace enlisted Babbage's help in a scheme to use the partly completed Difference Engine to beat the odds at the race track. The scheme appears to have carried on for a period of four years, until Lady Lovelace's death in 1852. It is a fact that the countess was heavily involved in betting on horses at the time, indeed, had pawned her jewels twice in order to pay her debts. Each time they were redeemed by her mother, Lady Noel Byron.

The surviving record indicates a blackmailer named Crosse was paid a small fortune by lawyers engaged by Lady Byron after her daughter's untimely death of cancer; the agreement stipulated that he was to destroy in the lawyers' presence "all of Lady Lovelace's letters and all her husband's, including 'the all-important letter from him to her, assenting to and authorizing her betting proceedings, and which letter she had handed over to Crosse.'"[162] Babbage had apparently acceded to Lady Lovelace's deathbed request that he make the payoff himself, but eventually agreed to the lawyers' involvement. One of Babbage's servants, Mary Wilson, had been mixed up in the scheme: Babbage had seconded her to the

Lovelace household where she appears to have acted as an accomplice, possibly a go-between for Lady Lovelace and the bookmakers who placed her bets. She was fired by an outraged Lady Byron when the whole messy story emerged with Ada's death. Babbage insisted Mary Wilson be given a settlement of £100 and he left her £3 a month for life in his will, the only person outside his immediate family to be recognized. The death of Ada had its own horrific overtones: her sickroom had to be padded with mattresses to prevent her from injuring herself when she was in the paroxysms of agony caused by her disease. She was just thirty-seven when she finally succumbed, the same age her father had been when he died.

Lady Lovelace's descriptions of Babbage's machines remain clear and readable today. The Difference Engine, she said, was to the Analytical Engine as arithmetic is to analysis. The Difference Engine could do nothing but add and problems had to be reduced to a series of additions. The Analytical Engine, on the other hand, could add, subtract, multiply and divide directly. The Difference Engine could only tabulate results; the Analytical Engine could develop more complex solutions by *feeding back* its results into the system for further computation. Because it was programmed using punched cards, it could be used for a wide variety of computations without altering its physical structure. The idea that his Analytical Engine could manipulate symbols other than numbers was well understood by Babbage, particularly in the case where the machine reached into its "store" to change its own program. Lady Lovelace spoke of the machine as "weaving algebraic patterns just as the Jacquard loom weaves flowers and leaves," and she envisioned it being used to write music and play chess.[163] Both knew they were tinkering on the fringes of artificial intelligence, as this particularly perceptive passage suggests:

> The bounds of *arithmetic* were, however, outstepped the moment the idea of applying the cards had occurred; and the Analytical Engine does not occupy common ground with mere "calculating machines." It holds a position wholly its own; and the considerations it suggests are most interesting in their nature. In enabling the mechanism to combine together *general* symbols in successions of unlimited variety and extent, a uniting link is established between

> the operations of matter and the abstract mental processes of the most abstract branch of mathematical science. A new, vast and powerful language is developed for the future use of analysis, in which to wield its truths so that these may become of more speedy and accurate practical application for the purposes of mankind than the means hitherto in our possession have rendered possible.[164]

As anyone who has endured a computer crash will know, a computer that does not function is a remarkably unimpressive device. Babbage's blueprints and half-finished pile of wheels and cogs[165] could not change the world. His vision was hopelessly beyond the reach of the technology of the day: in general, the clockwork machine he had blueprinted was just too complicated, and in particular, machine tools of the day could not produce gears to the fine tolerance required, certainly not at a cost that could be borne by Babbage and his backers.[166] He had demonstrated, however, that the great power of formal systems like mathematics and symbolic logic could be tapped by mechanical means—in effect, automated—and the implications were profound.

He died in his eightieth year in 1871, virtually forgotten; only one old friend outside his immediate family attended his funeral and the Royal Society of which he had long been a member did not think to print the customary obituary. He suffered a final indignity of having his brain removed and pickled for study. Of course, nothing was learned from it; the curious, however, can still see it for themselves at the Museum of the Royal College of Surgeons in London.

CHAPTER TWENTY

The Secrets of Automatic Formal Systems and the Enigmatic Alan Turing

Babbage was perhaps the first person to understand that the ultimate power of formal systems is realized through their automation. The formal systems of arithmetic and geometry can be used to abstract real-world problems (through symbolic logic) and, by slavishly following the rules of the formal system, provide reliable answers with relevance for the real world. The step-by-step methodology is tedious and no doubt mind-numbing, but it is at the same time very powerful due to the fact that it is error-free. That it is tedious in human terms doesn't really matter because formal systems can run on machines, whose "minds" presumably cannot know numbness.

As Babbage also knew, to his eternal frustration, to be of any practical value, an automatic formal system needs an appropriate medium—a machine of some sort—within which it can be executed. Babbage stretched the machine technology of his era to its limits and well beyond, without ever being able to realize his designs. The formal system called "symbolic logic" had to wait for the discoveries of Faraday and the other inventors of electronics, and in particular for Fleming's vacuum tube, before it would find a medium in which it could be embodied in a practical way, a means of construction appropriate to its capabilities. Two key conceptual notions were needed, as well, to bridge the gap between

Boole's theorems of symbolic logic and a device to manipulate them in useful ways. Both arrived just prior to World War II.

The young British mathematician Alan Turing was doing work on symbolic logic when, in one of this century's most impressive intellectual achievements, he provided the first of the needed concepts by working out on paper the complete details of how an electronic computer might function. In 1934 he was twenty-two years old, a newly installed graduate fellow at Cambridge with a £300-a-year stipend and free room and board. Although the fellowship involved no formal duties, he was expected to work on some mathematical problem of significance. He chose to take up where Kurt Gödel had left off, tying up a loose end left unresolved by Gödel's theorem, which was still causing consternation among mathematicians the world over (see Chapter 4, "The 'Invention' of the Electron"). The dangling question was: Is there any means by which it might be possible to distinguish between provable and nonprovable mathematical statements, that is, short of actually proving or disproving them (which Gödel had demonstrated was not possible). Might there be some mechanical formula or other that could be applied to give a "provable" or "not provable" answer?

The reader may well smile at the angels-on-the-head-of-a-pin nature of such a question. It would be reasonable to ask of what possible use could it be to spend one's time trying to answer it. A taxpayer might be prompted to demand whether a financial grant-in-aid of such a project was a sane way in which to spend public money. Turing's experience, though perhaps somewhat more dramatic than others, was nevertheless a prime example of how rarefied science can—in spite of its insistence on purity of motive—advance the cause of civilization in very concrete ways.

Turing tackled his problem by trying to conceive of a machine that could provide the required "provable/not provable" answers—a machine, because it would by definition operate without the intervention of human imagination or judgment to affect its decisions. To solve a problem mechanically, a machine must obviously be able to accomplish its goal within a finite number of steps. The finite operations used to solve a problem are collectively called an algorithm, and the question Turing was really asking is: Are there any kinds of problems in logic or mathematics that cannot be solved using algorithms? It is a question which could also be seen as a clarifying query about Boolean logic: what can and cannot be achieved with this kind of formal system? Turing was able to demonstrate that there was indeed no way to test for provability or nonprovability of mathematical statements, thus providing a definitive answer to the question he'd set out to answer. But much more important was what he'd learned about the machine he had invented as a by-product of his analysis. He could show conclusively that *any* computation that could be carried out by algorithms could be done on his "Turing Machine."

He found, in other words, that he had invented something quite miraculous: a machine that could do the work of *any other* machine. The Turing machine was capable of duplicating the operations of any formal system whatsoever, that is, any system that operated according to identifiable rules or algorithms—as do all machines, not to mention many natural systems. Put another way, if one could set out an algorithm for the operations of a process, that process could be simulated by a Turing machine. (Today's "Turing machines" in the form of supercomputers can, for example, accurately model the dynamics of a nuclear explosion or the growth of a cellular structure.) It was this brilliant achievement that led to the digital revolution that is in the process of changing our world. Turing's paper "On Computable Numbers," in 1936, described in complete detail the theoretical structure of all digital computers to the present day and is one of science's most significant documents.

The essential operations of the computer as understood by Turing are astonishingly simple. To accomplish its tasks all any digital computer does is write one of two symbols (1 or 0), one at a time. Which one it writes is determined by the existing state of the system as it finds it (or "reads" it) and a finite set of formal operating rules. Whether it is playing chess or crunching national tax accounts or navigating a space vehicle, any task being performed by a computer can be explained in terms of this simple, step-by-step reading/writing procedure. Strings of 1s and 0s are used to represent numbers, letters and other symbols, according to agreed-upon conventions or protocols (algorithms).[167] With different algorithms or sets of rules, different tasks can be done by the same machine.

As we've already observed, while this may seem a tedious way to go about solving a problem, it is also extremely precise and thus powerful: the straightforward engineering challenge is to find ways to make the machine perform its humdrum operations more and more quickly.

The second conceptual breakthrough that was required before the digital computer could become a reality was made by a master's student in electrical engineering at the Massachusetts Institute of Technology, at about the same time as Turing was writing his seminal paper. In what has been called the most influential master's thesis ever written, the American Claude Shannon showed how complex arrangements of electrical switches behave according to the rules of—Boolean logic! A switch, in being open or closed, is in effect saying "yes" or "no" to a current, or, in Boole's terms, "true" or "false." Therefore, any logical relationship that could be translated into Boolean algebra could also be expressed in an array of electrical switches.

The Boolean expression $a \times b = c$ (a or b is equal to c, where *or* is represented by $\times$ in Boolean notion) would be represented by two switches in parallel, either one of which could be closed to allow current to flow. This was termed an "or" circuit because it conformed to a Boolean expression in which two propositions are joined by the conjunction *or*: if either is true (if either switch is closed), then the sum is true (the current will flow). Two switches in series, on the other hand, is analogous to an *and* expression: both must be closed for current to flow and if one is open and the other closed, current will be blocked (since it must pass through both switches to reach c), just as in the Boolean expression $a + b = c$, both a and b must be true for c to be true (for current to flow). Shannon found that any conceivable arrangement of switches had a corresponding expression in Boolean logic. The converse was an even more powerful idea: any proposition that could be expressed in Boolean logic could also be expressed in an arrangement of electrical switches. Since arithmetic problems are problems in symbolic or Boolean logic, Shannon had also showed how electrical circuitry could be wired to solve arithmetical problems.

Minds like Shannon's and Turing's were not to be overlooked in pursuance of the war effort against Hitler. (In fact, the two mathematicians were to meet as senior technical advisors to their respective governments and discuss their mutual interests halfway through the conflict, when Turing crossed the U-boat-infested Atlantic on the *Queen Elizabeth* to inspect American efforts at code making and code breaking.) In 1939 Turing found himself attached to a motley group of tweedy classics scholars and unkempt chess champions housed in a series of cottages at the British Code and Cypher Unit in Bletchley Park, south of London. This was the modest beginning of what would become a very large and top-secret part of the war effort, one which was to play a decisive factor in the defeat of Germany. Throughout the war, Turing was a senior figure in astonishingly successful British efforts to break the German communications codes in all theaters of action.

The German cipher machine, which the Nazis believed with absolute assurance to be failsafe, was called "Enigma." The British effort to second-guess the machine was called "Ultra." Turing's insights into the logical workings of machines were literally invaluable, in that the Achilles heel of Enigma proved to be the very fact that it *was* a machine, and therefore its operations could be simulated by a Turing machine. By the end of the war, thousands of men and women were involved in the decoding and interpreting effort run out of Bletchley. And they were being assisted by super-secret, high-speed digital computers—at first electromechanical, using mechanical relays, but later fully electronic—in which Turing had played a key design role.

While his was undoubtedly one of the most important individual contributions to the winning of the war, it went unacknowledged for more than two decades because the work that had been accomplished at Bletchley was (and in some respects still is) highly secret. And at the height of espionage and Red paranoia excesses of the Cold War in the summer of 1954, Turing was found dead in the bedroom of his suburban Manchester home, a victim of cyanide poisoning. There was a hasty inquest, and the coroner pronounced it a "straightforward suicide."

The background to Turing's death suggests that final judgment on the coroner's verdict of 1954 should perhaps be reserved. Turing, an amiable, gentlemanly and thoroughly eccentric genius of scruffy professorial habits and with a passion for marathon running, was a homosexual whose innate honesty compelled him to inform any close friend of the fact. He was also in possession of the full mental catalogue of the secrets of Ultra, which Britain and the United States then regarded as of the highest strategic value. At Manchester University, where he spent his last years working, he had access to a Ferranti computer on which other scientists were doing work for the British atomic bomb program.

Two years before his death he was arrested and charged with gross indecency following a consensual homosexual encounter with a young man, in his home. (The police learned of it while investigating a burglary.) Homosexual relations between consenting couples were then illegal in Britain, as in most of North America, and the relaxed tolerance to homosexuality that Turing had found at prewar Cambridge had largely vanished after the war. Homosexuals were considered by the governments of Britain, the United States and Canada to be obvious security risks, unfit for military service of any kind and dangerously unsuitable for work of a strategically sensitive nature. In all three countries, loyal citizens with unimpeachable records were hounded from their jobs and sometimes to their deaths during the 1950s by government security agencies because they were homosexual.

Turing was tried in open court, convicted and put on one year's probation, during which time he was forced, as a condition of sentence, to take a drug that was supposed to help cure him of his "deviant" tendencies by causing temporary (it was believed) impotency. It also caused him to develop breasts. Shortly before his death he confided to a longtime friend that he had been providing continuing help to a government decrypting effort but had lately been told that there was no longer any place for homosexuals in such top-secret work.[168] Whatever these facts may say about Turing's death, in no sense does it seem just to call his suicide, if such it was, "straightforward." It was a disturbing and ignominious end for a person who deserves to be celebrated as one of the great men of the twentieth century.

CHAPTER TWENTY-ONE

The Computer Comes of Age

Turing's machine, no matter how brilliant in conception, would not have had much impact on society had it not found a physical embodiment appropriate to it, as poor Babbage discovered. Babbage died a disappointed man in 1871; in 1942, seventy-one years later, the world's first working analytical engine began operating at Harvard University. The IBM Automatic Sequence Calculator, Mark I, was a fully automatic digital computer that used mechanical switches operated electrically. Like Babbage's machine, it was controlled by punched media, in this case tape instead of cards. Its speed and capacity greatly exceeded anything achieved to that date by even the most sophisticated analogue calculating machines.

The impetus behind Mark I, as with so much of computer research and development, came from the military. One of the more serious problems of computation with which the United States Navy needed help was calculating the trajectory of various shells and missiles. This is a relatively straightforward task if the shell is to be fired in a vacuum; in real life, it becomes complicated to an entirely unmanageable degree by such considerations as atmospheric humidity, air density, the shape of the projectile and its velocity. Gunnery crews were forced to operate by trial and error, using approximations supplied by the best calculators then available. More precise calculations and faster compilation of tables were of important strategic value.

Britain was by now embarked on a desperate program to construct a fully electronic digital computer. Having broken the German cipher system at Bletchley Park, there was still an urgent need to speed the process of the computation involved. Banks of super-secret electromechanical machines, themselves far in advance of anything the world at large had ever heard of, were falling behind the flow of available information. Important messages were being decoded too late to be of use. Built between February and November 1943, "Colossus," as the electronic computer was called, was so deeply shrouded in military secrecy that it was an antique before the world knew it existed. It operated using binary arithmetic, which allowed it to get away with a relatively small complement of 1,500 vacuum tubes. Eleven of the machines were built during and immediately after the war, evidence of a truly staggering effort of technical skill, administrative determination and inventive genius.[169]

Across the pond at the University of Pennsylvania, hundreds of operators were being kept busy computing artillery trajectory tables, using the biggest and fastest of the last generation of electromechanical analogue calculating machines. Meanwhile, work had begun on another part of the campus, at the Moore School of Electrical Engineering, on an electronic version of the Mark I IBM machine, called the Electronic Numerical Integrator and Computer (ENIAC). It was not completed until the war was over, but it was by far the fastest computer in the world when it was.

ENIAC had a front panel 100 feet long, contained 17,468 vacuum tubes (it used the decimal rather than binary system, which would have called for fewer electronic switches), required a dedicated power line to supply its demand for 174 kilowatts of power, occupied 1,800 square feet of floor space and weighed thirty tons. It cost $800,000 in 1945 to build. Teams of engineers and technicians were required to operate it: replacement of tubes alone was a challenge; one burned out, on average, every few hours. It could do thirty-eight nine-digit divisions in a single second, which seemed incredible at the time.[170] In one of its first tests, it performed in twenty seconds an atomic energy calculation that was taking human-operated calculating machines of the Manhattan Project forty hours to complete. It was programmed by changing the wiring on a complex control panel that looked like an early telephone switchboard with its hundreds of patch cords. To change a program was a tedious and frustrating labor of many hours. It was made even more frustrating by cleaning staff who would occasionally accidentally knock one of the plugs out of the board and to avoid discovery stick it back into a hole at random.

As often happens, the makers of ENIAC made discoveries in the course of their work that made the machine obsolete even before it had been completed. With war still raging in 1944, they approached the army's Ballistics Research Laboratory for more funding, to produce a more advanced machine called EDVAC (Electronic Discrete Variable Automatic Computer). Then fate gave them the opportunity to recruit one of the brightest minds of twentieth-century science, a man whose stature would, they knew, add immeasurable prestige to their project. Mathematician John von Neumann was brought onto the team, to join ENIAC veterans J. Presper Eckert and John Mauchly. As a group, they developed a breakthrough in computer architecture which has since borne von Neumann's name.

John von Neumann had been a child prodigy in mathematics in Hungary. He received a degree in chemical engineering in Zurich and a Ph.D. in mathematics from the University of Budapest in his twenty-third year. In 1930 he joined the faculty of Princeton University, where he became a member of the Institute for Advanced Study along with Einstein; contemporaries placed them in the same intellectual class. (From 1936 to 1937 Alan Turing was also at the institute, on a visiting fellowship.) Von Neumann pursued an astonishing range of interests, publishing work on quantum theory, mathematical logic, continuous geometry, meteorology, aerodynamics, economics and the theory of hydrodynamics. It was this latter interest that would lead to his involvement in the Manhattan Project. There, he contributed the implosion method of detonation, which was used in the atomic bomb exploded over Nagasaki.

Von Neumann's initial exposure to advanced computers, we are told, took place on a railway platform in Aberdeen, Maryland, home of the army's Ballistics Research Laboratory, to which he was a consultant. There, on a crisp wartime evening, he met Herman Goldstine, who was the chief liaison officer between the Ballistics Lab and the ENIAC project at the University of Pennsylvania. Goldstine recognized the great mathematician and introduced himself. Though both men held top-level security clearance, their conversation was guarded at first, until Goldstine made reference to progress with ENIAC. He later recalled: "When it became clear . . . that I was concerned with the development of an electronic computer capable of 333 multiplications per second, the whole atmosphere of our conversation changed from one of relaxed good humor to one more like the oral examination for the doctor's degree in mathematics."[171] In September 1944, von Neumann visited the ENIAC project and soon after became a consultant.

His impact on the EDVAC proposal was profound. Leaving technical aspects of the machine to Eckert and Mauchly, he focused on the abstract problem of the relationships between the logical components. He wrote a paper on the team's developing ideas, which Goldstine circulated and which was immediately recognized as a breakthrough in computer architecture. Unfortunately, it bore only von Neumann's name, and the design it proposed became known as "von Neumann architecture." This did not endear him to Mauchly and Eckert, though in hindsight he appears to have been innocent of any intent to co-opt the team's work. There were major ramifications: the Moore team dissolved in acrimony and bitterness. Mauchly and Eckert had early on realized the commercial potential for the machine they were developing and were already embroiled in a dispute with Moore school officials over patent rights. Publication of von Neumann's paper scarcely helped matters for them: the Army argued it had placed EDVAC architecture in the public domain, making it unpatentable. Mauchly and Eckert angrily resigned their university positions, intending to develop a commercial version of EDVAC on their own. They filed for patent protection in 1947; Bell Telephone Laboratories and IBM vigorously intervened and the case dragged on and on until 1964, when the patent was finally issued.

By then, Mauchly and Eckert had sold their interests in the patent to Sperry Rand for $600,000. It was another bitter disappointment for them; they had hoped to carry on in business for themselves. But, though they had both been deeply involved in the top-secret military projects to build ENIAC and EDVAC, new military contracts were inexplicably denied them.

Finally, Mauchly learned that an army/FBI security check had found him to be "communistically inclined." The main basis for this allegation was an anonymous colleague's recollection that Mauchly had once signed a petition (along with several hundred other scientists) calling for civilian control of atomic energy in the United States. The petition had been sponsored by the American Association of Scientific Workers, which army intelligence regarded as having been a Communist-front organization. It was an egregious example of the McCarthyist excesses of the Cold War. Mauchly himself, not knowing the exact nature of the charges against him, thought he might have come under suspicion for having been a member of the Consumers Union, which the House Un-American Activities Committee had also identified as a Communist front.[172] It would take six long years for Mauchly to clear his name and regain high-level security clearance, but in the meantime his company had withered on the vine, forcing the sale to Sperry Rand.

In 1967, Honeywell Inc., with IBM watching with interest, brought suit against Sperry Rand, alleging that the EDVAC patent was invalid since the design had been appropriated from an obscure Iowa State University physicist whom Mauchly had met back in 1940. The physicist himself had long ago lost interest in computers and, until the idea was planted in his mind by Honeywell's lawyers, had never thought of Mauchly as having borrowed his ideas. He had never sought a patent on his own work, indeed, had never bothered to announce it in public. In the grand swashbuckling tradition of the Western Union suit against Bell, it was a thin thread on which to hang a lawsuit, but Honeywell's legal resources were formidable. Pretrial proceedings took three and a half years, during which thirty thousand exhibits, including the autobiography of Charles Babbage, were identified. The hearing itself ran from June 1971 to March 1973. Seven months later the court ruled that the ENIAC patent was invalid. It also ruled that Eckert and Mauchly had waited too long to apply for protection in the first place. Under U.S. law, inventors have one year after making an invention to apply for a patent; the judge ruled that ENIAC had been in "public use" since December 1945, but the patent hadn't been applied for until June, 1947. Whatever its other merits, the lawsuit had effectively demolished any serious patent barriers to ongoing development of the computer industry, which was plainly one of the goals of Honeywell and its collaborators. Sperry Rand, having sunk more than a million dollars in legal fees, declined to appeal. For them, it was strictly a financial decision.

John von Neumann did not live to see the resolution of the twenty years of legal wrangling. He died in 1957 when he was just fifty-four, of bone cancer probably resulting from exposure to radioactivity during his work on atomic weapons. Since 1955 he had been a key member of the U.S. Atomic Energy Commission, which at that time had as its main responsibility the development and stockpiling of nuclear weapons. Not long before his death, he crystallized his virulent anti-Communist views for an article in *Life* magazine in which he advocated preemptive warfare against the Soviet Union: "If you say why not bomb [the Soviets] tomorrow, I say, why not today. If you say today at five o'clock, I say why not one o'clock."[173]

The key to the success of von Neumann's legacy, the so-called von Neumann computer architecture which most computers employ to this day, lay in its use of a large internal memory that could be accessed directly. It was large enough to hold both data to be manipulated *and* operating instructions or "programming." The fact that the operating instructions could be accessed directly, and at electronic speed, rather than the machine having to read through punch cards or tape to find its next command, meant it ran much faster than ENIAC. It was also much more flexible, by virtue of its stored program feature. The fast, randomly accessible memory would come to be called "random-access memory," or RAM.

The world's first computer using stored programs in RAM, however, was put into operation at Cambridge University in Britain in 1949. The design, largely by Turing, took full advantage of his Bletchley Park experience, and as a result placed great emphasis on speed of operation. Pilot ACE (Automatic Computing Engine) was built to run at a "clock speed" of one megahertz (one million cycles per second), much faster than EDVAC. It also incorporated the idea of a large memory, an idea which was in fact implicit in Turing's original conception of the Universal Turing Machine: storage within ACE was in the order of 100 kilobits, an enormous sum for the time. Unfortunately, and tragically for Britain's computer industry, the development of ACE was placed in the hands of government bureaucrats, where it languished. Turing himself withdrew from the project in frustration in 1947. Britain's substantial head start in computer development was squandered, while nimbler private companies in the United States, operating with rich financial incentives from Washington, first caught up and then, by the early 1950s, established an irretrievable lead.

The first *commercial* computer with a stored program was made by the Remington Rand company (later Sperry Rand) on a design of Eckert and Mauchly, the Universal Automatic Computer (UNIVAC 1), and delivered to the U.S. Census Bureau in 1951.[174] UNIVAC represented a tenfold jump in speed over ENIAC, though it was still slower than ACE. It contained only 5,000 vacuum tubes shoehorned into a cabinet 14 feet long by 7½ feet wide by 9 feet tall, and operated at a clock rate of 2.5 megahertz. It used magnetic-tape storage instead of punch cards for its instructions and had about one kilobyte of RAM.

It was a fitting first sale for the first of the commercial main frame computers, because the Census Bureau had in a way been the first modern customer for practical mechanical means of calculation. Required by law to conduct a full census of the American population every ten years, by the 1880 census (when the U.S. population had reached fifty million) the bureau found to its dismay that it took more than seven years to process the data that had been collected. It was clear that unless some method was found to speed up the system, the bureau would be unable to fulfill its legal mandate—hence the department's continuing interest, beginning as early as 1890, in mechanical calculators. In that year it purchased a machine, built in Belgium and modeled directly on plans taken from an article written by Charles Babbage about his Difference Engine.

Eventually, forty-six UNIVAC machines were sold, despite the conventional wisdom of the time which insisted that there was enough demand in the United States for perhaps two or three computers of its power. A 1950 *Business Week* article said, "Salesmen will find the market limited. The UNIVAC is not the kind of machine that every office could use." Even more revealing of the challenge of selling the computers of the day is Lord Bowden of Chesterfield's description of the job in a comment to

British communications theorist Colin Cherry: "I had to sell the wretched things, if I was to earn my living and keep my firm in business. I decided that I had the most peculiar job in the world, until I met a man on the *Queen Mary* who sold lighthouses on commission and he told me about some of his problems!"[175]

The profound shock caused by the Soviet Union's test explosion of an atomic bomb in August 1949 provided the impetus for a new burst of creativity in computer design in the United States. The explosion, coupled with the knowledge that Soviet long-range bombers were capable of reaching U.S. targets via the Arctic, caused near-panic in the Truman administration's defense establishment. American air defenses were in no way capable of dealing with such a deadly threat, one which seemed palpably real at a time when China was falling to the Communists, the Berlin blockade had only just ended and insurgent nationalist, anti-colonial movements supported by the Soviets were springing up all over Asia and Africa.

The response to the Soviet atomic threat was a crash program to build a state-of-the-art network of early-warning radar stations around America's perimeters, scores of stations capable of picking up low-flying aircraft and tracking them until interceptors could be directed to them to shoot them down. Canada was willingly co-opted into the scheme through the North American Air Defence Command (NORAD), as many of the stations would have to be built in that country's high Arctic regions. What made the plan feasible was the emergent power of computers to coordinate information flooding in from the many radar stations.

The military turned once again to MIT, where it had been financing a major computer project called "Whirlwind." Initially envisioned as a program to develop a real-time flight simulator that could train pilots on a number of different aircraft, Whirlwind had evolved into much more than that. Its speed was such that it was capable of tracking several aircraft in real time and relaying intercept coordinates to attacking fighters. It grew into a huge project, employing seventy engineers and technicians and another hundred support staff, with a budget of more than a million dollars a year. Scientists Robert Everett and Jay Forrester were in charge; Forrester is credited with inventing, for Whirlwind, the first magnetic core memory systems, which further boosted the machine's speed by a factor of four.

With the promise of bottomless funding, MIT computer scientists turned to the development of SAGE, or Semiautomatic Ground Environment, the heart of the new computerized air defense system. Whirlwind became the prototype backbone computer, and Forrester and his team invented a whole new family of technologies to enable radar stations to communicate directly with the mainframes. In tests in 1953, SAGE, which weighed 250 tons, was able to successfully track as many as forty-eight targets simultaneously. Continually updated and refined with new computer technology (much of it codesigned by IBM) and new weapons of interception, SAGE remained in service until 1984. The system's eventual cost was $61 billion. But many of those who worked on SAGE now believe that despite its cost and sophistication, it would never have been capable of fending off a full-scale Soviet onslaught. It was the best system possible at the time, but it was excruciatingly vulnerable to being swamped by large numbers of attacking aircraft. To conceal its technical inadequacies, it was extensively promoted in public statements as virtually infallable. It was, in the words of Paul Edwards, "more than a weapons system: it was a dream, a myth, a metaphor for total defense."[176]

Whatever its shortcomings as a defensive weapon, SAGE provided a host of spinoffs that were of seminal importance, especially in the field of computer networking and communication. Robert Everett gave Whirlwind an interactive graphical interface: a technician could touch a blip on a radar screen with a light pen (invented by Everett) and information on the object would appear on the screen. Touching the dot and typing *T* on a keyboard would designate it as a target. So that the twenty-three SAGE control centers across the United States and in Canada could communicate, SAGE engineers developed the modem (modulator-demodulator), which translated the computer's digital information into analogue form for transmission through the telephone system, and then converted it back into binary form at the other end for computer consumption there. SAGE technology also formed the foundation for applications as diverse as air traffic control and airline ticket reservations systems. Contracts for its development put IBM solidly in the forefront of the international commercial computer industry. The original Whirlwind computer's sixteen-bit architecture also provided the foundation engineering for a new generation of smaller, faster, less expensive machines called minicomputers, which are the direct ancestors of today's ubiquitous personal computers.

IBM went on to design and market a range of business-oriented computers culminating in the Model 650, which rented for $3,000 a month and was tended to by a priesthood of clean-shaven, close-cropped, white-shirted, black-shoed IBM service personnel. Company officials felt they were taking an enormous risk with the new line and planned for a production run of only 250 machines; they were astonished when business snapped them up in the thousands. The Model 650 made IBM the world's undisputed leader in computer manufacturing.

The early notion that there was no widespread market for computers, and that only a handful could fulfill any nation's requirements, was widespread and deep seated. In 1951, the British physicist Douglas Hartree remarked: "We have a computer here at Cambridge; there is one in Manchester and one at the [National Physical Laboratory]. I suppose there ought to be one in Scotland, but that's about all."[177] In 1956 Howard Aitkin, builder of the pioneering Harvard Mark I electro-mechanical computer, told a German symposium on the future of computing: "If it should ever turn out that the basic logics of a machine designed for the numerical solution of differential equations coincide with the logics of a machine intended to make bills for a department store, I would regard this as the most amazing coincidence that I have ever encountered."

The failure of even the nascent industry's top postwar scientists to foresee the future that machines of their own design were helping to bring about can only be attributed to a failure to understand the implications of von Neumann's innovation, the large random-access memory, or RAM. Von Neumann of course understood that having enough internal memory to store both data to be manipulated and the operating instructions meant that a computer would be able to perform arithmetic on its own instructions and thereby modify them. He saw this as a way of making the machines operate faster and smarter; they would be able to tackle more complex problems with less overall memory. What he apparently did not see, and what none of his contemporaries appear to have seen, was that his architecture also allowed computers to write their own programs on the fly, and it was this ability that would lead to a world in which computers would become almost as commonplace as telephones and be used for a thousand purposes other than computation. It was this ability that permitted computers to be programmed—to be given their instructions—in "higher" languages, in assembly and compiler languages and interpreters that were easily understood by people, and that the computer itself translated into its own machine language of 1s and 0s. It was also this ability that allowed computer interfaces to be designed which would further simplify human-machine interaction through the use of menus, the mouse, joysticks, video display terminals, audio and even voice recognition.

The scientists, engineers, corporate strategists and military procurement experts failed to get beyond the narrow idea of the digital computer as a replacement for platoons of human computers working with calculating machines. It would take a generation of undisciplined student hackers, incorrigible phone phreaks, inveterate video game players and barely postpubescent entrepreneurs to begin to realize the true potential of the do-everything machine Allan Turing had invented. By simply *playing* with what was after all the most miraculous toy of all time, they uncovered its hidden cultural dimensions. It was this anarchic mob of unkempt social misfits that conceived of the personal computer and brought it into being.

CHAPTER TWENTY-TWO

Information, Chaos and Reality

While the air defense demands of World War II and the postwar era prodded computer hardware designers to ever greater efforts, they also led directly to one of the major theoretical insights of the century and a cornerstone discipline of the Information Age. It is called information theory, and as its name implies, it is central to the development of the various communications media because it is the discipline that deals with the problems of how to best transmit information in the presence of noise, or noninformation. It has been broadened, as well, to deal with more abstract problems, as indicated in this quotation from Norbert Weiner, the MIT mathematician who is credited with working out the major problems in the field:

> The commands through which we exercise our control over our environment are a kind of information which we impart to it. Like any form of information, these commands are subject to disorganization in transit. They generally come through in less coherent fashion and certainly not more coherently than they were sent. In control and communication we are always fighting nature's tendency to degrade the organized and to destroy the meaningful; the tendency . . . for entropy to increase.[178]

Information theory developed out of the military's requirement for better gun sights as the speed of air combat rose dramatically toward the end of World War II. The key to shooting down an enemy airplane (or a partridge!) is to "lead" the target so that the bullet or missile will intersect the flight path of the target. When the target is being piloted by a human, this becomes a difficult proposition, since it will normally take evasive action if it knows it is in danger. However, using position-indicating technology like radar, combined with statistical data on the evasive maneuvers permitted by aerodynamic, physical and psychological considerations that limit the enemy pilot's choices, useful predictions can be made. When the two sets of factors are blended and continually updated by a sufficiently powerful computer, it becomes possible to "lead" even an intelligent target with remarkable accuracy.

The key is in providing continuous "feedback" on the enemy's position to refresh the computer's calculations and sharpen its estimates. And the reason that works is that each choice made by the enemy pilot determines to some extent what his next maneuver will be. The choice of actions is limited by previous actions (he cannot, for instance, immediately turn west if he is flying east).

Norbert Weiner realized that his general theories developed during the war were mathematically related to one of the thorniest problems of modern electronic communications: how to deal with spurious interference to intelligent data, caused by noisy transmission channels. It turns out that the problem of how best to encode data in a noisy environment is closely related to the problem of predicting its value or content at some time in the future, a mainly statistical issue like that of "leading" an enemy aircraft. Telephone and telegraph engineers had been stymied by interference problems for decades; Weiner's work showed the way to solutions.[179]

Out of his wartime studies Weiner developed ideas of feedback and how it can be used to control technical systems, which he published in his great work *Cybernetics* in 1948. Closely related papers were published at about the same time by electrical engineer Dennis Gabor and Bell Labs researcher Claude Shannon. Together, the work of these three men founded the new science of information theory, sometimes called communication theory.

At its most basic level, the discipline recognizes that there is much redundancy in most communication and that *useful* communication exists exclusively in new or novel data contained in any transmission. Television pictures, for example, are routinely compressed for delivery to cable outlets, or "head-ends," using a digital process that transmits only elements of the picture that have changed, leaving the receiver to replicate those elements that have not changed. The difference between pictures compressed by ratios as high as ten to one is impossible to distinguish with the naked eye. Or perhaps a less technical application will make the concept clear: we all know that communication between friends goes more smoothly than between strangers, and that one can convey information to friends using just a few words which contain new information, whereas to get the same thought across to a stranger might take an elaborate explanation. This is obviously because friends have stored a great deal of information about us in their memories, and this stored information helps them in interpreting what we are saying to them. As Claude Shannon proposed in his ground-breaking book *Mathematical Theory of Communication* in 1948, information is really "the resolution of uncertainty."[180]

It is not much of a stretch to see how this notion relates to the problem of targeting an enemy aircraft. The more you know about the target, the better able you are to predict its next move. Or, put another way, the more you know about the target, the better able you are to reject spurious or inaccurate predictions of that move. Applied to the problem of noise on a telephone line, the theory states that the more you know about the data coming down the line, the more able you are to predict what the next character or string of bits is likely to be, which means that you are able to filter extraneous "noise" out of the system with increasing efficiency. Here we mean "know" not in the sense of knowing the message's content, but understanding its makeup from a statistical point of view. The closer you come to being able to predict the next character with complete accuracy—in other words, the closer you can come to rejecting 100 percent of the noise or interference—the closer you can come to completely error-free transmission. And in the case of the fighter pilot or antiaircraft gunner, the closer he will come to being able to shoot down the enemy with complete certainty.

Communications theory has two chief values, as described by John R. Pierce:

> One is rather like the value of the law of conservation of energy. [It] has made it possible for communications engineers to distinguish between what is possible and what is not possible. Communication theory has disposed of unworkable inventions that are akin to perpetual motion machines. It has directed the attention of engineers to real and soluble problems. It has given them a quantitative measure of the effectiveness of their systems. [It] has also inspired the invention of many error-correcting codes, by means of which one can attain error-free transmission over noisy communication channels.[181]

The error-correcting codes are developed on statistical models of the probability of a given bit of information being next in line to be received. To accomplish this, the theory ignores the content of a message, and treats information as a simple commodity, as a choice of one message from a set of possible messages. The choices occur with a certain level of probability; in other words, some are made more frequently than others. For example, an English sentence is a choice of the first word, then the choice of the second word, which is influenced by the first; then the choice of a third word, which is influenced by the preceding two choices, and so on.

It turns out that making predictions about transmissions of the English language is not as difficult as it might seem, because there is actually, on average, very little new information carried with each succeeding letter. This is because there is a wide range in the frequency with which letters occur in English usage: some letters, like *e* and *a* are very common, while others, like *z*, *q* and *x* are used only infrequently. Were all the letters to have the same incidence of use, prediction would be more difficult. If you know for certain what the next letter in a sequence is going to be, it actually carries no new information at all, because you knew in advance what it would be.

Information content is measured in bits: if all of the letters in the English alphabet had an equal probability of being used in a sequence, the information rate would be about 4.76 bits per letter. In fact, the information rate calculated for English is closer to one bit per letter. English is about 50 percent redundant, a fact which can be verified by observing vanity license plates on cars, in which information is often compressed while remaining readable: LT AGN; HVY METL; RLY RISR; UPN ATM.

Information theory had its genesis in the search for better weaponry, and it had obvious applications in both encrypting and decrypting wartime military messages. But its most important uses would come with the networking of computers, a development that grew out of the SAGE air defense project and moved into the civilian world of time-sharing on mainframes, and continued through to our own era and its proliferating on-line webs. On any computer network, in fact on any *information* network, bandwidth is the bedrock issue. That is, how much data can be crammed into a given transmission space, be it a television channel with bandwidth limited by government edict, or a length of copper wire with bandwidth limited by the laws of physics. Information theory is the tool used to get optimal results.

The code developed by Morse and Vail, known as Morse code, is a crude example of how information theory works to compress data. Morse code begins with an understanding of the frequency with which each letter of the alphabet is used in English text. To figure this out, Vail used the clever expedient of checking printers' type trays to see how many of each letter they kept in stock. The letter *e*, the most commonly used in the language, is thus coded as a single dot. The letter *t*, also frequently used, is coded as a single dash. The rarely used letter *z*, on the other hand, is coded as two dashes and two dots, a considerably longer (or bandwidth-hungry) sequence. Coding of data as complex as digitized television pictures or animated graphics obviously requires much more elaborate algorithms than Morse code, but the principle is the same.

In recent years there has emerged an interesting confluence of information theory with another brand-new scientific paradigm called Chaos theory. Chaos theory is the study of complex systems in nature, ones that cannot be adequately understood using classical Newtonian methods. Newtonian physics and its mathematical descriptions concentrate on systems that are predictable, that have clear cause-and-effect relationships; that can, in short, be described by linear equations or equations that can be plotted as straight lines or regular curves on a chart. A moment's reflection will show that most systems in nature are not regular in this way, but are much more chaotic. Weather systems are a good example: they are difficult, if not impossible, to describe mathematically and therefore cannot be predicted with much accuracy even using super computers; cause and effect are wildly disproportionate in that a small change in a variable in one part of the system can cause huge effects in another part of the system. It has been called the "butterfly effect": a butterfly flapping its wings in Hawaii can, in principle, change the weather in Toronto. This sometimes wildly disproportionate relationship between cause and effect is a prime characteristic of chaotic systems, which are full of surprises. Because they are so full of surprises—so full of waiting-to-be-resolved uncertainty—chaotic systems can be seen from an information theory point of view as being exceedingly rich in information.

Chaos theory has in a relatively short time had a profound effect on the way scientists and lay people alike look upon the world they inhabit. It is clear, with hindsight, that describing all nature in Newtonian terms was a lot like looking for your lost car keys under the lamppost (where the light is good) rather than in the ditch where they fell. Chaotic systems are not the exception in nature, but the rule. At the same time, chaos theory and experimental subdisciplines like fractal geometry also demonstrate that, surprisingly, there is order in chaos; that what may seem completely random patterns on the surface often have deeply buried consistencies or patterns of repetition called "strange attractors." These patterns, which emerged thanks to the powerful mapping capabilities of modern computers, display a beauty as remarkable as anything in nature—indeed, they might be said to be responsible for much of the beauty in nature.

Chaos theory has found wide application in the life sciences, where it is now recognized that it is the element of randomness in chaotic systems that is responsible for the robustness, the brilliance and diversity, of evolving life on the planet. Life is a process of self-organization, or bootstrapping. What was once seen as a spontaneous and unexplainable coming together of various elements in nature to produce living organisms is now understood to be a process which grows out of the fundamental dynamics of chaotic, information-rich systems. Chaos theory has demonstrated, principally through computer simulations, that chaotic or information-rich systems facilitate rather than impede self-organization. The element of surprise or unpredictability in these systems is what produces the variants in development, or mutations, that are necessary to evolution via natural selection. Indeed, as one chaos scientist has said, "Evolution is chaos plus feedback." It might equally well be said that evolution is dynamic information plus feedback. Life, it might be added, is a creature of information density, and is therefore most likely to be found in chaotic, information-rich surroundings. Chaos may be seen as a condition of maximum opportunity. Rather than a black void, it is an inexhaustible sea of information.

In this modern conception of chaos can be found resonances with some of the earliest recorded metaphysical thinking. Early Greek cosmology designated Chaos as the void that existed before things came into being, but it also identified it with Tartarus, the underworld. Ovid and later thinkers saw it as the formless, disordered mass from which the universe was created. Many world religions, primitive and sophisticated, deal with eschatology or the ending of things by predicting a return to chaotic beginnings, because in chaos there was a kind of purity in which salvation may be found. Out of chaos comes cosmos (order) and then decay sets in; an endgame involving a return to chaos allows for renewal and a fresh start. Christian theology on this theme is well known. Perhaps some idea of its universal application can be sensed from the fact that the Semang pygmies of Malacca believe that in the beginning there was chaos (a void) into which the ultimate being, Ja Pedeu, blew wind and storms, creating stars, trees, water and everything else in the material universe. At the end of time, when everything is dead, Ja Pedeu will destroy her creation with her storms. A great flood will submerge the world and the bones of humankind will swim together. Finally, the bones will rise again. Chaos, for both the Semang pygmy and the Bell Labs' Ph.D., is the information soup out of which the world is made.

The link between chaos and information can be defined in terms of something called Shannon's theorem. Among the contributions of Claude Shannon to information theory is the cornerstone theorem of the discipline, which may be formally stated this way: for any level of interference whatsoever, there exists a way to code a message such that, when it is decoded after transmission, it will be as free of distortion as you may wish it to be. In other words, the key to distortion-free reception of messages lies in the way they are coded for transmission. This is a very general theorem, and it applies to any and all kinds of messages. It has many practical applications in everything from advanced radar to cryptography to compression of television images. It also has powerful implications for the way we see the world in the Information Age. Like Newtonian physics, Einsteinian relativity and the uncertainties of quantum mechanics, Shannon's theorem is a scientific idea which can be expressed in the language of mathematics, but which also finds its way into our most basic thinking about ourselves and about the nature of reality. This seepage from mathematics to metaphysics is worth briefly pursuing here because it provides further evidence of why the technologies of the Information Age are fundamentally different and potentially far more powerful than those of earlier technological eras.

Let us suppose that the sensory information we use to shape our perceptions of nature is actually a series of "messages" about reality. Nature as we perceive it, in other words, may be thought of as an isomorphism for an underlying reality, in the same way as a vinyl record is an isomorphism for the music held in its jittery grooves, or a portrait in oils is an isomorphism for its human subject. Can information theory help us to discover the underlying reality behind the message?

Mathematician J. M. Jauch provides a clue which suggests that it can. In an invented dialogue using characters borrowed from Galileo, Jauch sets out the following proposition in his book *Are Quanta Real?*:

> SALVIATI: Suppose I give you two sequences of numbers, such as
> 7 8 5 3 9 8 1 6 3 3 9 7 4 4 8 3 0 9 6 1 5 6 6 0 8 4 . . .
> and
> $1, -\frac{1}{2}, +\frac{1}{5}, -\frac{1}{7}, +\frac{1}{9}, -\frac{1}{11}, +\frac{1}{13}, -\frac{1}{15} \ldots$
> If I asked you, Simplicio, what the next number of the first sequence is, what would you say?
> SIMPLICIO: I could not tell you. I think it is a random sequence and that there is no law in it.

SALVIATI: And for the second sequence?
SIMPLICIO: That would be easy. It must be $+1/17$.
SALVIATI: Right. But what would you say if I told you that the first sequence is also constructed by a law and this law is in fact identical with the one you have just discovered for the second sequence?
SIMPLICIO: That does not seem probable to me.
SALVIATI: But it is indeed so, since the first sequence is simply the beginning of the decimal fraction [expansion] of the sum of the second. Its value is $\pi/5$.
SIMPLICIO: You are full of such mathematical tricks. But I do not see what this has to do with abstraction and reality.
SALVIATI: The relationship with abstraction is easy to see. The first sequence looks random unless one has developed through a process of abstraction the kind of filter which sees a simple structure behind the apparent randomness.

It is exactly in this manner that laws of nature are discovered. Nature presents us with a host of phenomena which appear mostly as chaotic randomness until we select some significant events, and abstract from their particular, irrelevant circumstances so that they become idealized. Only then can they exhibit their true structure in full splendor.[182]

The idea is that nature presents itself to us in coded messages which appear to be random and chaotic, but which have underlying patterns. The patterns can be discovered, and the message interpreted, by applying the correct code or filter. The code is a way abstracting from the messages only the essential information, in other words, of filtering out the irrelevant noise or interference. We can thus begin to see the link to information theory. Information theory is all about distilling or abstracting information from communication, i.e., messages. Shannon's theorem demonstrates that there is indeed a way of filtering out the interference in the messages we receive from nature in order to get at the hidden reality behind them, and it even provides the mathematical tools to do so. In a very broad sense, it could be said that mathematics is the filter through which we connect with the reality of nature, and information theory is a mathematical tool kit for building filters to be applied to specific questions or problems in nature, filters which allow us to see the meaning behind the apparent chaos on the surface.[183]

If we pull the threads of this chapter's arguments together, we have convincing support for our claim that Information Age technology is unique in not only its scientific but its metaphysical implications. Information theory and chaos theory show us that there is important meaning in complexity and truth behind apparent chaos. These are insights we could not have explored—indeed, which may not have occurred to us in the first place—without digital computers and their immense capacities for manipulating raw information. We might say that digital computers are really ocular devices which transcend the telescope and the microscope to offer us the possibility of a metascope, the comic book X-ray glasses of the Information Age. Computers and their associated technologies allow us for the first time to investigate the information-rich world of nonlinear or "chaotic" equations, which, it can be argued, are the filters that allow us to perceive the reality behind our sensory perceptions. At no other time in history have tools of similar power been available, let alone distributed worldwide in their tens of millions and linked by high-speed communications networks. There is no guaranteeing that all of this potential will be realized—certainly no one would expect millions of chaos hackers to spring up worldwide and reveal the keys to the universe. Nor is there any way of knowing how that realization might manifest itself in either pure science or in society, though the very fact that the tools of this power of insight have been so widely distributed leads to interesting speculation. It is as if we have discovered a new planet which, with baited breath, we are setting out to explore.

CHAPTER TWENTY-THREE

The Prodigal Semiconductor

The Bell Laboratories in New Jersey were humming in 1948. It was the year Claude Shannon published his first, ground-breaking work on information theory, and it was also the year of the announcement of the first transistor, an invention for which John Bardeen, Walter Brattain and William Shockley of the Bell Labs were to be granted the Nobel prize eight years later. It was with the transistor that warp-speed development in the electronics industry got under way. The little device could do everything the vacuum tube could, but it was much smaller, much faster, much more reliable, used very little electricity and generated very little heat.

The transistor is one of a whole family of electronic components that developed out of continuing research into materials called "semiconductors." These are minerals whose ability to carry an electric current lies somewhere between that of a good conductor and an effective insulator. Semiconductors and their peculiar electrical properties had been known since 1906, and as we've seen, some of the earliest practical radio receivers, crystal sets, made use of the rectifying ability of lead sulphide (galena) crystals tickled by a fine silver wire or "cat's whisker." Advances in understanding of the subatomic world brought about by continuing developments in quantum theory made possible great strides in the practical application of these puzzling devices. Behind the scientific curiosity was the compelling impetus of an urgent need for something to replace the vacuum tube, which World War II had pushed to the limits of its practical applications. A typical large aircraft of the period, for example, contained as many as three thousand vacuum tubes, a serious weight and power drain consideration, and, on the civilian side, telephone switchboards using vacuum tubes for improved switching speed presented major reliability headaches due to the limited life span of the tubes.

At the Bell Labs, Shockley and his colleagues were fascinated by the mysterious electrical phenomena occurring at the junction where the "cat's whisker" made contact with the semiconductor surface. They probed the area with a second and then a third wire, discovering that the current between wires 1 and 2 varied with the voltage applied to wire 3. The semiconductor thus had a lot in common with Fleming's vacuum tube. Not only would it act as a rectifier, but it could be made to act as an amplifier, just as the three-element vacuum tube did: a weak incoming current or signal was reproduced at much greater strength in the output current.

The U.S. military was immediately interested in furthering the development of the transistor and funded it lavishly through contracts for the Atlas intercontinental ballistic missile, airborne computers for the air force and improved radar-tracking equipment for the navy.

What exactly *was* it that Shockley, Bardeen and Brattain had found so fascinating about the point where the cat's-whisker wire touches the surface of the semiconductor crystal? Something mysterious was going on at that fragile interface, something that allowed the crystal to act as a detector of radio waves and a rectifier of alternating currents.

What Shockley and his colleagues discovered was this: the crystal structure of silicon (and other semiconducting solids) is such that even at room temperature there is enough heat energy to allow some of the electrons in the individual atoms' outer orbits to break loose from their electrical bonds and roam freely. Where those electrons have departed the crystal structure, a vacant space is left. Since electrons carry a negative charge, the newly vacant slot has the appearance, to nearby electrons, of a positively charged space, and the Bell Labs group called it a "hole," for lack of a more descriptive term. A "hole" is best imagined as a bubble in a pot of simmering water: the bubble is actually an absence of water in that small region. By virtue of their positive charge, holes are able to attract nearby electrons. When an electron succumbs to that attraction and jumps to fill the hole, it creates a new hole in the atom from which it jumped. Consequently, it appears as if the hole is migrating from one atom to another, while the electron is seen to be moving in the opposite direction. Imagine two atoms, side by side, the one on the left with a loosely bound electron, the other with a hole. When the electron jumps from left to right to fill the hole, the hole appears to jump from right to left, showing up in the spot vacated by the electron. Electrical current, in other words, is flowing in both directions: positive current in the direction of the holes and negative current in the direction of the electrons.

Holes do not exist in most metals because they are a feature of the crystalline structure of minerals. It is their existence that make semiconductors strange and unique and allows for their many useful applications. Prior to their discovery it was believed that only electrons conducted current, which, as a consequence, flowed only in one direction. The new notion of two-way electrical flows proved crucial to the understanding of semiconductor devices.

Having discovered this much, the Bell group speculated about what would happen if the crystal were deliberately contaminated with impurities, to change its electrical properties by giving it more holes, or more free electrons. Could a "supercrystal" be made? Pure silicon crystals were melted in furnaces and "doped" with tiny amounts of arsenic and antimony, gallium and indium. Two different effects were noted: in the case of arsenic and antimony, doping added to the number of free electrons available in the crystal, and since electrons are negatively charged, this was called n-type silicon. In the case of gallium and indium doping, the effect on the crystal bonding within the silicon was to increase the number of holes, since the impurities robbed the silicon of electrons. Because holes are positively charged, this was called p-type silicon.

And this is where the experiments got exciting. If a thin slice of p-type silicon were sandwiched with a slice of n-type silicon, it was discovered that a current would flow only in one direction through the unit. It was an entirely solid-state semiconductor that had the electrical properties of a diode or two-element vacuum tube. It could therefore serve as a rectifier of alternating current, or a detector in a radio receiver. The explanation for what was happening was that when the two doped wafers were sandwiched together, an orgy of attraction between oppositely charged holes and electrons occurred at the junction region. This immediately created a band of space on both sides of the junction where electrons and holes had neutralized one another, an area where both positive and negative charges were depleted. Common sense might have expected the transfer of charges to continue between the two wafers until they became electrically neutralized, but this was not what the scientists observed. The process proved to be self-limiting, with only a narrow band on either side of the junction being affected. The researchers called this the "depletion zone," and it turned out to be just wide enough to make a stable barrier against any further transfer of charges from one silicon wafer to the other. There was nothing left in the depletion zone to attract either electrons or holes.

When a battery or other source of electrical current is connected across the diode formed by the two silicon wafers, the equilibrium in the device is upset and holes and electrons will now have enough energy to bull their way across the neutral depletion zone. But here is the critical distinction: if the battery is connected one way, current will flow; but if the battery connections are reversed, current will *not* flow. The device is either a conductor or an insulator, depending on the direction of current flow in the circuit into which it is wired. The reason for this is not difficult to explain in terms of holes and electrons and their properties. Connect the battery one way, with its negative terminal in contact with the *n* side of the silicon sandwich, and holes will be attracted from the *p* side toward the diode's depletion zone and through it. At the same time the battery's positive terminal, wired to the *p* side of the device, will attract electrons from the *n* side toward the depletion zone and across it, going the other way, and in this way a current will flow in both directions. If the battery connection is reversed, its negative pole (now hooked up to the *p* side) will attract positively charged holes *away* from the depletion zone to one end of the diode, and its positive pole will attract negatively charged electrons *away* from the depletion zone to the other end of the device. The depletion zone has now become so wide a barrier that the diode is effectively neutralized and will act as an insulator (or very high resistance); no current will flow in this direction.[184] It is as if the depletion zone in the transistor becomes a powerful suction device when the battery is hooked up one way, drawing current through itself in both directions, and behaves like an equally powerful blower when the battery terminals are reversed, preventing current from leaving the battery to enter the transistor.

Shockley's team had created a practical, rugged, mass-producible replacement for Fleming's thermionic valve or vacuum-tube diode, which had been the workhorse of the electronics industry for a generation. Not only was the new solid-state diode more reliable, it was much, much smaller and lighter, and it did away with the power-consuming and heat-producing filament required by the vacuum tube.

More excited experimentation demonstrated a phenomenon that seemed as magical as alchemy. If a wafer of either p- or n-type silicon were placed between two wafers of the other kind (i.e., *p-n-p* or *n-p-n*), the resulting device would amplify an electric current presented to it! This tiny sandwich of doped silicon crystal was capable of taking a weak current such as a faint radio or audio signal and reproducing it in amplified form, just as the three-element vacuum tube of Fleming and De Forest had done.[185] As if that were not amazing enough, the device could be made to conduct current or not, by changing a tiny voltage applied to the filling in the sandwich. It was a solid-state switch that could be turned on or off *in a picosecond* by remote control! A device worthy of the wizardry of the emerging technological revolution, it was ten times easier to manufacture than the vacuum tube; it was about a hundred times smaller, it used a thousand times less energy and it would become ten thousand times more reliable. And, of course, as a switch it was orders of magnitude faster than anything else on the planet. The device was dubbed a "transistor," because the inherent job done is *trans*ferring a current across a res*istor*.

With the development of the transistor and the avalanche of other solid-state devices that quickly followed, the race to miniaturization and subminiaturization was under way; fifty years later it has not ended. Equally as important as the triple virtues of reduced size, reduced power consumption and reduced weight were the gains made in the speed of operation of electronic circuits due to the unbelievable rapidity with which these devices were able to switch their state from conductor to nonconductor and back again.

The transistor went in to commercial use in hearing aids in 1953 and in portable radios the following year. The first transistorized digital computer was assembled in 1965 at the Massachusetts Institute of Technology's Digital Computer Laboratory, with the support of IBM, on a military contract. Its job was to replace a 55,000-vacuum-tube machine that was the cornerstone of the contemporary air defense system. By the decade's end, transistorized computers were available in the civilian market: the Control Data 1604 of that era contained 25,000 transistors in its central processing unit.

In 1955 Shockley left the scene of his triumphs at Bell Laboratories to set up his own company in the Santa Clara Valley near Palo Alto, California, about halfway between San Francisco and San Jose. Nearby Mountain View was his hometown, and Stanford University was close at hand. His plan, in the spirit of Marconi and Bell and Edison, was to manufacture products based on his own inventions, to become an amalgam of scientist, engineer and entrepreneur, and, not incidentally, to get rich in the process. He set up shop in a concrete block building in an industrial park and hired a dozen Ph.D. electrical engineers and a handful of clerical and administrative staff. They set to work melting silicon in electric furnaces, doping it with traces of exotic minerals and testing the results for improved semiconductor characteristics.

Genius though he was at the business of research, Shockley was by all accounts an insensitive and overbearing calamity as a manager of people. In the summer of 1957, less than two years after he'd founded Shockley Semiconductors and less than a year after he'd received his Nobel prize in physics, eight of his key engineers mutinied. They got together to form a company of their own: it had occurred to them that all they needed to succeed in this new business was their own intellectual capital—their brain power and experience—and some money to rent another empty manufacturing space. Fairchild Camera and Instrument Company staked the start-up money in return for an option to buy back the new company, Fairchild Semiconductor, for $3 million any time during the next eight years. They would eagerly exercise that option just two years later.

The defectors had picked a propitious time to go it alone. IBM, Hewlett-Packard and others were preparing designs for electronic computers employing semiconductors rather than vacuum tubes, and demand for transistors from the radio industry was beginning to boom. Nineteen fifty-seven was also the fateful year of Sputnik I. The news that the Soviet Union had placed an artificial earth satellite in orbit rattled the American nation to its entrepreneurial roots and galvanized its political, scientific and educational institutions into desperate action to catch up, to close the gap in what immediately was seen as "the space race." And the transistor was the foundation on which new space technologies, including smaller and more powerful computers, would be built.

Before Fairchild Semiconductor was two years old, its research director, Robert Noyce (one of the Shockley defectors), had designed an entirely new way to fabricate circuits that did away with the need to wire individual components together on circuit boards.[186] It takes a nanosecond for an electric current to travel through a foot of copper wire: at the computational speeds computers were achieving, circuit slowdowns due to connection length were already becoming a real problem. Noyce's idea was to etch both the circuitry and the transistors and other electronic components such as resistors and capacitors directly onto fingernail-sized wafers of silicon, using a process similar to photolithography. The significance of the new process for producing these "integrated circuits" or "microchips" was clear to everyone, including Fairchild Camera and Instrument which immediately exercised its option to purchase, making Noyce and his fellow defectors wealthy men. Noyce was just thirty.

Within a decade of its formation, Fairchild Semiconductor had grown from the original group of ex-Shockley employees and a handful of support staff to employ some twelve thousand people. Revenues soared to $130 million. By the early 1960s, competing microchip businesses were springing up like mushrooms after a spring rain, most of them next door to Fairchild in the Santa Clara Valley. So many of them were there that people began calling it "Silicon Valley." And most of the start-ups were formed by defectors from—where else?—Noyce's own Fairchild Semiconductor, getting into business on their own with their intellectual capital and a bit of money from venture financing. About fifty of them launched their own firms to produce microchips, and they came to be called the "Fairchildren." Thus was an industry born of sand and scientific insight. Without it, the communications revolution could not have happened.

Shockley, the pioneer, never lost his bitterness over the defections from his company. It struggled thereafter and was eventually bought out by Clevite Transistor. He had been teaching at Stanford, and was elevated to an endowed professorship of engineering and applied science in 1963. In the late 1970s, he figured importantly in the counterculture turmoil of the time, unfortunately as a focus of antiestablishment enmity. He had begun campaigning about the threat of something he called "dysgenics," by which he meant the retrograde evolution of the human species caused by the unfettered freedom of people of subnormal intelligence to have children. It was an idea Samuel Morse might have found appealing; many saw it as closet racism. Shockley proposed a private foundation which would reward hemophiliacs, epileptics and people with low IQ who would agree to be sterilized. He sued the Atlanta *Constitution* for a million dollars when the newspaper likened his ideas to Nazi eugenics and was awarded a token one-dollar payment, which amounted to a stinging moral rebuke. Undaunted, he entered the Republican primary of 1982 seeking a California Senate seat as a single-issue candidate warning of dysgenics; he placed eighth with eight thousand votes. He continued thereafter to be more interested in talking about his bizarre theories of race and intelligence than semiconductors, for which he eventually held more than ninety patents.

CHAPTER TWENTY-FOUR

Robert Noyce and the Integrated Circuit

In 1968, Apollo 8 circled the moon with three American astronauts on board and returned safely to Earth, an extraordinary feat of navigation and piloting made possible by the integrated circuits Robert Noyce had pioneered.[187] The on-board computer was more powerful and far faster than ENIAC, but was about the size of a suitcase. In the same year, Noyce, now forty-one, caused a sensation in Silicon Valley by quitting Fairchild Semiconductors to set up his own company. With another of the original Shockley defectors, Gordon Moore, he rounded up about $2.5 million in capital (including half a million of their own money), rented another nondescript Silicon Valley multipurpose concrete building, and launched Intel Corporation.

Intel would specialize in manufacturing memory chips. The company's first commercial product, the 1103 chip, was about the size of the *o* on a computer keyboard and contained four thousand transistors. Intel's workforce leaped from forty-two at start-up to more than a thousand four years later, in 1972. The next year, 1973, the workforce ballooned to more than twenty-five hundred. Sales in 1972 were $23.4 million; in 1973, they hit $66 million.

There was more to come. A thirty-two-year-old Intel engineer named Marcian E. (Ted) Hoff, Jr., fresh out of Stanford, invented the "microprocessor" which, amazingly, put all of the logic and arithmetic functions of a complete computer on a single silicon chip. Intel had been commissioned by the Japanese company ETI to construct a programmable desktop calculator, but Hoff was unhappy with the design they'd submitted. It called for thirteen separate chips containing different logic functions. In a leap of imagination, Hoff saw that it should be possible to create a chip that would be a complete programmable computer, and link it with a few memory chips that would contain its operating instructions. The computer-on-a-chip would be a generic product that could be mass-produced and would be usable in all sorts of applications from traffic lights to cash registers and video games to nuclear missiles and telephone switchboards. After all, as Turing had shown, all computers are essentially the same machine, so why design complicated, special one-off devices for every application that came along? It didn't make sense. It was much more efficient to program a generic processor chip with software, which was a lot less expensive to produce than custom hardware. The programming would go on a separate memory chip.

The calculator that Intel eventually produced for ETI (marketed as Busicom) had just four chips, mounted on a circuit board. The processor chip itself contained 22,000 transistors.[188] Intel began marketing the new chip, designated 4004, late in 1971. It was the size of a fingernail, sold for less than $100 and matched the performance of IBM machines of the early 1960s that had sold for upwards of $300,000.

Industry historian Stan Augarten notes:

> Although Intel did not realize it at first, the company was sitting on the device that would become the universal motor of electronics, a miniature analytical engine that could take the place of gears and axles and other forms of mechanical control. It could be placed inexpensively and unobtrusively in all sorts of devices—a washing machine, a gas pump, a butcher's scale, a jukebox, a typewriter, a doorbell, a thermostat. . . . Almost any machine that manipulated information or controlled a process could benefit from a microprocessor.[189]

Just four months after the 4004's release, in April 1972, Intel announced the 8008 chip,[190] which was followed a year later by the 8080. The 8080 was twenty times faster than the 4004, a robust and flexible platform that would form the foundation of the nascent microcomputer industry.

Intel stock tripled in value. By 1982 the company's sales had grown to almost a billion dollars a year. Along the way, Noyce had attracted renewed attention, not as a scientist and inventor, but as an influential business manager, whose corporate style was decidedly informal and (superficially) democratic—Noyce was by any definition by now a plutocrat. The management hierarchies Noyce set up at Intel were as thin as the silicon wafers he manufactured. The Noyce style was lean and efficient, well adapted to constant change, and it managed to wring maximum effort and commitment out of the young engineers it employed. The rate of burnout, divorce and family breakdown in Silicon Valley would become legendary.

In a colorful biographical essay on Noyce, writer Tom Wolfe pointed out that the principal players in the microchip revolution that began in the 1960s had remarkably similar backgrounds: most had grown up and gone to college in small towns in the American Middle West and the West.

Wolfe attributes their successes to the shared background of dissenting Protestantism in these small American towns and universities and the egalitarian, hard-working approach it fostered to getting things done. Although generations removed from the pious men and women who founded the small towns they came from, these homesteaders in Silicon Valley had unaccountably carried with them the Protestant work ethic that valued achievement through labor and eschewed ostentatious displays of wealth and position (though not wealth and position themselves).

Wolfe concludes:

> Surely the moral capital of the nineteenth century is by now all but completely spent. . . . And yet out in Silicon Valley some sort of light shines still. People who run even the newest companies in the Valley repeat Noycisms with conviction and relish. The young CEOs all say: "Datadyne is not a corporation, it's a culture," or "Cybernetek is not a corporation, it's a society," or "Honey Bear's assets"—the latest vogue is for down home non tech names—"Honey Bear's assets aren't hardware, they're the software of three thousand souls who work here."[191]

Wolfe might with equal persuasiveness and perhaps greater accuracy have assigned Silicon Valley's successes to the less pious engineering ethic that had developed through the late nineteenth and early twentieth century, in which engineering is explicitly and heroically at the service of capitalist enterprise (see Chapter 7). This would account for the glaringly anomalous fact that the private lives and social intercourse of Silicon Valley denizens are, famously, anything but morally righteous in nineteenth-century Protestant terms. Corporate capitalism is not concerned with moral issues, or even with private lives. Even Noyce's vaunted collegial approach to management has less to do with moral righteousness than with engineering efficiencies: it is simply the most productive and profitable way to organize creative people. At the manufacturing level in the semiconductor industry, it should be noted, collegiality is as hard to find as it is in a Taiwanese garment factory.

Working out the manufacturing processes for integrated circuits deserves recognition as one of the all-time triumphs of engineering in any field. Where the engineers of the mechanical era strove to test the limits of scale with the biggest ships, the longest canals and tunnels, the tallest iron structures, the deepest underwater cables, the most powerful steam locomotives and the longest railways, the new field of electrical engineering focused more and more on the other end of the continuum, on the world of microscopic and then submicroscopic structures and processes. The structures they have succeeded in building are so tiny as to be invisible to the naked eye and are also the most complex artifacts ever made by humans.

At Bell Labs, where the transistor was invented, the idea of integrated circuits was investigated and dropped. They had reasoned that if the circuit is to function, all of the transistors, capacitors, resistors and other electronic components within it must work. If a circuit containing even as few as twenty transistors was to have a 50 percent probability of working, the probability of each individual component working would have to be 96.6 percent. This, Bell scientists believed, was an impossible success rate to achieve.[192] They were, of course, wrong, as Robert Noyce and others were to demonstrate in such dramatic fashion. Miniaturization in integrated circuits quickly proceeded to a state where, as early as the

1970s, researchers had begun to examine the fundamental limits to size and power dissipation imposed by the laws of thermodynamics and the new science of information theory. Though chips were by then so small that a microscope was required to see individual components, it appeared they were still many orders of magnitude away from the absolute physical limits to size. Since 1968, the density of transistors on a single chip has quadrupled every three years. Intel's flagship product in the late 1980s, the 386 chip, contained 375,000 transistors. In 1989 it was superseded by the 1486 chip, which contains more than one million transistors. In 1990, Motorola (maker of microprocessors for Macintosh computers, among others) announced a new chip containing four million transistors. Today, the number of transistors being routinely fabricated on a single chip is approaching *one hundred million*. The size of individual components has shrunk at a comparable rate, from about 10 microns (a micron is one-hundredth of a centimeter) in 1968 to 0.35 microns and less today.

Much of the early impetus for this amazing progress came from persistent and pressing demand from the military for ever more sophisticated computing and communications devices for increasingly complex Cold-War armaments and defense systems. The space race provided added push. But the economics of integrated circuit fabrication have a logic of their own that promotes miniaturization. Each wafer of silicon must go through a series of steps in the manufacturing process that are the same whether the wafer is to contain one or one thousand components. Obviously, the unit cost of components drops dramatically with the number that can be squeezed onto each wafer. Likewise, there is a very strong natural economic incentive for manufacturers to find ways to achieve very high levels of reliability in individual chips, in other words, to minimize the number of defects creeping into the manufacturing process. In practice, wafers of silicon are commonly about 200 millimeters in diameter, and each is divided into many small chips containing identical circuitry. If each wafer contains 100 chips which sell for $50, and if the rate of discarded chips due to defects is 70 percent (a typical figure in the late 1960s and 1970s), then the factory is producing gross income of $75 million a year. If the defect rate can be reduced to 50 percent, the factory can increase its earnings to $125 million a year, and almost all of the gain will be pure profit.[193]

A modern integrated circuit, when viewed under an electron-scanning microscope, has an eerily half-organic, half-mechanical look to it, with its many layers of oxide and metal deposits, its etched-in geometric shapes and its gently curving slopes. Seen in cross-section, it resembles a profile of geological strata laid open by a highway engineer's road cut, with overlapping layers of sediment and rock formations, lava flows and mineral deposits. In truth, it can be hard to believe it is a human artifact.

The manufacturing process takes place in stages, and is nowadays almost completely automated. The first step is to design the circuit, a process facilitated by design software. Then the circuit must be translated into a series of screens, as many as twenty or thirty for a typical integrated circuit, which will be used in the photoengraving process. This extremely complex drafting process, too, is highly automated through the use of computer-assisted design (CAD) software that can produce detailed drawings based on input of various design and fabrication criteria. These screens are the "blueprint" which the engineer hands over to the manufacturer. They represent about a third of the total cost of chip manufacturing, but fortunately can be used over and over to pattern many thousands of wafers.

The actual fabrication involves growing large silicon crystals of exceptional purity in a furnace within a sealed chamber filled with argon or some other inert gas. The electronic-grade silicon melted in the crucible is 99.999999999 percent pure and is refined from common quartz, one of the earth's most abundant minerals. A seed crystal of silicon is carefully lowered into the melt, and then slowly pulled back out as a much larger crystal grows. The product of this computer-controlled process is a crystal of pure silicon that may be as much as thirty centimeters in diameter and one to two meters long; it looks a lot like a big sausage.

Next, the crystal is sliced into thin wafers with a steel saw blade that has been dusted with diamond chips. Once a wafer has been washed and cleaned and polished, it is ready for the photographic process. Typically, a layer of doped silicon is applied to the wafer, followed by a thin layer of a "photo resist" material that will dissolve in an etching liquid wherever it has been exposed to light. Here is where the screens come in. The first of them is placed over the chip, and the chip is exposed to a flash of bright light. The screen is removed, and the chip is washed to dissolve away the exposed photo resist. A layer of doped silicon is applied, filling the exposed areas. Then, another layer of photo resist and another screen. Many succeeding layers of doped silicon, quartz insulation and various metals are applied until the process has been completed, and the individual chips can be separated, tested and packaged. All of this takes place in some of the most expensive manufacturing plants on earth, which are also among the cleanest environments on the planet, to limit the defects caused by stray dust particles. Technicians, clad from head to toe in special dust-free fabric, quietly monitor computer-controlled machines of incredible sophistication and complexity beneath their unassuming metal cabinetry.

The challenge currently facing integrated circuit manufacturers is the approaching limits of the optical lithography processes, in which the wafers are exposed to light through the various screens so that the surface can be etched. Component size has become so vanishingly small that it is close to matching the wavelength of light used to expose the resist material. The search is on for a new process using nonoptical techniques for etching. Electron-beam lithography has been used, and it has the advantage of being able to produce feature sizes of about 100 angstroms, which is far smaller than will be needed for the foreseeable future. But the process is a very slow and, therefore, very expensive one. X-ray lithography seems more promising as it works much like optical lithography and at comparable throughput rates, though with its shorter wavelength it can produce much smaller features.

While chip technology has been evolving, similarly spectacular advances have been made in the capacity and speed of access for long-term data storage in computers. UNIVAC made a major step forward with the use of magnetic tape, an enormous advance over punched cards, but data still had to be sought out in a linear process that involved moving the tape back and forth over reading and writing heads. The magnetic disk, in which data is recorded in concentric rings on a constantly rotating platter, was introduced in the 1950s. The read/write head skimmed the surface of the rotating disk, dropping to the surface to do its job as required. The physical distances involved in accessing data were greatly reduced, speeding the process significantly once again. The now-familiar "hard drive" is a stack of such disks mounted in a hard shell or box, with read/write heads for each surface. As with integrated circuits, no limit to advances in this area of computer technology has yet appeared.

Two well-known, rather fanciful benchmarks for progress in computer technology are perhaps worth quoting here:

> If the aircraft industry had evolved as spectacularly as the computer industry over the past 25 years, a Boeing 767 would cost $500 today, and it would circle the globe in 20 minutes on five gallons of fuel.[194]

> If the automobile business had developed like the computer business, a Rolls-Royce would now cost $2.75 and run 3 million miles on a gallon of gas.[195]

Astonishing as those claims may seem, they are based on accurate statistical evidence of the power/cost ratios experienced in the computer industry. They are also, as this is written, nearly seventeen years out of date, making them conservative by a factor of perhaps 100. That Rolls-Royce would now cost about twenty-seven cents and run 300 million miles on a gallon of gas.

CHAPTER TWENTY-FIVE

The Personal Computer

It was just one pleasant surprise after another at Intel. The first integrated circuits had brought the company overnight wealth almost beyond comprehending; now its 8080 microprocessor, the "computer on a chip," was about to launch an industry that would spark a revolution in communications that would in turn change the world.

We've seen that, thanks to the exigencies of World War II and the interminable Cold War that followed, the development of the electronic computer up to and including the 8080 chip had been financed and supervised largely by the military in Britain and the United States. No technology is "value-free" or "value-neutral," least of all communications technologies; they carry within them the values of the social structures and institutions out of which they develop. Take the humble family washing machine. Electric and automatic, it is such a commonplace that it may seem value-empty. However, to someone from another culture, say, an Eretrian bedouin, it is as value-laden as a mechanized Tibetan prayer wheel is to us. The washing machine has built into it all the values of a modern, industrial, consumer society. It tells us that cleanliness is important, that clothes should be washed in the home rather than outside, that they should be washed by a family member rather than a servant (otherwise why bother to make it so easy), that menial work is something to be avoided if at all possible (or why have the machine at all) and so on. As Michael Shallis observes in *The Silicon Idol,* "The [washing] machine requires electricity, piped water and drainage services. It is dependent on a detergent industry and systems of transportation. . . . It has also been produced with a certain attitude to work built into it. It is not neutral at all; to accept a washing machine is to accept a specific attitude to nature, to man's place in the world."[196]

As a child of the military, the computer circa 1975 bore unmistakable traits of military organization and priorities. A 1978 study of the U.S. computer industry estimated that between 1958 and 1974 the U.S. federal government, through the Department of Defense and NASA, had funded computer development to the extent of one billion dollars. By then, IBM, the main commercial beneficiary of this funding, boasted revenues of more than eleven billion dollars: the computer industry at that time was characterized by the wryly descriptive name of "IBM and the Seven Dwarves."[197] The computers the military bought and paid for, whether made by IBM or, infrequently, by other companies such as Sperry Rand, were massive, hugely expensive, highly secret in their operations, protected by elaborate security measures, pampered by air-conditioning and attended by an anointed priesthood of white-coated minions who mediated between actual users and the machine itself, taking delivery of the batches of punch cards that constituted programs and placing them reverently in the machine's maw. Mainframe computers of the era, whether military or commercial, were a reflection of the military's obsession with hierarchical structure, control and order. They could be nothing else.

But that was not the whole story. A computer counterculture was emerging among people who believed that computers could and ought to be placed in the hands of ordinary people. Its members were mainly very young and almost exclusively male. Most were more or less loosely affiliated with university engineering schools where they had privileged access to the few computers then in existence; they were influenced both by their shared fascination with electronic technology and by the wider "flower power" counterculture movement swirling around them. Stephen Levy has documented their story admirably in *Hackers: Heroes of the Computer Revolution*. Although he warns that it is essentially uncodifiable and is best understood by observing its adherents, he nevertheless lists the salient elements of the "hacker ethic": access to computers, and anything which might teach you something about how the world works, should be unlimited; all information should be free; authority should not be trusted, decentralization promoted; hackers (people) should be judged by their hacking (accomplishments), not bogus criteria such as degrees, age, race or position; art and beauty can be created on a computer and, finally, computers can change life for the better.[198]

It has been noted by more than one jaundiced observer that the hacker ethic sounds a lot like plain old-fashioned youthful rebelliousness, idealism and irresponsibility, something apparently programmed in DNA. That, however, is an observation of the ethic's derivation, not a judgment of its value or utility. While much of it may seem trite today, the ideas behind the ethic were anything but commonplace when they evolved, during the era of the mainframe computer, the primacy and infallibility of IBM and the dominance of military funding. Computers, at that time, were an explicit extension of the traditional social control mechanisms fostered by the institutions responsible for their development.

Hackers simply believed that this was not the most appropriate use of the technology, that computers ought to be placed in the service of changing society. The hacker ethic states, in essence, that by providing access to information and the power to manipulate it, computers can be powerful tools for liberation of the innate creativity of the individual in society. It is a view that owes a great deal to the German-American philosopher Herbert Marcuse, whose 1964 *One-Dimensional Man* was much discussed on university campuses everywhere. Marcuse argued that automation could ultimately lead to egalitarianism, by allowing workers to wrest control of the means of production from the corporate power structure. The proletariat in both capitalist and socialist countries had been so thoroughly co-opted by materialism and philistinism that it would require the leadership of far-sighted individuals on the fringes of society to bring about this shift in social structures. Marcuse saw the hippie counterculture and the antiwar protest movement of the time as playing this catalytic role, and might have included the hackers, had they not been so nearly invisible. Once ordinary people had assumed control of the means of production, their natural goal would be to liberate the human spirit rather than simply maximize profits. In more mundane terms, the productively self-employed were bound to have other priorities beyond simply making money. True freedom, he believed, can emerge from automation, though not before corporate capitalism and bureaucratic communism both had done their worst.[199]

The Intel 8080 microprocessor, an artifact of total, unfathomable mystery to the intellectual humanist establishment in 1975, was excitedly seized upon by hackers as the means to realize their humanist goals. It was the tool that would put the enormous power of computers into the hands of the ordinary individual: it was the antiauthoritarian silver bullet. While the hippies sought to promote change through passive resistance, grass-roots organization, art and good vibrations and the more militant yippies fought pitched battles all over North America and Europe with authority as represented by the police and the military, the hackers, virtually unnoticed at the time, were busily working toward similar social goals through management of technology. It remains debatable which of these groups of avowed revolutionaries ultimately had the most impact. Hippie icon Stuart Brand, founder of the *Whole Earth Catalogue* and, more recently, the Electronic Frontier Foundation, has this to say:

> I think that hackers—dedicated, innovative, irreverent computer programmers—are the most interesting and effective body of intellectuals since the framers of the U.S. Constitution. . . . No other group that I know of has set out to liberate a technology and succeeded. They not only did so against the active disinterest of corporate America, their success forced corporate America to adopt their style in the end. In reorganizing the Information Age around the individual, via personal computers, the hackers may well have saved the American economy. . . . The quietest of all the '60s subcultures has emerged as the most innovative and powerful.[200]

In the late 1960s the computer industry began to diversify in a small way, producing a range of machines that were much more compact than the big IBM-style mainframes, and also less powerful. Often, however, they made up for their lack of brute strength with extra speed and flexibility. These were called "minicomputers": produced by companies such as DEC and Data General, they typically sold for less than fifteen or twenty thousand dollars and were used mainly in business and industry, for process control and accounting. They also found their way into university computer labs, where they were inexpensive enough to be made directly accessible to students. It was hands-on experience with minicomputers such as Digital's PDP line that inspired many hackers to get interested in the possibilities of the truly personal computer, if only as an abstract concept.

The notion took on concrete potential when, in January 1975, the American hobbyist's magazine *Popular Electronics* published a cover story about a shoe-box-sized computer called the Altair 8800. Built around the Intel 8080 chip, it was available either in kit form or complete from a tiny Albuquerque, New Mexico, electronics hobby firm called Micro Instrumentation and Telemetry Systems, or MITS. Fully assembled, the MITS computer sold for $650; in kit form it cost just $395. In its editorial that month the magazine said: "For many years, we've been reading and hearing about how computers will one day be a household item. Therefore, we're especially proud to present in this issue the first commercial type of minicomputer project ever published that's priced within the range of many households. . . ." It was, in fact, the first personal computer.

The president of MITS and the Altair's designer was Edward Roberts, a former U.S. Air Force captain who had run MITS at first out of his garage and later from a couple of rooms in a nearby shopping mall.[201] Prior to publication of the magazine article, MITS was $300,000 in debt and on the ropes; two years later, Roberts sold it to a large computer peripherals manufacturer for $6.5 million, and retired to a farm in Georgia.

The Altair was a machine only a dedicated computer aficionado could love. To get it to do anything, you had to enter a program bit by bit using a series of front-panel toggle switches. Its internal memory (RAM) was just 256 bytes, and it had no external memory (or ROM), so that whatever program had been laboriously loaded into it would vanish whenever it was turned off. There was no display screen, just a few rows of tiny red lights, by which a user was able to divine what was going on inside the box. It sold like hotcakes. MITS was overwhelmed by demand, much of it in the form of prepaid orders. Shipments fell weeks, then months, behind. Promised peripheral equipment such as memory boards, terminals and paper-tape readers for entering data wasn't available for almost a year, and then in only limited supply.

Much of the extraordinary demand came from members of computer hobbyist's clubs, principally in the United States, the most famous of which is the San Francisco Bay area's Homebrew Computer Club. Its meetings were attended by well over a hundred enthusiasts, some of whom had taken delivery of Altairs, many more of whom were impatiently waiting for their orders to be filled. They showed each other programs they had written that performed such miraculous feats as adding together two digits. One member brought the entire club to its feet in wild applause by programming his Altair to play the song "Daisy" by causing raspy noises on a portable radio.

Homebrew Computer Club members helped each other with advice, components and peripherals. But what was most intensely lusted after by the majority of them was a BASIC interpreter for the machine. BASIC was an "assembly language" that would reside within the computer memory: its purpose was to translate plain English instructions, entered via a keyboard or punched paper tape, into machine language of 1s and 0s. That would eliminate the tedious, time-consuming and error-prone procedure of programming directly in machine language using the front-panel toggle switches.

MITS had such a BASIC interpreter for the Altair, but the chaos within the overburdened company was such that it, too, failed to ship on time. It had been written by a computer-savvy Harvard Law School freshman and computer prodigy named William Henry Gates, III, and his programmer friend, Paul Allen. They, too, had seen the *Popular Electronics* article and realized immediately that there was an opportunity for anyone who could put together a BASIC tiny enough to fit into the Altair's minuscule memory. Six weeks later, they had finished the job and Allen flew to Albuquerque to show Ed Roberts what they'd achieved. MITS bought the program and made Allen the company's software director. Bill Gates dropped out of school and began programming full time.

The knowledge that an Altair BASIC existed but was not being made available was enough to drive the average hacker crazy, and it led to the inevitable: the program was pirated at one of MITS's marketing demonstrations. It was copied and distributed at computer club meetings and through the mail all over the world. To the hobbyists involved, it seemed like simple justice; after all, many of them had already paid MITS for their copies and hadn't received them. And, in any case, software heretofore had always been free. It was part of the hacker ethic: if you wrote a program, you distributed it free to others in the expectation that they would work on it themselves and, one hoped, improve it. At worst, you might let people use it and suggest to them that if they liked it and found it useful, they might send you a few dollars just to cover your costs. The idea of making software specifically to sell at a profit . . . well, there was something *not right* about it.

For Bill Gates, though, piracy was theft, pure and simple. As president and co-owner (with Allen) of the fledgling Microsoft Corporation, he wrote an open letter saying as much, and it was widely published. "As the majority of hobbyists must be aware, most of you steal your software," he wrote. "Hardware must be paid for, but software is something to share. Who cares if the people who worked on it get paid?" It was perfectly in character for Gates, who, it would soon develop, had the business instincts of a wolverine. But reviews were scathing: one computer club in California even threatened to sue Gates for calling hobbyists thieves. It was the beginning of Gates's career as the Richard Nixon of the computer industry, the ambitious overachiever whom people love to hate. Some have compared him with radio pioneer David Sarnoff, another man whose overweening ambition and cutthroat competitiveness often put him on the wrong side of U.S. federal antitrust regulators bent on keeping the industry competitive. Gates, like Sarnoff before him, had the brilliance to see that the real money to be made in his industry was not in hardware, but in software. For Sarnoff, that meant networks and radio programs; for Gates, it meant computer applications. It would make him the richest individual in the world.

Other personal computer companies sprang up, many in California garages and vacant warehouses; by 1977 there were thirty or more products on the market, including Heathkit, Cromemco, IMSAI, Radio Shack and Commodore. And there was Apple Computer, the creature of

Homebrew Computer Club enthusiast Steve Wozniak and his chum Steve Jobs. Their company and its products were destined to have a profound impact on the budding industry.

Where Ed Roberts of MITS had been more or less in the mold of the classic American small businessman and entrepreneur right down to the overstuffed, short-sleeved cotton-polyester shirt with pocket protector, and Allen and Gates had been typical owl-eyed, skinny computer nerds, Wozniak and Jobs were anything *but* conventional. Jobs, a self-described "freak" with shoulder-length hair, sought spiritual insight in soft drugs, vegetarianism and Eastern mystical philosophy. He preferred bare feet to shoes and drove a Volkswagen minivan. Wozniak, "Woz" to his friends, had a penchant for corny practical jokes and cultivated wildly erratic work habits. As a scruffy college dropout who was a spectacularly good programmer, he managed, but only just, to hold down a steady job in the calculator division of Hewlett-Packard in Palo Alto, California. Jobs and Wozniak had met in 1971 when Jobs was sixteen and Wozniak twenty-one. Despite the age difference, they had much in common, including the same high school alma mater. That year they could be found lurking in the halls of UC Berkeley, selling Wozniak-designed "blue boxes," gadgets that allowed you to make free long-distance phone calls.

A fellow student from that era recalls:

> I was happy to be at UC, Berkeley, in 1972, but unlucky enough to be a resident of one of the dormitories, filled with freshmen away from home for the first time. One night, acting on a lead from a mutual friend, two young men known by the names "Hans" and "Gribble" came to my room for a visit. ["Gribble" also went under the name Oaf Tobar.] They were really named Jobs and Wozniak, and they were selling blue boxes.[202]

When he finished high school, Jobs spent two years at Reed College in Portland, Oregon, before dropping out, returning home and landing a job as a programmer with a start-up computer games company called Atari. A few months later he took what savings he had been able to accumulate and headed for India as one of the generation's "dharma bums," returning in 1974, just about when the Homebrew Computer Club was getting organized.

Jobs and Wozniak formed the Apple Computer Company in 1975 to sell a computer kit "motherboard" Wozniak had designed around an integrated circuit chip being produced by a company called MOS Technology. It was less powerful than some of the competing products from Atari and IMSAI and others, but it had the virtues of a low price ($666.66) and the built-in ability to be hooked up to a television monitor. Jobs landed an order with a hobby shop for a hundred of the boards at $500 each, and the two suddenly had a viable business. To raise operating capital, Jobs sold his VW van and Wozniak sold his HP calculator. That brought in $1,350 and they were able to borrow $5,000 from a friend. They got the components they needed on thirty-day credit.

In all, Apple sold about 175 of the boards, enough to convince the two Steves that they should continue on in business. While Wozniak turned to designing a new and better computer that would become the famous Apple II (he hung on to his day job at Hewlett-Packard), Jobs looked to organizing the business end of things. He asked one of Silicon Valley's newly minted millionaires, a thirty-two-year-old former Intel marketing manager named A. C. Markkula, to help him draft a business plan. Before he had gotten very far into the project, Markkula realized that Apple was sitting on a potential gold mine, and bought a third of the company for $91,000. Then he arranged a $250,000 line of credit with a bank and raised another $700,000 in venture capital.

On a firm financial footing, and with professional management on tap, Apple was positioned to handle the wild success that its Apple II machine became. It was the Volkswagen beetle and the Austin mini of the personal computer industry, a fun, friendly machine selling for an affordable $1,195. It sported a floppy disk drive when many of its competitors were still using slow and unreliable cassette tapes for memory storage. It would support a color monitor. Most important of all, it left open expansion slots so that third-party inventors could come up with hardware to enhance the machine. The young company also made a point of working with software developers, sharing the machine specifications needed to develop programs. This openness was one of the keys to the rapid early development of the personal computer industry and was a dramatic departure from the policy in place at IBM and other established manufacturers, which preferred a strategy of locking in market share to their closed, proprietary systems.

Jobs had hired an industrial designer to do the plastic case for the Apple II—not just any designer, but Harmuth Esslinger, creator of the Porsche 928 fuselage—and had seen to it that the machine had a warm and friendly look and feel about it. Wozniak, an inveterate computer gamer, had made it a great game machine. It was also the only computer of its time that ran the breakthrough business spreadsheet program called VisiCalc. From the perspective of our current, supersophisiticated software world, VisiCalc now looks amusingly antique and feeble. But it was authentically revolutionary in its time in opening the public's eyes to the potential of these little machines. Jean-Louis Gassé, who would go on to become product-development manager for Apple Computer, has written of his first encounter with the program in 1981:

> So VisiCalc offered itself to me on the screen: a sheet of ruled paper with rows and columns. Little by little I noticed that this program, which looked like nothing in particular, allowed three budget simulations (something every company head needs but never has the time or courage to do) to be executed in two steps . . . a single item changes and everything is recalculated. . . . I couldn't believe my eyes. . . . That was the day I realized that you didn't have to be a programmer anymore to use a computer.[203]

The Apple II was just right for the market when it was introduced in 1977, a near-perfect match between consumers' dreams and builder's vision. Apple became the fastest-growing corporation in the United States, its sales soaring from $775,000 in 1977 to $35 million in 1981 and nearly a billion dollars in 1983. When the company went public in 1980, its stock hit a market value of $1.2 billion on the first day it was offered. Markkula had parlayed his $91,000 investment into $154 million; Jobs was now worth $165 million and Wozniak $88 million.

In a classic case of hubris in large corporations, IBM ignored the personal computer market for three long years after the Apple II and its competitors from Commodore, Radio Shack, Atari and others were introduced. But when it decided to move, it surprised nearly everyone with the sophistication of its entry. The IBM PC was startling—astonishing even, considering the reputation of its maker for secrecy and proprietary technologies—because the company had decided to open up the computer's architecture. Open architecture meant that any third-party developer could get a copy of the computer specs and go to work on designing add-ons, peripherals, software, or even clone the motherboard itself. IBM PC clones did in fact begin to appear almost immediately. The company had made it possible for small ventures to compete with its PC on an equal footing, and that led to rapid development of the product line.

When it came to writing an operating system for its personal computer, IBM turned to Bill Gates's Microsoft. It was, for Gates, the opportunity of a lifetime: whatever operating system the IBM PC used was sure to become the industry standard, which would mean that most personal computers sold around the world would use it, and that was bound to make Gates a very, very wealthy man. The operating system Microsoft designed for the IBM PC was called, simply, MS DOS, for Microsoft Disk Operating System. It was robust, flexible and workmanlike, but it was decidedly unfriendly to anyone not already familiar with computers and their languages. Nothing about it was intuitive; it came with thick manuals that needed to be studied before the simplest operations could be performed. The IBM PC was a gray, buttoned-down accountant's tool kit. It offered the power of the microchip to a broad public at a reasonable price, but on *its* terms, not theirs. Despite its open architecture, it was very much a product of top-down thinking.

Thus, while the IBM PC was built for, one might even say imposed on, the market, the Apple II had grown out of its market. The distinction was apparent in virtually every aspect of the two machines, from operating systems—especially operating systems—to cabinet design to clarity of monitor presentation to the size of their respective owner's manuals (Apple's was slim and graphically appealing), right down to the way the motherboards were designed. Its open architecture was the IBM product's eventual salvation: it became more friendly as third-party designers and software developers gradually changed it to suit the needs of ordinary human beings. But it was a long, slow process.

Whatever its initial shortcomings when compared to the Apple II, the IBM PC served to legitimize the personal computer industry in the minds of many ordinary users and business people, and to take it into the mainstream in the same way that Western Union's entry into the telephone business had almost overnight greatly expanded the market initially developed by the then-unknown Bell Telephone. The IBM product's introduction was a key step in making the personal computer a commonplace appliance in homes, and in businesses of all kinds.

In 1984 Apple introduced a new model that was to make the distinction between the IBM-compatible PC and its own products crystal clear. The Macintosh, a product of the vision of Steve Jobs and his messianic desire to make computing accessible to everyone, introduced two innovations that radically simplified the interface between user and machine: the now-ubiquitous "mouse" and the icon-based "desktop." The Mac was a further development of Apple's earlier and unsuccessful "Lisa" business machine, which failed principally because it was aimed at the wrong market. Corporate business purchasers were not ready for a machine that made computing fun, and the home-business computer market had not yet developed to the degree that it could support a product as expensive and sophisticated as Lisa.

The Lisa interface was developed over a period of nearly three years, beginning in 1978, by Apple designers who were working with an explicit set of goals that detailed Jobs' vision.[204] Instead of an inscrutable blank screen and a flashing cursor, the new interface presented the user with a picture of an electronic desktop with icons the user manipulated to tell the computer what to do. A menu bar across the top of the screen offered further options with pull-down submenus. Documents were displayed in windows which could be sized to suit the user. The general layout of the interface could be altered to suit a user's individual tastes and preferences. Some of the guidelines which the Apple designers worked with were listed in the project's Marketing Requirements Document (1980), which opened with the declaration:

> Lisa must be fun to use. It will not be a system that is used by someone "because it is part of the job" or "because the boss told them to." Lisa will be designed to require extremely minimal user training and hand holding. The system will provide one standard method of interacting with a user in handling text, numbers, and graphics. . . .[205]

The Macintosh transferred these guidelines to their logical environment, the personal computer market, and the result was a radical machine had been designed from the ground up with the idea that computers ought to be easy to use, playable, in fact, like a musical instrument, by anyone who knew a few rudimentary rules. You could use it for spreadsheets and word processing, but you could also draw with it, make music with it, design page layouts. The hardware itself was a giant step forward, with its all-in-one plug-and-play format and attention to design aesthetics, small footprint, auto-eject floppy disk drives and loudspeaker:

> Even the packaging showed amazing creativity and passion; do any of you remember unpacking an original 128K Mac? The Mac, the unpacking instructions, the profusely illustrated and beautifully written manuals, and the animated practice program with audio cassette were packaged together tastefully in a cardboard box with Picasso-style graphics on the side. Never before had a computer been delivered with so much attention to detail and the customer's needs.[206]

This was a computer designed to be an extension of its user, in direct contrast to the IBM product which treated its operators as living peripherals who were required to know pages of machinelike DOS command codes and obey machine, rather than human, logic. Anyone who has been puzzled over the years by the fierceness of the loyalty of Macintosh users will do well to keep in mind this crucial distinction. If Bill Gates was the Richard Nixon of software developers, then the IBM PC was Nixon as cyborg: it had the power, it had the authority, it had the market success, but nobody *loved* it. It is also true that there was a certain macho appeal to the very complexity of MS DOS in the business and technical communities. It gave those who had mastered it a special cachet and improved their job security. The DOS-literate had a vested interest in opposing anything that would reduce the economic scarcity of their special knowledge. They had reason to like the fact that DOS was difficult.

By the 1990s, Apple Computer had seen the MS DOS operating system occupy the lion's share of the personal computer market thanks in part to the availability of bargain-priced IBM clones. What was even more worrying than the gap in hardware prices, however, was Microsoft's success in emulating the very features of the Macintosh operating system that had accounted for its strong appeal. With each succeeding iteration of its Windows operating system, Microsoft drew closer to mimicking the Macintosh interface. A long and expensive lawsuit charging Microsoft with infringement of copyright had failed. By the mid-1990s Apple seemed to have lost its nerve, and its direction. It toyed with licensing its hardware, and licensed Macintosh clones made a brief appearance in 1997–98. By then, in attempting to be all things to all people, the rudderless company had introduced a bewildering plethora of models designed to compete head-to-head in the Windows/IBM marketplace, in which the core virtues of the Macintosh were increasingly obscured. Work on a "revolutionary" new Mac operating system dragged on endlessly; meanwhile, Macintosh computers were promoted as being capable of running the Windows operating system under emulation. The distinction between the two was becoming perilously blurred to all but the most knowledgeable users, and yet Macs remained relatively expensive. By 1997 Apple was facing the dangerous prospect of steadily shrinking market share combined with a loss of market identity.

In desperation, the company asked Steve Jobs to return to the helm. (He had been squeezed out in the early years of expansion in a boardroom putsch, a not-uncommon fate among the industry's founding geniuses.) In one of the more remarkable business achievements of the decade, within a year of his return Jobs had put Apple firmly back on the rails, principally by refocusing its efforts on its traditional areas of strength: elegant operating systems, intuitive software, brilliant hardware design within a limited product line and technical superiority. To this list he was able to add, for the first time, strongly competitive pricing. With the Internet and World Wide Web markets extending Macintosh's acknowledged superiority as a communications and publishing platform into the vast new world of on-line interactivity, the company's prospects seemed once again secure.

Perhaps more than any other single event, the introduction of the Macintosh helped to clarify the future direction of personal computing, helped to make clear the possibilities latent in the phenomenon of computing power widely distributed among the general populace. It had been designed to be used as a communications tool, and it was eagerly adopted as such. While it did poorly in the mainstream business market, it was snapped up by writers, artists, musicians, educators, architects, mathematicians—creative people in general and communicators in particular.

The phenomenon of desktop publishing was a case in point. With its strong built-in graphics capabilities, the computer quickly became the cornerstone of a new cottage industry in which it was possible to publish virtually anything from newspapers to handbills to books, at relatively low cost. Text and images could be prepared, edited and even typeset on the computer. With the introduction of the laser printer, high-quality printing, too, became a desktop operation. Every aspect of the traditional publishing industry had been made accessible, except the mechanisms of distribution, which remained largely in the hands of bookstore chains and their buyers and periodical distributors. With the advent of the World Wide Web in the mid-1990s, this final hurdle to personal publishing would be overcome as well, giving any individual with a computer and a modem, in principle at least, access to the communications power of a newspaper editor or network television producer. The democratic revolution in textual communication, launched but only partly realized by the printing press, was taken a giant step closer to fruition by the personal computer.

What was true of publishing was also true in the fields of music production, sound editing and the graphic arts and, eventually, video editing and production. In each case, the personal computer dramatically extended access to the field by drastically reducing both capital costs and the cost of materials. In television, for example, lightweight digital cameras costing a tiny fraction of the price of a standard professional video camera, matched with low-cost desktop editing software, made it feasible for the freelancer to enter the field of TV news and documentary production. Until the computer chip made these products available, extremely high equipment and operating costs had maintained this as an exclusive preserve of the corporate media businesses and the networks. As with desktop publishing, however, distribution remained a problem for freelance video producers; traditional televison news and current affairs operations have historically been loath to accept outside contributions for reasons related mainly to professional pride, union contracts, consistency of style and legal liability. And, of course, setting up one's own TV station was out of the question for all but the very wealthiest in the community. The World Wide Web, once again, has begun to offer accessible, low-cost alternatives for distribution and these will proliferate as Internet bandwidth expands to support full-motion color video. On the Web, there is no reason in principle why a small, independent group of news producers could not compete effectively with CNN or NBC; why independent actors and producers could not make and market their own dramas.

It was with the advent of the computer network, then, that the personal computer's ultimate destiny as a powerful communications appliance began to crystallize. Like the telephone before it, the personal computer had been thought to be essentially a business tool; the Macintosh computer itself was dismissed by business as a toy, just as Bell's telephone had been dismissed a century earlier by the president of Western Union as a "scientific toy" not worth serious development and certainly not worth purchasing the rights to manufacture.

Today's personal computers are vastly more powerful and orders-of-magnitude faster than the early prototypes of the breed. A typical machine of today boasts more computing horsepower than was available to the entire U.S. defense establishment at the height of the great air defense panic of the 1950s. And that power is being amplified enormously, and with unknown consequences, as the world's computers are linked in their tens of millions on worldwide networks like the Internet. It is to the Internet that we turn our attention next.

CHAPTER TWENTY-SIX

A Digital Mardi Gras

The Internet is a technology without precedent. It was not invented so much as imagined. It was not built—it just grew, as if instructed by some deeply embedded coding. It is not so much the product of individual minds as the realization of the collective vision of small groups of men and women, mainly academics and computer enthusiasts, all over the world. It was born into the deeply psychotic world of Cold War nuclear gamesmanship, yet transcended it magnificently. It mirrors human needs and aspirations, playfulness and genius, creativity and depravity better than any other technology ever devised. A visitor from another galaxy could learn just about all there is to know about humanity from the Internet. Science fiction writer Bruce Sterling said of the Net, in a wonderful image:

> No one really planned it this way. Its users made the Internet that way, because they had the courage to use the network to support their own values, to bend the technology to their own purposes. To serve their own liberty. Their own convenience, their own amusement, even their own idle pleasure. When I look at the Internet . . . I see something astounding and delightful. It's as if some grim fallout shelter had burst open and a full-scale Mardi Gras parade had come out.[207]

The implications have been concisely described this way:

> It took a hundred years and billions of dollars to wire the world into a switched telecommunication network. It took half a century and billions of dollars to create computers you could afford to put on your desk. A ten-year-old kid with a hundred dollars can plug those two technologies together today and have every major university library on earth, a bully pulpit, and a world full of co-conspirators at her fingertips.[208]

It had begun in the 1960s, when schoolchildren all over the Northern Hemisphere were receiving regular instruction on how to survive an atomic attack by sheltering under their desks in a fetal crouch, and when building contractors were doing a land-office business in backyard bomb shelters. The RAND Corporation, Washington's best and most frightening Cold War think tank, applied its formidable brainpower to what had been identified as a pressing strategic problem: how could some semblance of civil and military authority be maintained in the United States after nuclear bombs and warheads had laid waste to the country?

It was well understood that no conventional communications network, wired or radio, no matter how well fortified and armored, could survive an all-out nuclear attack. And presuming that an Armageddon-proof network *could* be built, how would it be organized and managed? Any central authority hunkered in concrete-hardened headquarters would be an obvious, high-priority target for enemy missiles, and no bunker could hope to survive a direct hit from a hydrogen bomb. It would be the first place to be obliterated. The think tank wrestled with this grotesque puzzle, and arrived at an audacious solution.

RAND's Paul Baran, in the landmark report "On Distributed Communications," eventually made public in 1962, proposed a network that would have no central authority. Furthermore, it would be built from the ground up to operate while in shambles.[209] It was to be a system so full of redundancies that it could survive horrible mutilation and still keep functioning. What's more, it would be able to grow back dismembered limbs.

The principles were simple but revolutionary. The network would be assumed to be unreliable at all times, and would be designed to work around its own breakdowns. All the nodes in the network would have equal status, each with its own authority to originate, relay and receive messages. There was no limit, in principle, to the number of these nodes. The messages themselves would be divided into packets of binary code, each packet equipped with the "address" of its destination. Packets would speed their way through the network on an individual basis, beginning at a source node and ending at a specified receiving node, the one in the address. The particular route taken would be unimportant: each packet would be bounced like a pinball from node to node to node, in the general direction of its destination, until it ended up in the right place. Once there, it would be reassembled with other packets to reconstitute the original message. If whole chunks of the network had been vaporized, that wouldn't matter. The packets would find their way via whatever nodes happened to survive. It was, in retrospect, a solution that seemed to rely as much on the lessons of biology as physics. New high-speed computers made it feasible.

The Los Angeles freeway system provides a useful metaphor for what goes on in a packet network. Just like a packet, each vehicle on the freeway knows where it is going (or at least its driver does) and where it has come from, and if one exit ramp is blocked by traffic or construction, police posted there will wave it on to the next, allowing the vehicle to find a detour by which to reach its destination. Vehicles of many kinds originating from many different locations and with as many different destinations can share the same freeway at the same time. (On a standard telephone network, the analogous situation would be for authorities to open the most direct "freeway route" from sender to receiver to only one vehicle at a time, in other words, to create a dedicated circuit.) As traffic thickens, the vehicle's speed decreases, but it will always (barring mechanical breakdowns and drive-by shootings!) eventually get to its destination. In fact, packet networks are more efficient than the freeway system because they have a way to handle the equivalent of an L.A. drive-by, which is the occasional corrupted or damaged packet. Error-correcting protocols report "bad" packets back to the sender and they are immediately retransmitted. It is also important to note that the more complex the freeway system becomes, the more on- and off-ramps and interchanges it has, the more likely it is that a given vehicle will be able to get where it wants to go despite widespread damage to the system. There are simply more options for detours in a more complex system.

Packetized data has other advantages: packets may be compressed in accordance with the rules of information theory, to increase transmission speed and take maximum advantage of limited network bandwidth, and they can also be individually encrypted for security.

Nevertheless, it seemed a horribly inefficient, almost chaotic system by traditional engineering standards and the whole idea of it drove telephone company engineers crazy just to contemplate. They were most comfortable in a straightforward world of dedicated circuits carrying analogue signals. AT&T personnel conscripted to the project did their best to kill it before it was ever made public. Their objections carried weight, because it was over the telephone network, that intricate, switched web that had been built and paid for over the preceding hundred years at a cost of billions of dollars, that the packet network planned to piggyback. Trunks here and there would need to be beefed up to provide a reliable, high-capacity network backbone, but the plan was feasible mainly because so many millions of miles of switched telephone lines were already in place. In the end, though, there was just no denying that the switched packet network idea perfectly matched the specified military requirements in that it was about as close to an indestructible piece of communications machinery as could be imagined.

At the same time as Paul Baran's RAND report was being released in 1962, the U.S. Department of Defense's Advance Research Projects Agency (ARPA)—part of the fallout from the trauma caused in American military and scientific circles by the Soviet Union's successful Sputnik satellite launch in 1957—was asked to organize research into how best to take advantage of the nation's growing inventory of computers, in particular those advanced machines that had been developed for air defense and other command and control functions. This project, like Baran's doomsday system, also involved high-security computer networking, and so it was rolled in with the RAND research, broadening its scope and objectives beyond strictly military applications, into pure research on intercomputer communication. Dr. J. C. R. Licklider, known universally as "Lick," was appointed to head up the project in October of that year. It was a felicitous choice: Lick was exactly the right person for the job, his background in physiological psychology and engineering giving him a rare degree of insight into the potential of computers and a refreshing openness in his handling of research associates.

Among Licklider's first initiatives was to shift the project's research contracts away from the corporate world into the best university computer research programs. ARPA's farsighted mandate was to fund research that was likely to result in advances of at least an order of magnitude over the current state of development and his assessment of the industry at that stage was that further major gains in computer applications would be made only by "out-of-the-box" thinking, that is, research into areas that had no obvious or immediate commercial applications. According to the completion report filed on the ARPANET project in 1977,

> [t]he computer industry, in the main, still thinks of the computer as an arithmetic engine. Their heritage is reflected even in current designs of their communication systems. They have an economic and psychological commitment to the arithmetic engine model, and it can die only slowly. . . . Even universities, or at least parts of them, are held in the grasp of the arithmetic engine concept . . . [whereas] the ARPA theme is that the promise offered by the computer as a communication medium between people, dwarfs into relative insignificance the historical beginnings of the computer as an arithmetic engine.[210]

Licklider was especially interested in the area of interactive computing, as opposed to the "batch processing" methods that were then in vogue. He was, in short, a hacker at heart. From the beginning, he seems to have understood that communications was computing's destiny. In a moment of prophetic whimsy he dubbed his team of computer specialists the "Intergalactic Network," and he focused research on developing software that would allow computers to talk to one another, and thus allow humans to talk to one another through computers. His insight had arisen out of his observations of a phenomenon connected with time-sharing on mainframe computers, that is, allowing several users access to the same computer at the same time rather than scheduling batch processing of stacks of punch cards:

> Lick was among the first to perceive the spirit of community created among the users of the first time-sharing systems. . . . In pointing out the community phenomena created, in part, by the sharing of resources in one time-sharing system, Lick made it easy to think about interconnecting the communities, the interconnection of interactive, on line communities of people. . . .[211]

By 1967 the basic conceptual structure of what would become ARPANET, the predecessor to the Internet, had been established by the team. The problem of communication between computers running on different operating systems (i.e., speaking different languages) would be handled by constructing a network to which entry would be gained through nodes consisting of minicomputers programmed to handle the translation issues. The idea was that these gateway computers would translate incoming data from users and their polyglot computers into a universal Net language, and would also translate data stored on the Net back into the language being used by the accessing computer. This strategy has survived in the current system of using Internet Service Providers for dial-up access to the Internet: ISP's now constitute a sizable industry with subscriber revenues in the hundreds of millions of dollars.[212]

In December 1969, there were four nodes on the infant American network, which was named ARPANET, after its sponsor. (These were all at universities with large defense research establishments: University of California at Los Angeles, University of California at Santa Barbara, Stanford Research Institute and University of Utah.) Scientists and researchers were now able share one another's computer facilities by long-distance telephone lines.[213] By 1971 there were fifteen nodes in ARPANET; by 1972, thirty-seven.

As early as its second year of operation, the technicians who minded the ARPANET noticed it was changing. ARPANET's users had transformed the computer-sharing network into a kind of electronic post office. The main traffic on the network was no longer long-distance computing. Instead, it was news and personal messages. Researchers were using ARPANET to collaborate on projects, to trade notes on their work and, increasingly, to just chat about whatever interested them. Clever software routines were being devised almost daily to make this informal communication easier and more efficient. User surveys reported great enthusiasm for these nonregulation network services with their automated mailing lists, newsgroups, digests and other nifty, time-saving wrinkles—far more enthusiasm than they felt for long-distance computing, which in any case was fast being made obsolete by plummeting hardware prices, and the increasing power of minicomputers and the new PCs.

Next to its speed and efficiency, it was the Net's colloquial informality that most endeared it to its users. Licklider noted:

> One of the advantages of the message system over letter mail was that, in an ARPANET message, one could write tersely and type imperfectly, even to an older person in a superior position and even to a person one did not know very well, and the recipient took no offense. The formality and perfection that most people expect in a typed letter did not become associated with network messages, probably because the network was so much faster, so much more like the telephone. . . . Among the advantages of the network message services over the telephone were the facts that one could proceed immediately to the point without having to engage in small talk first, that the message services produced a preservable record, and that the sender and receiver did not have to be available at the same time.[214]

In its early days, ARPANET use was confined to those students and faculty lucky enough to be affiliated with one of the host universities in the United States. However, those who did not share that good fortune, but nevertheless understood something of the communications potential of computers, did not sit idly by. In 1970 the first of many so-called store-and-forward nets was launched. These were essentially e-mail discussion groups in which letters or "postings" were available to anyone who was a member of the group. Nowadays the process is called "conferencing" and the interest groups involved are called "newsgroups." Early store-and-forward nets tended to be clumsy to use and slow, but by the late 1970s and early 1980s these had evolved into much more friendly environments based on user-developed software with names like Usenet, BITNET and Fidonet, and they were now capable of catering to the burgeoning numbers of personal computers. They would comprise the "network of networks" that would become the Internet.

Fidonet proved to be the precursor of the BBS or bulletin board, of which there are today tens of thousands, including some very big ones, like Prodigy, CompuServe and America Online—about which more in the following chapter. These are so-called client/server arrangements, in which all of the data is stored on a single computer and those wishing access (the clients) dial in to that computer (the server) via telephone lines, using PCs and modems. Some BBSs are run for profit and some are set up in the public interest. With low-cost or free software like FidoBBS, almost anyone who owned a computer of moderate capabilities could set up a BBS and begin selling or giving away memberships and passwords. Many catered to special interests, including, inevitably, pornography. Others attempted to offer a selection of topics catering to subscriber interests, along with on-line conferencing or "chat" on different subjects.

Other forms of networking sprang up. Usenet, short for Unix User Network, was based on a software protocol that allowed computers running the Unix operating system popular on university campuses to communicate with one another. It is another example of a client/server arrangement: a user connects to a computer, which connects to a main server on which are stored all Usenet postings, ready for retrieval. These are downloaded from the server to the client computer, in accordance with the user's requests. The user can peruse the selected postings and reply directly to an author via e-mail or by posting a message of his or her own on the server. (Usenet software was eventually revised and adapted to the IBM and Macintosh personal computer operating systems, making it much more widely accessible.) In its earliest incarnation, Usenet was used mainly by university students and faculty as an area for discussion of academic issues and sharing of new information about everything from research grants and papers to conference dates and proceedings. Soon, however, thriving discussion groups, or newsgroups, sprouted around such nonacademic topics as science fiction, movies and food.

Usenet was a haywire, ad hoc electronic post office that had grown up without any planning, and to try to put some order in the mounting chaos of proliferating newsgroups, its early volunteer administrators had established two categories, "mod" for those newsgroups which used a moderator (who filtered out spurious postings and generally kept the discussion on track), and "net" for those groups which were unmoderated.

By the summer of 1986, Usenet had become so big and rambunctious it seemed at times like a wild-and-woolly frontier town. There were newsgroups on hundreds of subjects, and postings were becoming less polite, orderly and disciplined, though at the same time more diverse and in many ways more interesting, if one could afford the time to separate the wheat from the chaff. The administrators decided a radical restructuring was needed. Seven new subject hierarchies were proposed: *comp*uters, *misc*, *news*, *rec*reation, *sci*ence, *soc*iety and *talk*. The last of the groups, "talk," was designed as a repository for all the unsavory, salacious, politically incorrect and socially psychotic newsgroups that had appeared like banana slugs among the Usenet flora. If they were all confined to a single hierarchy, they would be easier for network administrators to remove from their Usenet feeds, should they wish to do so. There was, significantly, no serious talk of attempting to ban or in any way censor the groups, beyond making them easy to avoid.

The reorganization is remembered in Net lore as the "Great Renaming" and it caused a flame war—a spate of e-mail argument and vituperation—of unprecedented proportions. Everybody seemed to have a strong opinion on whether it was necessary or unnecessary, a good or a bad thing, high-handed and autocratic or just businesslike. It was clear as never before that the Net's users regarded it as their community property, and were actively hostile to any attempt to impose a hierarchy of management upon it. The volunteer administrators who had organized Usenet and its backbone of servers and who carried out the Great Renaming were vilified as the "Backbone Cabal."

The pot boiled over when a gadfly named Richard Sexton proposed two deliberately provocative new discussion groups under the "rec." umbrella: they were "rec.sex" and "rec.drugs." Following normal protocol, Usenet denizens were given a chance to cast e-mail votes on whether the groups should be officially listed. The vote passed, but in a decision that became a watershed in the cultural history of the Net, the Backbone Cabal refused to create the groups or to carry them on backbone machines. What happened next does much to explain both how and why the Net has remained an ungoverned, unstructured, organic entity. First "drugs" and then "sex" were granted status in an alternative Usenet setup using routings that were separate from the official backbone. "Alt.drugs" along with "alt.gourmand" and a handful of other alternative discussion groups were created by student Brian Reid, who had become an unhappy camper when the Backbone Cabal had commanded him to change the name of his own "rec.gourmand" recipe exchange to "rec.food.recipes." The cheek! Surveying his new alternative creations and noticing an oversight, Reid sent the following impertinent message to Usenet administrators:

> To end the suspense, I have just created alt.sex. That meant that the alt. network now carried alt.sex and alt.drugs. It was therefore artistically necessary to create alt.rock-n-roll, which I have also done. I have no idea what sort of traffic it will carry. If the bizzaroids take it over I will . . . moderate it; otherwise I will let it be.[215]

In reminiscing on-line about his exploits, Reid would confess five years later:

> At the time I sent that message I didn't yet realize that alt. groups were immortal and couldn't be killed by anyone. In retrospect, this is the joy of the alt. network: you create a group, and nobody can kill it. It can only die when people stop reading it. No artificial death, only natural death.

He added, in an insightful aside:

> I don't wish to offer an opinion about how the Net should be run; that's like offering an opinion about how salamanders should grow: nobody has any control over it, regardless of what opinions they might have.[216]

As early as 1970, computer networks similar to Usenet had begun emerging all over the world, and it was recognized from the start that if a way to interconnect them could be found, it would be a boon to academic life and communication. The Internet, the global network of networks, was formally proposed in 1972. In October of that year, the first International Conference on Computer Communications was held in Washington, D.C. A public demonstration of ARPANET as a working model of how the international net might work was given using forty computers set up in one of the conference halls. Scientists from Canada, France, Japan, Norway, Sweden, Great Britain and the United States discussed the need for agreed-upon protocols that would allow computer networks in various countries to be tied together. An InterNetwork Working Group (INWG) was created to shepherd the protocol project and Vinton Cerf, who was involved with UCLA's ARPANET node, was chosen as the first chairman. The vision proposed for the architecture of the network of networks was, as Cerf would later recall, "a mess of independent, autonomous networks interconnected by gateways, just as independent circuits of ARPANET are interconnected by IMPs [Information Message Processors—gateway computers]." Figuring out how to program those magic Internet gateways was more easily said than done. It would be two more years before Vinton Cerf and UCLA colleagues completed writing and testing some code they called TCP/IP.

In an act of profound pragmatic and symbolic significance, the protocol was placed immediately in the public domain by its inventors, freely available to everyone. The tradition of public disclosure had been firmly established by Licklider and the graduate students involved in the ARPANET project from the beginning, in the best hacker tradition. All ARPANET developments, including beta or "trial" versions of software, had been posted and commented upon via the Net itself in a system called Requests for Comment, or RFC's. By dramatically broadening the range of people involved in software development, the project had extended the hacker ideal of "bottom-up" development into the realm of government-sponsored, high-level research, and the results were to be spectacular. It was, moreover, a distinct departure from the prevailing mode of secrecy in research as practiced by corporate laboratories and the university research establishments they sponsored, and it reflected the mood of antiestablishment radicalism on university campuses as dramatized in the "Free Speech" movement and the occupations by students of university administrative offices on several campuses. That mood is reflected in an early RFC posted by graduate student Steve Crocker, a member of the team working on the ARPANET:

> The content of a note [RFC] may be any thought, suggestion, etc., related to the [development of gateway software] or other aspect of the network. Notes are encouraged to be timely rather than polished. Philosophical positions without examples or other specifics, specific suggestions or implementation techniques without introductory or background explication, and explicit questions without any attempted answers are all acceptable. The minimum length for a note is one sentence. . . . These standards (or lack of them) are stated explicitly for two reasons. First, there is a tendency to view a written statement as ipso facto authoritative, and we hope to promote the exchange and discussion of considerably less than authoritative ideas. Second, there is a natural hesitancy to publish something unpolished, and we hope to ease this inhibition.[217]

Robert Braden, another student participant in both the ARPANET and Internet projects, reflected years later:

> For me, participation in the development of the ARPANET and the Internet protocols has been very exciting. One important reason it worked, I believe, is that there were a lot of very bright people all working more or less in the same direction, led by some very wise people in the funding agency. The result was to create a community

> of network researchers who believed strongly that collaboration is more powerful than competition among researchers. I don't think any other model would have gotten us where we are today.[218]

TCP/IP is, arguably, one of history's most important linguistic developments. TCP, or Transmission Control Protocol, codes messages into streams of packets at the source, then reassembles them at their destination. IP, or "Internet Protocol," handles the addressing, seeing to it that packets can be routed across multiple nodes and networks using computers of many different makes and models.

Still, getting it adopted by the Internet (which had been launched with early ARPANET software called Network Control Protocol or NCP) took some finessing by Cerf and other ARPANET developers. By now, habitués of the Net had taken a proprietary, even chauvinistic, interest in it and were reluctant to see it changed. However, without a universal protocol, the Net would be effectively hobbled and prevented from reaching its full international potential.

In the end, as Cerf recalls, the new standard was simply imposed by force majeure:

> In the middle of 1982, we turned off the ability of the network to transmit NCP for one day. This caused a lot of hubbub unless you happened to be running TCP/IP. It wasn't completely convincing that we were [serious about the changeover], so toward the middle of fall we turned off NCP for two days; then on January 1, 1983, it was turned off permanently.[219]

The Internet grew. The network's decentralized or "distributed" structure made expansion as easy as building with Tinkertoys. The original Usenet and others like it in the United States had gradually migrated to the ARPANET backbone throughout the 1980s, and in other countries a similar consolidation under national backbones had been long under way. From a handful of hosts in 1983, the Internet grew to 2.5 million computers linked to 10,000 hosts in 1993, to 30–50 million computers and 4.5 million hosts in 1995. (A "host" is the equivalent of what we have until now been calling for sake of clarity a "gateway" computer. It is a computer with a direct connection to the Internet trunk lines provided by telephone companies worldwide. Hosts are normally operated by a commercial or academic Internet service provider. A host computer may serve hundreds, even thousands, of "dial-up" users of Internet services, the group into which most households fall.) As the century closed there were estimated to be about 160 million computers with Internet access worldwide.

Cerf, looking back, is bemused by his success:

> I had certain technical ambitions when this project started, but they were all oriented toward highly flexible, dynamic communication for military application, insensitive to differences in technology below the level of the routers. I have been extremely pleased with the robustness of the system and its ability to adapt to new communications technology. . . . But I didn't have a clue that we would end up with anything like the scale of what we have now, let alone the scale that it's likely to reach by the end of the decade.[220]

In a remarkable development in April 1995, the U.S. government withdrew its support of the country's Internet trunks, the high-capacity backbones of the Net that had been built on government subsidy. In a carefully planned transition, the Internet in the United States switched seamlessly over to commercially operated backbones—with no noticeable interruption to service, despite the fact that the main load-carrier, the National Science Foundation's NSFNET, was at the time handling nearly twenty terabytes of data per month (a terabyte is a thousand million bytes). The Net, in the United States at least, had untied the maternal apron strings and now stood alone and independent.

CHAPTER TWENTY-SEVEN

Hypertext, Search Engines and Browsers

How the Net Became a Mass Medium

There are three milestones in the evolution of the Internet that are so significant as to merit examination as breakthrough technologies in their own right: the World Wide Web, the graphical-interface Web browser and the smart search engine. The first gave the Net a usable index with hypertext links when it had previously struggled along with only a table of contents; the second provided a colorful, friendly, point-and-click interface to replace bleak text and arcane keyboard commands; the third automated the search process and made manageable the mountains of information available on the Net. Together, they made of the Internet a mass medium in the truest sense—a medium of, by and for ordinary people. And in doing this, they played a critical role in shunting aside attempts to control the medium for purely corporate purposes, as had happened with radio and television. When commercial interests did come to the Net, they found it necessary to adapt to an implacably embedded technical environment and protocol regime that had been designed to give priority to the interests of the individual user. The story of how all of this came about is the tale of two competing paradigms for on-line services, one which views users as consumers, and the other which sees them as citizens.

Nineteen seventy-four, the year that TCP/IP was released to the public, was also the year in which another new form of computer-based communications emerged with the first of what are now called Bulletin Board Systems, or BBSs. At first mainly academic or interest-related and non-commercial, the idea was quickly taken up as a business opportunity by commercial BBS operators. CompuServe, which would become one of the giants of this new industry, went on-line in that year.

The idea of the commercial BBS is to provide fee-paying subscribers with services such as e-mail, databases of various kinds (depending on the interests of subscribers) and forums for real-time chat on all manner of topics. By the end of the decade, the conspicuous success of CompuServe had attracted attention and giant commercial interests began moving into the field, offering continent-wide and eventually worldwide access to a greatly expanded menu of databases and services, including on-line shopping. The accounting firm H&R Block purchased CompuServe; General Electric launched GEnie; IBM, Sears and CBS got together to establish Trintex, later to become Prodigy; and America Online or AOL was put together by a group of young entrepreneurs who soon made it the biggest and most successful of the lot.

Until about 1990, the general public remained largely ignorant of the Internet itself, although the notion of on-line services of the kinds provided by BBSs had begun to receive some attention in the traditional media. The larger BBSs had by now established substantial subscriber bases thanks to heavy promotional expenditures, and they had begun to exploit new revenue opportunities by offering on-line advertising in the form of a variation of highway billboards, which would be seen by their subscribers as they signed on and perused databases. The commercial BBSs had largely co-opted consumer on-line communication and made it their own, by virtue of their easy-to-use interfaces to databases geared to satisfying business customers and providing a wide range of entertainment and information services. E-mail and on-line chat[221] also developed into major selling points for potential new subscribers. Chat, which tended to devolve into on-line flirting, helped to bring down the median age of those using the services, while significantly broadening the subscriber base beyond its early business and computer aficionado stalwarts.

Throughout this period of evolution, the telephone industry had been slowly supplementing its worldwide copper-wire infrastructure with optical fiber, linked by satellite radio transponders of very wide bandwidth. Staggering amounts of data can be moved down fiber optic pipelines—millions of simultaneous phone calls, hundreds of TV channels.[222] The telephone companies soon found themselves, in principle, with virtually unlimited bandwidth on their hands, an embarrassment of riches. All that was required to realize the potential was heavy investment in infrastructure to bring the bandwidth now available on intra- and intercity trunks right to the customer's doorstep. With that done, almost any conceivable mix of programming and content could be delivered. The telcos began to show an active interest in fields that before 1990 had been the exclusive preserve of television. In short, they saw new revenue opportunities in owning and providing what they called "content" or "video dial tone" and what the television industry calls programming. At the same time, cable television had reached levels of household penetration throughout much of the industrialized world that made it possible for cable companies to entertain thoughts of providing telephone service in competition with the telcos. The trend toward deregulation in industrial economies everywhere was making it possible for those corporate dreams to be realized, first in Britain, then in the United States and Canada, where most legislative restrictions on competition between cablecos and telcos were removed in the 1990s.

By the middle of the decade, a major turf war had erupted with cablecos and telcos jockeying for position in what was thought by industry mavens to be the new frontier of the multibillion-dollar home-entertainment business, the so-called information highway. This was to be a high-bandwidth delivery system to homes, for what was envisaged in corporate boardrooms as digital, pay-per-view television with an interactive component. The interactivity would be confined mainly to games and to searchable databases for news and various kinds of information, including financial services and shopping. It was perceived from the start as a television-based rather than PC-based enterprise; almost no thought was given to adapting the innovations in interactivity already in use on the Internet. Indeed, the number of senior executives in either the cable or telephone industry who had had Internet experience was vanishingly small.

On the proposed information highway the "brainpower" needed to make television sets digital and interactive was to be added in the form of set-top boxes containing microprocessors. Sale, rental and support of these devices would provide one more revenue stream for the service providers. In competing for this perceived information-highway market, cable companies had the advantage of an installed subscriber base served by high-bandwidth coaxial cable, capable of carrying hundreds of digitized and compressed television channels. However, these cables had been installed with millions of one-way traps and amplifiers, which would prevent data from being sent back down the line. This was not thought to be a serious handicap at the time: the only return data the cable companies were interested in receiving was pay-per-view orders and payments, and, if need be, these could be handled by telephone links. The telephone companies, although they had built intercontinental fiber trunks of enormous capacity, had to contend with the fact that individual subscribers were still linked to the system with twisted pairs of copper wires which, even with the best of digital compression, could only carry a single television channel along with some data and one or two voice channels. On the plus side was the fact that the telephone network was truly symmetrical, allowing data to be sent in both directions, to the capacity of the lines. However, this was an asset that was all but ignored in the early stages of the broadband competition.

Major capital investments would be required by both the telephone and cable industries if they were to begin delivering the kind of bandwidth-consuming, "interactive" video fare each believed held the cherished key to the home-entertainment vault. Both industries sponsored expensive trials of so-called video-on-demand in Britain, the United States and Canada, and these received widespread journalistic coverage. Typically, a demographically correct subdivision or urban neighborhood would be rewired with high capacity, two-way cable connected to a bank of computer servers. The computer databases would contain dozens of digitized movies along with other video entertainment, video games, news outlets and on-line shopping services. It was, in reality, an attempt to adapt commercial broadcast television to the digital environment in a way that would maintain TV's one-way control structure while exploiting new revenue opportunities offered by digital media and limited interactivity.

But a funny thing happened on the way to the bank. Lucky consumers who had been given free or highly subsidized broadband services in these market tests fooled around with them for a few days, and then lost interest. On-line movies, which the cableco and telco planners had expected to provide a revenue bonanza, were a bust. People preferred to make their selections from the bigger catalogues available at video stores, and they were unwilling to increase the time previously devoted to watching movies at home. The news services, which were inferior to conventional TV news, were ignored. In 1996, the last of these market tests threw in the towel and wrapped up operations. By then, the on-line world had changed forever in ways that had taken the cablecos and telcos utterly by surprise.

In one of the classic miscalculations of modern business history, both the entertainment and communications industries had failed to grasp the important values of digital, computer-mediated communications networks, had failed to understand the fundamental nature of their appeal. They conceived of interactivity only narrowly, in terms of the ability to select and pay for content. They saw their so-called interactive broadband services as enhanced broadcasting, rather than an entirely new medium in which every subscriber is both a consumer and a provider of content. They failed to understand the lesson of the telephone, which was that people, when given the opportunity, will chose not simply to *get* content, but to *be* content as well, or the broader historical lesson that it is dangerous in the extreme to offer people a little freedom where none had previously existed: given a little, they will quickly rise up and demand a lot.

The place where those lessons *were* understood was on the Internet. The Net had, after all, been built from the bottom up by its users, built to serve their interests exclusively, in a network environment that resists top-down management. The Net viewed its users as citizens (or "netizens" in the argot), while the information highway sponsors viewed users as consumers. Citizens have rights and expectations that consumers do not share. And chief among these is the right to have one's say, and to be heard. The Net was structured in its very hardware and software to respect that right; the broadband video-on-demand systems were set up in ways which denied it, their content being the exclusive purview of their owners.

It was sometime in 1995 that the highly publicized and excruciatingly expensive wars for on-line dominance between the telcos and cablecos became irrelevant, consigned to history by a broad flanking movement on the part of the Internet, a sudden and surprising maneuver which settled the issue definitively by crushing both contenders. Even AOL and the other hugely successful BBSs were shaken to their foundations.

Until the spring of 1993, the Net had experienced continuous, modest growth of about 10 percent a month, which reflected both the increasing richness of resources available on the one hand, and the intimidating nature of its interface on the other. Already an enormous, even overwhelming, resource, it required expert knowledge to reveal its treasures, despite the distribution by university computer science labs of several rudimentary search tools with such whimsical names as ARCHIE and Veronica. Researchers at the European Particle Physics Lab (CERN, for Conseil Européen pour la Recherche Nucléaire) were among those who found that the volumes of valuable information available to them were getting to be hopelessly unmanageable. Oxford graduate Tim Berners-Lee and some colleagues at CERN tackled the problem in an imaginative way by creating universal standards for data to be posted in their network, and a universal addressing system with which to retrieve it. They called the network the World Wide Web, and in the tradition of the Net, they made the protocols freely available at no charge. Any Internet document formatted according to World Wide Web protocols was ipso facto a part of the Web.

The World Wide Web is an innovation that can be compared to the invention of the alphabetical index for books, or the development of the various indexing systems for libraries that allow us to zero in on the material we're looking for and quickly retrieve it. But it is also much more than that, thanks to its employment of hypertext linking. Any word or passage on a Web page can be highlighted (usually by underlining) and the coding behind the page allows the creation of a link between that word or phrase and any other word, phrase or document on the Web. Simply clicking on the underlined text will activate the link and carry the reader to the linked data automatically, wherever in the world it might be. The idea was to connect relevant material in a way which would facilitate both vertical and lateral exploration of an area of information. From its earliest implementation, the Web spawned a new recreation called "surfing," which is somewhat akin to browsing in an encyclopedia (if one can imagine an encyclopedia of 100 million or more pages!). Surfers found seemingly limitless pleasure and stimulation in aimlessly following World Wide Web hyperlinks wherever they might take them in the labyrinthian world of Web-enabled databases.

To use the Web required a software program called a "browser." The earliest browsers were efficient but uninspiring and somewhat complicated to use. Very soon, however, the synergy between developers of the Web infrastructure and designers of browsing tools led to a stage of evolution in which Web pages were presented in color and a variety of typefaces, complete with photographs and graphic images and even rudimentary animation. And then came live audio and full-motion video: the Web was clearly destined to become a highly sophisticated multimedia environment, awaiting only adequate bandwidth to deliver its limitless potential.

The remarkable phenomenon of the "personal home page" developed early and spontaneously, as Web users succumbed in their thousands and eventually millions to the temptation to add their own content to the network. It was a simple matter for anyone with a computer and modem to program a Web page in the standard HTML (HyperText Markup Language), add some digitally scanned photographs, and publish a calling card that announced his or her existence and uniqueness to the world. Most contained hyperlinks to favorite Web sites. These personal home pages are a unique feature of the Web that is endlessly moving as an expression of human diversity and the universal desire to share information.

Despite all of these advances, and however popular they may have been with the initiated, the Internet and the World Wide Web remained well below the horizon of public consciousness. Then, in August 1993,

Marc Andreessen, a greenhorn programmer working for peon's wages with the U.S. National Center for Supercomputing Applications (NCSA) in Urbana-Champaign, Illinois, wrote a Web browser program called Mosaic. Of course, it was made available for downloading on the Net, free of charge. Thanks to its elegant interface and point-and-click ease of use, Mosaic was an instant hit and its impact was every bit as significant for the Net as VisiCalc's had been for the personal computer industry. Anyone who spent half an hour browsing with Mosaic came away from the experience understanding that a new and important medium of mass communication had been born in the World Wide Web. The impact was every bit as powerful and intuitively exciting as had been the first public demonstrations of the telegraph or the telephone or the motion picture: it was the reaction one might imagine of someone whose sense of hearing or vision had suddenly become vastly more acute, opening up the potential of greatly enhanced interaction with the world. It was, in a word, thrilling.

With the release of Mosaic in January 1993, the World Wide Web began a period of breathtaking, exponential growth that was to change the communications landscape, confounding the plans of some of the world's biggest telecommunications and cable television companies in the process. It was the Web, more than anything else, that had made the Net a giant-killer and doomed to oblivion the monopolistic dream of a lucrative, broadcast-style information highway.

For Andreessen and a handful of young programmers attached to the Mosaic project, success was not an unmixed blessing. With the soaring numbers of Mosaic downloads on the Net came an equally impressive number of requests for help in implementing the program, at all times of the day and night, from eager users all over the world. And then there were increasingly insistent offers from investors who saw potential in Mosaic and wanted to either buy it outright or license it. Andreessen was in way over his head, as he confessed later in an interview, and he abruptly quit his NCSA job intending to drop the Mosaic project and return to a quiet life of programming: "At the NCSA, the deputy director suggested that we should start a company, but we didn't know how. We had no clue. How do you start something like that? How do you raise the money?"[223] he said in an interview a year later.

The answer was, if your idea is good enough, the money will raise you. In March 1994, Andreessen was tracked down at a small California software company by a savvy Silicon Valley veteran on the rebound from an unpleasant breakup with Silicon Graphics, the computer animation powerhouse he'd founded and where he'd been chairman of the board. Jim Clark understood Mosaic's commercial potential, and he also knew how to set up and run a business. Clark and Andreessen formed Mosaic Communications Inc., flew back to Illinois and in a single afternoon hired all of the key Mosaic developers away from NCSA. To avoid paying licensing fees to NCSA for Mosaic, they set about writing a new browser incorporating Mosaic's most popular features, and improving on them. It was called Netscape, and within weeks of its release it had taken over more than 80 percent of the worldwide browser market. The business strategy was unusual, and its success led to later adoption by many other software developers: Netscape was made available free for downloading on the Net, and the company made its money licensing it to corporate users and selling Internet service providers (the companies that offer Internet hookups to subscribers for a monthly fee) the software needed to implement its features.

There was a deeper significance to Netscape's successful game plan than initially met the eye. By the time of the program's release, the Web had evolved through several generations of increasing sophistication, and some farsighted observers were beginning to have inklings that it had the potential to become much more than a way to construct, index and link databases. Software was being developed by Sun Microsystems and other companies that allowed Web pages to behave in interesting ways by rapidly downloading small applications to the browser's computer. These so-called applets (called *Java*, in Sun's case) might, for example, animate a graphic, or cause lettering to scroll, or any number of other things. It seemed a small step to extend this technique to larger applications, and before long Net visionaries were talking about the Web of the future as a gigantic "hard drive" (ROM) full of applications, available for instant downloading whenever, wherever and by whomever they were needed. The Net would indeed become the computer, and users would plug into it with cheap appliances that needed only enough memory (RAM) to run a browser.

In this scenario, the browser was seen to take on new significance, because it had the potential of becoming the operating system for the Net-as-computer. In other words, it would be the vehicle through which the operator is able to both access and manipulate information. Microsoft had proved with MS DOS how crucial market domination in operating systems can be in terms of controlling the industry and its standards. Whoever dominated the Net browser market would be in a position to shape the development of this new medium while effectively keeping a lid on competitive threats.

The rise of Netscape did in fact force the mammoth Microsoft Corporation to radically rethink its corporate strategy during the early months of 1996, in order to avoid being leapfrogged into the anticipated new generation of Web-based applications and Web-centered computing. It did this with surprising alacrity, accelerating development work on its own Web browser, Microsoft Internet Explorer, and adapting many of its standard software application packages for easy Web access.

Another highly visible and highly significant change made by Microsoft was to drastically revise its plans for the Microsoft Network, which had been initiated as a vehicle to compete with the likes of America Online, Prodigy and CompuServe, the big commercial BBSs. In making the changes, Microsoft had finally come to accept what had been foreseen by others in the industry months or even years earlier—that the new, graphic-enhanced, user-friendly World Wide Web was a serious if not fatal threat to the proprietary BBSs. As the Web continued to expand at a rate of 20 percent a month, there was very soon almost nothing offered by the BBSs that could not be had for free on the Web, and access to the Net was significantly less expensive than typical BBS subscriber fees. Furthermore, even in early 1996 there was a staggering volume of information available on the Web that was not to be had on any BBS as the Web went from a few thousands to tens of millions of sites, worldwide, in astonishingly short order. In terms of volume and range of content, there was simply no way any BBS could compete with the Web. The reason was structural, in that the BBSs typically paid for their content from "content providers" (who in turn returned a portion of any earnings to the BBS), while on the Web, content was provided free and, with only a tiny number of exceptions, was available to users equally free. Since no one owned the Web, no one could control its content (except on individual sites) or charge for it en bloc. But the very fact that it was accessible to hundreds of millions of people all over the world, twenty-four hours a day, made it of immense strategic interest to business, and to social institutions of all kinds. (More on this in succeeding chapters.) Content of every conceivable description was being placed on the Web in stunning, almost incomprehensible quantities.

Internet "newbies," or new users, are invariably astounded by the fact that so much genuinely valuable information is available free of charge. It is as if a department store in their neighborhood had decided to stop charging for its merchandise! How can it all work if virtually everything is free? Answers to that question can be offered at several different levels. It can be said, for instance, that information is provided free because it is an extremely inexpensive commodity in and of itself: the expense associated with it is usually connected with its packaging and delivery. On the Net, these costs are minimal. Low cost thus provides a healthy incentive to publish on the Net. Secondly, information is unlike other commercial products in that distributing it does not deprive the provider of its future use. When you part with your automobile, it is gone. When you share some information, you still have it. Information is thus replicated, rather than consumed. This fact also helps to keep its cost down. Finally, as the hacker ethic says, information wants to be free. There is at any given time a great deal of information, both public and proprietary, that would be made freely available if there was no cost attached to the distribution process. The Net has dramatically lowered the publication cost threshold, thus releasing enormous quantities of information that was ready and waiting for public distribution. This includes everything from public domain literary masterpieces to product information to scientific databases to gallery contents to government documents. Finally, as the Web matured, it came to be understood that value could be created out of thin air, as it were, simply by attracting visitors to Web sites in large numbers, just as population densities and traffic patterns determined the value of real estate in the traditional economy. The way to attract them was to provide useful or entertaining content free of charge. All those visitors were potential sources of information, or potential customers, or potential voters, or potential supporters of a cause, or potential travellers to your country or city, and so on; they constituted value greater than the cost of attracting them in the first place. As Kevin Kelly notes in *New Rules for the New Economy*:

> In the first 1,000 days of the web's life, several hundred thousand webmasters created over 450,000 web sites, thousands of virtual communities, and 150 million pages of intellectual property, primarily for free. And these protocommercial sites were visited by 30 million people around the world, with 50% of them visiting daily, staying for an average of 10 minutes per day.[224]

The commercial BBSs, with their hefty subscriber fees, were left facing what appeared to be a gaunt future. In response to the Web's popularity, they dramatically lowered subscriber fees, and then reluctantly began offering access to the Web through their own facilities, hoping to hang on to subscribers by providing a friendly interface through which to explore the Net, along with access to their familiar and popular proprietary features such as on-line versions of newspapers and magazines. To add insult to injury, however, many of those proprietary features withdrew from BBSs to set up shop independently on the Web where they had complete autonomy over their sites and where they did not have to return tithes on whatever income they might generate to the BBS owner. This was the situation facing the industry in 1996 as Microsoft prepared to launch its lavishly funded Microsoft Network BBS (MSN, for short). The plan was withdrawn, revised and relaunched as an elaborate Web site, charging no subscription fees. The hope—and it proved to be justified—was that by producing an engaging and informative "portal" site,[225] the company would be able eventually to capitalize on large numbers of visits by selling advertising. It was one more indication that the Net was likely to become the metamedium of the coming century. A year or two later only America Online remained of the once-flourishing consumer BBS industry, having successfully transformed itself into an Internet

Service Provider while swallowing up its only surviving competitors (including the mammoth CompuServe).

In a dramatic wave of large-scale consolidation of Internet business in late 1998, AOL purchased Netscape in a $4.2 billion stock swap that also involved Sun Microsystems, aiming to become the world's dominant ISP, and positioning itself to take advantage of emerging markets for hand-held Internet access devices and other new technologies on the horizon. Netscape's long feud with Microsoft appeared to be at an end: AOL announced it would continue to use the Microsoft Internet Explorer software as its proprietary browser, which it had agreed to do originally as the price Microsoft had demanded in exchange for a clickable AOL icon on the "most valuable real estate in the world," the Microsoft operating system's desktop display. But whether Microsoft would be able to claim victory was a question that would only be answered with the outcome of the U.S. Department of Justice's protracted antitrust case against the software giant, a suit reminiscent of Washington's ultimately successful drive to break up AT&T into a number of regional "Baby Bells." In any case, the most immediate opportunities for profit on the Net now seemed to be in capturing users for Internet "portals"—Netscape's principal attraction for America Online lay in its very successful *Netcenter* site. And industry observers were also quick to note that the AOL-Netscape-Sun alliance had brought into a single camp three of Microsoft's most dangerous and capable opponents, each of which had at one time or another had occasion to express public outrage at Microsoft's high-handed approach to market dominance.

The AOL-Netscape merger may have signposted a decisive turning point in the history of Microsoft. The company had clawed its way to overwhelming dominance in the computer software industry primarily by leveraging its operating system; in other words, it had first seized market dominance with MS DOS and Windows, and then sold much of the software written to run on that platform. As a rule, Microsoft had not developed the software itself, but simply purchased independent development companies who had produced smart, successful products. In some cases, these software tools were incorporated right into the Windows operating system. But by late 1998 the issue of operating systems was undergoing a radical rethinking. On the one hand, the reality of the Net-as-computer inched ever closer with the continuing development of cross-platform or universal computer languages like Java and Jini. The idea was that in the future, personal computers would be a lot smaller and simpler, built to function as tools to access the Net, where the software and the horsepower to operate it would reside. Operating systems that were designed for a world in which each individual's computer held within its memory all the software its user required seemed destined for obsolesence. Future operating systems would likely look a lot more like Web browsers and would be much less complex. The computers they would be running on would in any case be incapable of running gargantuan programs like Windows. The heavy-duty computing power a user might require from time to time would be resident in the Web-based applications accessed on an as-needed basis. Microsoft's operating system dominance began to look like a future liability, an albatross around the company's neck, because it committed the firm to continually updating an obsolete product base, namely PC-based software, in a world where the real money had moved onto Net-based software. And on the Net, the best software is often distributed free. W. Brian Arthur of the Santa Fe Institute observed:

> Once you've locked in a huge user base—DOS, Windows, whatever—you have to keep the technology backward-compatible. That slows you down in terms of innovation. It's like Napoleon trying to take Moscow: the further you go, the longer your supply lines get. A fast-moving start-up with a new technology doesn't have to worry. For Microsoft the problem gets worse every day.[226]

As the century closed, one fact must have kept many a Microsoft executive awake at night: nobody needed Windows to run the "killer applications" of 1999—e-mail and Web surfing.

If the explosion of interest in the Web is strongly reminiscent of the great radio gold rush of the 1920s, there is an additional parallel in the fact that those pioneer businesses eagerly staking claims in the cyberspace of the Internet seventy years later had no more idea how to make those claims pay off in a commercial sense than did the early radio entrepreneurs. The Big Question was exactly the same: Where will the money come from to pay for the content? Charging fees for Web access beyond the basic hookup tariffs was not feasible for reasons we've already discussed (how can you charge for access to something you don't own?), and it was soon made painfully clear to several high-profile Web sites that subscriber fees to individual sites were unacceptable to the vast majority of Web users, if for no other reason than there was such a vast wealth of directly competing material available free of charge. As with radio, the business model that quickly gained favor was commercial sponsorship. But on the Web, sponsorship was a whole different ball game for both the sponsor and the user.

For a start, advertising was not among the "approved uses" for the Internet permitted by its early sponsor in the United States, the National Science Foundation (NSF), which provided major funding for the American Internet backbone. But by the mid-1990s this was a rule obeyed more in the breach than the observance, and it seems clear that the NSF recognized that advertising could no more be censored on the Net than pornography or political extremism or anything else. Beyond that, there was good reason to allow experimentation with commercial applications to this new medium, particularly since government sponsorship was understood to be only an interim arrangement designed to assist in start-up operations. When the NSF eventually did withdraw its support of the Internet in 1995, the issue of advertising and other "permitted uses" became moot.

As a nonlinear, or random-access, interactive medium, the Web operates on the principle that content is selected by users, rather than being broadcast at them. This means that commercial messages can be skirted, ignored or eliminated in any number of ways, both manual and automatic. For example, a Net surfer can simply turn off the graphic enabler of her Internet browser, and advertisements, which invariably use a graphical format, will no longer be visible. There is, in fact, a positive incentive to do this as a matter of routine, since text-only Web pages load far more quickly than those with pictures or animation. And the pictures remain accessible, just a mouse click away (on the "broken image" icon that replaces the missing picture). For an advertisement to be effective in an interactive medium like the Web, it must be of some use; otherwise it will be ignored, since it cannot intrude or impose itself on the user in the way a radio or television commercial can.

At first confounded by a medium so seemingly advertising-averse, businesses soon began to see a silver lining, in that the Web was a two-way street where sponsorship was concerned. While it was true that users could and would ignore advertisements that held no interest for them, advertising sites could also count on those who did visit to have a genuine interest in their product (or else, why would they visit?). Virtually every visitor was a sales lead. Advertising as it evolved on the Web was thus tightly targeted and highly informational, to a degree where it could be argued that it provided useful content to the Web as opposed to littering it with noisesome clutter. It would prove a sucessful strategy: surveys in 1999 indicated that most American Internet users spent about 70 percent of their time on-line searching for product information related to planned purchases.

Advertising on the Net was a fundamentally different phenomenon than it had been in the linear broadcast media of radio and television, and posed no inherent threat to the quality or integrity of content, simply because the Net is endlessly expandable, whereas the radio and TV spectrum had been strictly limited. In the broadcast media, advertisements displace and intrude into "noncommercial" content. On the Net, there is unlimited room for both commercial and noncommercial sites, and users can choose among them freely. Commercial and noncommercial content exist, as it were, in the same space but in different dimensions.[227]

Having said that, it needs to be acknowledged that advertisers have been trying very hard to bend the Net to their needs. Many large Web sites now carry banner ads designed to attract attention, and studies show that in terms of simple brand recognition they are as memorable as television commercials even when there is no "click through" by the user. The Web's technical flexibility in this regard led overstimulated imaginations to come up with a strategy impossible in any other medium: when a user sought a particular Web site by entering its name in a search engine, the engines "returns" page would feature an automatically inserted banner ad for a related product or service along with the standard list of Web addresses. Advertisers had to rethink the idea when they were sued for copyright infringement by several companies, notably Playboy and Estèe Lauder. Playboy objected to its name being juxtaposed with banner ads for X-rated porn sites, and Estèe Lauder executives hit the roof when they discovered that searching for their company's name brought up ads for Fragrance Counter. Teething pains.

Another attempt to adapt tried-and-true broadcast strategies to the Net was responsible for the huge investments being made in portal sites like Netcenter, Microsoft, Yahoo!, SEEK and SNAP at the century's close. Just as radio station owners invented the broadcast network to accumulate the large audiences wanted by sponsors, portal sites attempt to persuade large numbers of users to make the portal the page that opens first when the user's browser is launched, and leadership in the portal sweepstakes translates directly into enormous advertising revenue. The means of persuasion vary: Microsoft places the button linking users to its portal prominently on the Windows desktop; Netscape browser software defaults to Netcenter unless the user enters a custom default page; NBC relies on cross-media advertising to publicize SEEK; Yahoo! publishes its own print magazine. Doubtless, many more advertising strategies will be tried and discarded, or adopted as part of the marketing repertoire. One thing is certain, however: the Net "audience" can never be considered captive or passive in the way audiences have been captive to the limited numbers of outlets in the broadcast media. The Net surfer, unlike the television viewer, will always be able to tailor his experience on the Net to his own tastes.[228]

From the World Wide Web's earliest days, the most-visited sites were the search services that sprang up in bewildering profusion before settling into a series of consolidations. Their popularity quickly made them the most valuable sites on the Web. For if the Net's staggering resources are its principal attraction, they are also its Achilles heel. Any information repository as diverse and dynamic as the Internet is only as good as its index, and the Net does not have an index, at least, not in the ordinary sense. What it does have is search engines and catalogues.

The best-known and most widely used catalogue, Yahoo!, is the creation of a pair of Stanford University graduate students named Jerry Yang and David Filo. With the help of artificial intelligence specialist Srinja Srinivasan, they set up an operation in which Web sites are catalogued according to a very large and continuously expanding hierarchical classification system that strives to cover the entire field of human knowledge. Addresses of new sites arrive at Yahoo! headquarters in California in the form of e-mail requests for registration from Web site developers, and from the Yahoo!'s robotic Web crawler or "spider," a software program that automatically roams the Web from link to link, searching for new sites. A staff of dozens of classifiers looks at each site and decides on the appropriate slot for it within the system. As catalogues do, Yahoo! provides a context for every Web site it lists, and this is its major virtue. Looking through the Yahoo! catalogue is akin to browsing a library's stacks, where books are shelved side by side according to topic. Its drawback is that, given the Web's rate of growth, there is little or no hope that Yahoo! will be able to keep its catalogue either up-to-the-minute or comprehensive. That does not mean it will become irrelevant, anymore than a year-old set of *Encyclopedia Britannica* is useless. It does suggest that Yahoo! and other catalogues will eventually find a role as trusted and familiar sources of advice rather than know-it-all indexes: Yahoo! has been able to parlay its initial successes into becoming one of the top three portal sites in much of the world.

In the know-it-all index field, a number of contenders for the biggest, fastest and best Web search engine were to emerge as the Web matured. They were initially launched for one of two reasons: because there are tough technical problems to be experimented with, in the process showing off state-of-the-art computer hardware and software (for which there is a strong corporate intranet market), or because they held the potential of earning their makers a lot of money. As the most visited sites on the Web, search engines were soon handling millions of requests a day—and charging premium rates for advertising banners. In the great consolidations of 1999, several of the most successful search engines were purchased for incorporation into portal sites.

The search engines—Lycos, Excite!, Alta Vista, Inktomi, Infoseek and HotBot, to name just a handful—are more like indexes than catalogues, and are generally fully automated. The basic principle behind their operations is an information grid in which rows list Web pages and columns list words. Some engines list every word on a page; others list only key words. The pages themselves are located by automated Web spiders. When a user initiates a keyword search, the engine scans its grid for intersections where there is a match between keyword and Web page. The addresses to those pages, and sometimes a brief description, are retrieved and presented to the searcher as clickable hyperlinks.

One challenge facing search engine developers is the sheer size of the grid needed to cover all Web pages and their contents. This has been tackled in a variety of ways, from scalable linkups of many computers, to various data-compression techniques which shrink the grid's size to manageable proportions. A more difficult problem is the librarian's curse of synonyms and homonyms. Synonyms are a problem because keyword searches only work if the correct keyword is used: if the information required is filed under a synonym for that word, the search will fail. A search for data related to automobiles will not succeed if that information is filed under cars, for example. Homonyms, or words that are spelled the same but have different meanings, are an even stickier challenge. How does the computer know the difference between battery as in gun battery and in flashlight battery? One solution is to check other words used in relation to the keyword in each document: the presence of energy, electricity, charger, voltage or similar words in proximity to the word battery will indicate to the search engine that it is looking at a reference to an electricity source and not a tactical unit of artillery. Given enough computing power, this can become a remarkably complex and sophisticated process, resulting in uncanny accuracy.[229]

At this writing there are an estimated 120 million indexed Web sites with 320 million pages, and 1.5 million new pages are being added every day. The number of Web sites doubles every eight months. The Web, catalogued, indexed and fully searchable, has become a resource whose breadth, depth and value are almost beyond comprehension.

CHAPTER TWENTY-EIGHT

Anarchy, Business and Public Space

The Net Grows Up

The Internet is destined to become the world's most important communications medium, gobbling up traditional analogue media as quickly as they can be switched over to digital technology. Private commercial and financial communications have already gone digital and migrated on-line more or less en masse; public communications services are quickly following.[230]

Digital television, for example, has replaced analogue delivery on a growing number of cable systems and on direct broadcast satellite systems, where it was introduced to save valuable bandwidth and allow systems to offer more channels. Over-the-air television is planning a switch to picture-perfect high-definition technologies during the next few years, and, in North America, the system of choice is digital. Radio is preparing for a phased-in switch to digital transmission in both Europe and North America. The Canadian Broadcasting Corporation's AM and FM national radio networks, the world's largest, were simulcasting all of their programming on the Net as early as 1996, as were many American commercial stations and broadcasters throughout Europe. At about the same time NBC formed an active alliance with Microsoft to migrate content to the Net; the Discovery Channel scrapped plans for new cable offerings and put the money instead into an enhanced World Wide Web site. Cnet is a successful TV/Internet hybrid that is available in analogue, linear version on television and digital, nonlinear form on the Web. CNN, the cable TV news network, created a small sensation (and a tidy new advertising income) with an interactive Web site featuring video news clips along with more conventional text reportage. BBC, ABC and other major news organizations followed suit. Popular television shows by the score developed elaborate Web sites with exhaustive biographical data about their stars, synopses of past and future episodes and, more and more frequently, video clips. Major special events on TV, like the Olympics, made a routine practice of offering detailed scheduling, results and other information available on Web sites. Major movie releases routinely involved elaborate Web sites. Some Web site developers began offering TV-like content, ranging from alternative reporting from world trouble spots to interactive soap operas.

The lines between the media are getting increasingly blurred. Recognizing this, in 1996 consumer electronics manufacturers introduced the first Internet "toasters." These were black boxes containing a telephone input jack and enough computer muscle to allow a conventional television set to become a window on the Internet. In the same year, the technology that promises to supersede the Net toaster made its first appearance: the combination TV set-Internet appliance offered the Net as simply one more "channel," all but eliminating the distinction between the two media. The days when viewers have to watch their favorite shows by appointment, at a time selected by the program provider, are clearly numbered. On-demand, view-when-you-wish capabilities offered by digital, nonlinear technology will be insisted upon by the public just as earlier generations shouted down expert technical opinion and other vested interests to demand, and get, the penny post and universal home telephone service. It also seems at least possible that linear programming on the model of the TV sitcom or drama will be challenged, if not superseded, as the dominant home-entertainment format as users become more and more accustomed to interactivity in their entertainment.

By 1998, new Internet compression technologies had made music of all kinds available for downloading, sometimes from authorized music company sites, but more often from "pirates" who simply posted their favourite songs on their personal Web sites for others to enjoy, free of charge. The phenomenon seemed irresistible, heralding a sea change in music industry distribution techniques, perhaps on the pattern of the software and shareware industries.

The trends are unmistakable: as early as 1995, North Americans were spending significantly less time in front of their television sets and more and more on the Net; in mid-1995 it was reliably reported that North Americans were spending more time using the Net than they were watching rented videos. The demographics of the Net, which had for years matched those of a mosh pit at a Pearl Jam concert, changed dramatically with the advent of the World Wide Web, to the point where there was a healthy representation of all age groups and both genders. By mid-1997, the male-to-female ratio was approaching unity in North America (Europe and Japan lagged in this respect); by the end of the century in most mature Internet markets, slightly more women than men were active on-line. As they had with the telephone, women discovered in the Net a compelling new way to maintain personal and business contacts and seek out information, and were clearly not intimidated by the technology.

Telephone services seem destined for a full-scale migration to the Internet as well, led by user demand for toll-free long-distance calls made possible by packet technology which encodes chunks of sound in digital form and sends them over the Net in the same way as text is sent. (The process allows many more conversations to be squeezed into a given pipeline, since the packets can be intermingled; a dedicated circuit for each conversation is unnecessary.) Since the Internet is composed of telephone links, Net telephony will mean mainly a change in the nature of the end-use appliance, from the conventional telephone to the personal computer. One can expect future computers or other Internet-access appliances to be equipped with telephone-like hand-sets, since these afford privacy that users appreciate in the telephone.

Late in 1996 Telecom Finland Ltd. became the first national long-distance carrier to offer Internet telephony service to its customers. The company plans to make money on the service in several different ways. As with current, still rather primitive Internet telephony software available free on the Web, users pay only for the normal local call to their Internet service provider; there are no long-distance charges. Right now, however,

Telecom Finland, like other long-distance carriers worldwide, pays hefty interconnect fees to the local telcos that provide the "last mile" of the long-distance connection, from the nearest international cable landing or satellite earth station to the home. With Internet telephony, these charges will be eliminated. The company also plans to charge users for value-added services similar to those sold so profitably on the regular telephone services: voice mail, call answering, directory assistance and so on. Said U.S. Sprint Corporation's director of Internet telephone applications, "We share a similar vision with Telecom Finland. The public-switched telephone network and Internet are starting to merge. Ultimately, we will have only one network.[231]

This may be the place to remind ourselves that whatever can be done in the way of data transmission over telephone links can be duplicated over the air using radio technology. There are currently entire cable television systems operating via radio rather than conventional coaxial cables: ultrahigh-frequency radio signals carry scores of digital television channels simultaneously. The area of coverage is limited by the power of the transmitter and the height of the antenna: satellites of course provide very high antennas but currently cannot pump out much real broadcasting power, and parabolic antennas the size of trash can lids are required to reliably pick up their signals.

But the larger, more intractable problem with radio as a carrier of digital networks is that Internet hookups which use satellite access for "downstream" or incoming data must use telephone hookups for "upstream" data sent from the user. Amateur radio operators around the world long ago demonstrated the feasibility of two-way digital packet communication by radio, and even have several amateur radio earth satellites aloft, capable of relaying Internet data in both directions in high volumes. But a publicly accessible system of two-way Internet connections via radio will have to wait for complete digital conversion of cellular telephone networks, and completion of the planned (inhale deeply . . .) low-earth-orbit-satellite-based global digital cellular telephone systems. These are systems which place scores of small satellites into orbit to act as the equivalent of earthbound telephone relay cells. The intention is to make cellular telephony available everywhere on the planet. The first of them, called Iridium, began operations late in 1998 with sixty-six low-earth-orbit satellites. Iridium's competitor, Globestar, which will use fifty-six satellites, was slated to become operational the following year.[232]

Print, as a medium, seems likely to hold out somewhat longer than the purely electronic media against the pressure to migrate to the Net. For one thing, books, magazines and newspapers are already packaged in a digital, nonlinear way that is extremely easy to access. The eye can skip down a newspaper column in search of relevant information; the hand is free to flip a magazine page to avoid a tasteless advertisement. A book can be opened anywhere to provide enjoyment, or it can be addressed more systematically with the help of its table of contents or index. Books and periodicals have the great advantage over most computer-mediated technologies of being highly portable and, in most cases, disposable. It will be some time before electronic devices can match them in this respect.[233]

Furthermore, it is not inconceivable that books in future will contain pressure-sensitive chips to supply audio clips or graphics to supplement written text. For several years now it has been possible to purchase greeting cards containing recorded music and voice messages on tiny, paper-thin microchips, and children's books incorporating this feature are already on the market. If an appropriate viewing appliance were to be attached to a chip-impregnated book, it is also quite conceivable that video could be implanted in the printed pages in the same way. One can easily imagine the bound book replacing the computer for many purposes, rather than vice versa!

The newspaper, as a designed-for-disposal product, is unlikely ever to evolve along the lines of the computer-enhanced book as we've imagined it here. On the other hand, so perfectly suited to its purpose in its present form, there seems little likelihood that it will be replaced by electronic media in any configuration. Reading tablets currently undergoing development by U.S. newspaper chains display newspaper-like pages on thin, flat screens, but are bulky and fragile (and expensive) and seem unlikely to find a market beyond a few specialty niches, such as applications currently being served by even bulkier interactive kiosks. (Data would need to be downloaded into them periodically by an Internet connection.)

As we've seen, commercial aspects of the Internet began receiving major attention in the mid-1990s. Current predictions for worldwide Internet commerce by the year 2003 range from a low of US$1.4 trillion to US$3 trillion and beyond.[234] A large majority of businesses in the United States expected sales on the Web to account for more than 20 percent of their revenues in the short to medium term, and most predicted it will have either a "huge" or a "significant" impact on their sales processes.[235] On-line retail sales in the United States had reached US$7.8 billion by 1998, and predictions were for that to soar to $108 billion by 2003. In Europe, sales figures for 1998 were $5.6 billion, and similar growth was expected. In some areas, such as travel reservations and computer sales, analysts predict that half of total business volume will be conducted on the Web by 2003.

It would be a mistake to classify the Net as simply another tool of commerce like the Macadam highway, the railway, the steamship or even the telegraph, because the Internet is uniquely capable of creating commercial commodities in and of itself. A seemingly trivial example is the e-mail greeting card, a whimsical product that drove the market value of one home-grown Web site into the stratosphere. And advertising potential is enormous: in 1998 on-line advertising on the Net generated revenues of nearly US$2 billion, projected to grow to US$7.7 billion by 2002.[236] However, even as a simple communications aid to commerce, the Net has capabilities that suggest its business impact will exceed that of earlier innovations in transport and communications. As an illustration one need only cite the case of Amazon.com, the Internet's biggest bookseller. Beyond having access to a worldwide pool of book buyers, Amazon has devised an ingenious way to develop a worldwide network of sales agents, all working on behalf of Amazon. The company offers official, contractual "associate" status to any Web site that wants to sell books—any books.[237] An electronic order form is supplied free of charge to the associate site, and that order form is linked by hypertext to Amazon's own order desk. The purchaser fills out the order form and clicks on a "submit" command which sends it instantly to Amazon. Amazon fulfills the order and gives the associate a small commission on sales over a certain threshold. The associate, for its part, adds value to its

Web site by offering visitors a chance to buy relevant books it recommends. A Web site devoted to dog breeding, for example, might review and recommend several volumes on that subject, and provide a direct-purchase form. Everybody wins: book purchasers have more outlets from which to choose, and benefit from informed reviews posted by associate sellers; the associate sellers are able to improve and add value to their Web sites by offering books for sale, without having to get into the book business themselves, and they stand to make a modest commission as well; Amazon reaps the rewards of an infinitely expandable commissioned sales force selling books it knows and understands and is enthusiastic about. Only on the Net is such an eminently sensible commercial arrangement possible. Innovation like this is destined to change the way we do business worldwide. On the strength of this kind of market innovation Amazon.com had become a $4.6-billion property by 1999.

As interesting and important as Internet commerce has become, the social changes wrought by the new computer network technologies can be expected to outweigh in significance the economic innovations they inspire, as has been the case with other breakthrough communications innovations. Even in the narrow sense of a tool offering fast, convenient access to information, the Net cannot help but bring about significant changes in the way society operates. Consider for a moment how much of current day-to-day life is bound up in the quest for information. Commuting, for most of us, is just a way to get to where information related to our jobs is located. Shopping is another time-consuming quest for information that the Net can largely eliminate. Most ordinary purchases such as clothing and groceries require visits to shops and supermarkets not because that is where the merchandise is located—merchandise can be delivered, after all—but because that is where the *information* we require to make an intelligent selection is located. In a rich multimedia, network-based shopping environment, we would not need to travel to the grocery store to know whether asparagus were in stock and fresh. We visit the Gap because we need a rich information environment to make choices about color and fit: we need to try the clothing on, or at least handle it. We could, perhaps, order from a conventional mail-order catalogue, but that involves a long wait.[238] On-line shopping can offer a much richer information environment than print catalogues, point-and-click ordering and overnight delivery via courier.[239] Face-to-face interactions with checkout clerks and sales staff will of course be eliminated in on-line transactions (though on-line shopping can involve video conferencing); for most of us, in most circumstances, this will be more boon than sacrifice and the time we've saved can be spent with people we really want to see. Few of us would wish to return to the days when all banking had to be done in person in front of a teller's wicket at a bank that was open from 10:00 A.M. to 3:00 P.M. weekdays only. The ATM or cash machine combined with the Net's reach into private homes and rich multimedia environment provides a sound conceptual model for many other businesses. The same principles apply to most forms of entertainment, other than live performances.

A further point needs to be made about the loss of face-to-face social contact. The fact is that in our consumer society, genuine social interaction has to a large degree been subverted to commercial ends: our agoras are for the most part shopping malls and not true meeting places. A family outing is as often as not just a "shopping opportunity." To the extent that we actually do value this kind of social interaction, we do so because it is all we have time for and it is better than nothing. We may rest assured that people will continue to find excuses to get together in groups both large and small whether or not in-person shopping continues to be a prime motivator for visits to public (or quasi-public) places. People may, indeed, find opportunities and environments that are much more convivial.

Journalists often refer to the Net as "anarchic" and that's not an inappropriate description, although, clearly, the Internet does not exist in a political vacuum: the laws of nations apply on the Net as they do in ordinary life. In Canada, for example, there have been criminal convictions of BBS system operators for purveying child pornography on-line, and in the United States, authorities have brought successful fraud prosecutions against Net businesses.

My dictionary has two definitions for the word *anarchy*, one negative and one positive: "a state of lawlessness or political disorder due to the absence of governmental authority," and "a Utopian society having no government and made up of individuals who enjoy complete freedom."

But the sense in which *anarchy* applies to the Net is tied up in the history of anarchism, as a vision of an alternative libertarian society based on cooperation as opposed to coercion. Readers old enough to remember the "flower power" era of the late 1960s and 1970s will recall that wherever hippies congregated in any number, there were sure to be "Diggers" who specialized in supplying free food, clothing and lodging. *The Whole Earth Catalogue* (with its motto "access to tools") grew out of the Digger phenomenon, and it is no surprise to find Stuart Brand and others who were prominent in the movement involved in today's Electronic Frontier Foundation, which attempts to ensure universal access and freedom of expression on the Net. The Foundation can credibly claim to speak for the dominant attitudes among Net users when it comes to the politics of control and authority or the "Net ethic."

The first Digger was in fact the first English anarchist, Gerrard Winstanley, a dissenting Christian who founded the Digger movement in 1649, just after the English Civil War. He held the now familiar views that power corrupts, that property is incompatible with freedom, that authority and property are the causes of crime and that happiness is possible only in a society free of rulers, where work and its products are shared and where people act according to their consciences and not laws imposed from above.

He and a little group of his followers set up a commune on some uncultivated hillside land in southern England, began growing crops and used passive resistance to fend off hostile local landowners. They were sadly unsuccessful in this latter endeavor. The commune was dispersed and Winstanley vanished into oblivion. But his ideas lived on in the English socialist tradition and in modern libertarianism. (It was the mad Russians Mikhail Bakunin and Peter Kropotkin who steered the anarchist movement disastrously into assassination and terrorism in the middle of the last century, thereby destroying its right to legitimacy in civilized political discourse.)

It is interesting how much the anarchist outlook resembles the views of present-day futurists who claim no particular political allegiance and who see computer-mediated communication and telecommuting as auguring fundamental changes in work and society. Pierre-Joseph

Proudhon, the great French anarchist, in the mid-1800s envisioned a networked society composed of independent farmers and artisans, where factories would be run by associations of workers. Instead of a central state authority, governance would be by contracts among autonomous local communities and industrial associations. The broadly similar futurist viewpoint of today sees production in a knowledge-based economy coming from "virtual corporations" of dispersed, networked and mostly autonomous workers. Politics reverts from the current representative democracy to a more direct democracy in which polling and voting are done continuously via the Net. Early anarchists like Proudhon and Winstanley would have found this an attractive prospect.

The Net is public space that is shared by millions of "citizens" but lacks a government.[240] It survives and flourishes thanks to the fact that the vast majority of its users mind their manners and obey the Golden Rule. It owes much of its staggering resources to simple altruism and goodwill. As an experiment in anarchism, it can only be described as an encouraging success.

"Public space" as it relates to the Net is an idea worth exploring a little more deeply.[241] It is first and foremost an architectural concept. In the official lexicon of urban designers, space devoted to streets, parks, squares, boulevards and so on is public space. But so is a mall concourse or the lobby of a hotel, in that they are ordinarily accessible to the public. This ambiguity makes it a controversial definition. At architecture schools, students bemoan the loss to private interests of what once was public space, as happens when construction of a shopping mall shifts the focus of a town from the central square or main street to the enclosed commercial spaces of the mall. While the main concourse of a mall may *seem* to be public space, it is not—pamphleteers, panhandlers, buskers and boisterous teenagers all learn this lesson definitively when they are given the bum's rush by private security guards. Furthermore, the public has no control over the amenities or lack of same in such pseudo-public spaces.

The on-line environment in which the Net operates is, and must be seen to be, public space analogous to the air waves that carry broadcast signals—*real* public space, and not the pseudo, shopping mall variety. There is nothing in this recognition that precludes the physical structures of the Net—the copper and optical-fiber lines and the switching devices and digital-coding machinery—from being owned and operated for profit by private enterprise. It is the space that is created when those lines are humming with data, the cyberspace, that has to be acknowledged as public space. The Net exists in that space and is, by definition, owned and controlled by its millions of users. It was designed and built to be that way, and the design works. The public nature of the Internet is lodged deep in its defining technologies.

If we can begin to think of the Net as a new kind of "built environment" or human living space, then we'll begin to better understand how to deal with it and profit from it. We might be moved to ask ourselves, for instance: What is the role of government in running a city? And the answer is to provide the amenities that make people's lives easier and more meaningful, to help people do the things they want to do safely and conveniently. It is *not* to tell people how to live, or what to say or do. There is an analogous role for government in the on-line environment. Government should facilitate access, enforce the criminal law and provide public amenities not made available by private enterprise. There is a role here for a CBC or BBC or PBS as well, in providing useful on-line resources that would otherwise not be there, and all three public networks have become heavily involved in Web site development, with the BBC's portal garnering nearly half of Britain's Web users (while its television audience share continued to slide). There is also a clear on-line role for business within this metaphor. It is *not* to own or control the public spaces of cyberspace. It is to provide services the public wants and needs (at a profit) and to generally behave like good citizens, refraining from despoiling the landscape or ripping-off the burghers. It also means sharing the wealth by reinvesting a portion of profits in public and quasi-public amenities like useful databases, navigation aids and entertainment sites.

CHAPTER TWENTY-NINE

Can the Net Think?

If we see evolution in communications and information management as consisting of a series of progressively more advanced technologies, from the telegraph through to the worldwide network of computers, then the next logical step in this progression is artificial intelligence embodied in computers and their networks. Just as steam and electricity gave us the horsepower we needed to bring the machine age to full fruition, artificial intelligence could provide the tools we need to productively manage, at every level, an economy based largely on information. Artificial intelligence could be the key to automating the processes of adding value in the information economy, just as cybernetic robots have done in machine-age processes such as automobile assembly.

In some senses, widespread application of artificial intelligence is a frightening thought if we consider to what degree even today's computers have displaced human workers and set them adrift. On the other hand, a world in which wealth is created in abundance by automated, intelligent processes could be a utopia if we could but devise humane mechanisms of wealth distribution. All of which begs the fundamental question: How smart can a machine be? Is machine intelligence even possible?

Back when the twentieth century and radio were both young and Alan Turing was still in knee pants, a comic-strip named *Buck Rogers in the 25th Century* was enormously popular. It involved a World War II airman who conks out in combat only to regain consciousness five hundred years later. The world of the twenty-fifth century as revealed in *Buck Rogers* is full of amazing technological achievements. There are television and radar, rocket ships and space suits, disintegrator guns and recoilless energy projectors—even a substance called Intertron, which has antigravity properties and is so cheap to produce it is as common as cardboard.

What was not imagined in *Buck Rogers*, or in any other contemporary sci-fi writing, was the intelligent machine—the thinking computer. When Buck Rogers faced a problem that was too difficult for his human mental powers, he turned not to a machine but to the eggheaded Professor Huer and his prodigious brain. It was not, in fact, until Arthur Clarke's *2001: A Space Odyssey* introduced Hal to the world that the idea of machine intelligence made an impact in the mainstream of science fiction writing. Well into the 1980s, even in *Star Trek* plots that were otherwise wildly inventive, the idea of conscious, thinking machines was strangely absent. Computers calculated, monitored, retrieved data and projected; only Spock displayed superior mental abilities, and although he was not human, neither was he a machine. It wasn't until *Star Trek: The Next Generation* introduced the android Data as a leading character that the idea of machine intelligence became commonplace in mainstream science fiction.[242]

In science itself there is continuing, deep controversy over whether artificial intelligence is possible. After nearly half a century of debate, during which computers have come to exhibit increasingly impressive capabilities, the principal naysayers have narrowed their arguments to focus on the distinction between "simulating" intelligence and "duplicating" it. Philosopher John Searle puts the argument against machine intelligence succinctly in those terms:

> No one supposes that computer simulation of a five-alarm fire will burn the neighborhood down or that a computer simulation of a rainstorm will leave us all drenched. Why on earth would anyone suppose that a computer simulation of understanding actually understands anything? It is sometimes said that it would be frightfully hard to get computers to feel pain or fall in love, but love and pain are neither harder nor easier than cognition or anything

else. For simulation, all you need is the right input and output and a program in the middle that transforms the former into the latter. That is all the computer has for anything it does. To confuse simulation with duplication is the same mistake, whether it is pain, love, cognition, fires or rainstorms.[243]

Arguments like this hinge ultimately upon the issue of consciousness, as do virtually all strong arguments against artificial intelligence. What Searle is really saying is that even a perfect computer simulation of human intelligence is not intelligence at all. Why? Because the machine is not conscious, and consciousness as it exists in humans is a prerequisite for true intelligence.

But is the premise correct, or might it be possible for intelligence to exist in the absence of consciousness? The answer turns on what is meant by "intelligence," which is an interesting question. The *Shorter Oxford* defines *intelligence* as the faculty of understanding. Turn to *understand* and you find it defined as "to comprehend." Flip to *comprehend* and you find "to grasp with the mind, to take in." Turn to *mind*, the key word in that last definition, and you find it defined as . . . eureka!: "consciousness." So, intelligence, strictly defined, demands consciousness, just as Searle and others have argued.

But if we take the hunt for literal meanings a little further, the issue becomes murkier. What does it mean to be conscious? The dictionary says it means "to be aware of what one is doing or intends to do . . . aware of one's existence and thoughts, and of external objects." We might then say that intelligence demands "awareness." Unfortunately, there doesn't seem to be any way to define "awareness" without resorting to the idea of consciousness. We have encountered the impasse that ultimately confronts any dictionary search for the real "meaning" of a word: all dictionary definitions are ultimately circular, because words in a dictionary can only be defined in terms of other words in the dictionary. There is a gap between words and the things they express or represent that cannot be jumped. A language is a formal system that maps onto experience, but the map is not the territory. Mired in this pointless "aware = conscious, conscious = aware" tautology, we are left with the frustrating fact that consciousness is undefinable in any useful way, at least at this stage in the evolution of human knowledge and understanding.

But if consciousness cannot be defined satisfactorily, then the appeal to consciousness as a defining characteristic of intelligence is meaningless. Worse, it is arbitrarily limiting, a mental straitjacket.

Two dictionary definitions of *intelligence* manage to side-step the consciousness issue: "the ability to adapt to new situations and to learn from experience," and "the inherent ability to seize the essential factors of a complex matter." Under these definitions, a computer running any of a myriad of highly complex programs in common use today would have to be considered an "intelligent" machine.

However, accepting either of these definitions backs most skeptics into a corner where they are forced to show their true colors. At bottom, they simply are unwilling to accept that innate intelligence can be anything other than a human attribute. Oh, to be sure, a dog or a parrot can exhibit intelligence too, but, skeptics would insist, the issue at hand is the higher, *human* form of intelligence responsible for art, invention, civilization, and whether *it* can exist in a machine. And, they will insist, it cannot. The pursuit of this argument, however, leads into another unproductive cul-de-sac. The problem is that it implies that human intelligence is in some inherent and ultimately mysterious way superior to any other kind of intelligence. Where is the evidence for that?

All of this raises yet another, ultimately much more interesting question: If a machine can exhibit intelligence, does this mean that human intelligence is in some way machinelike? In other words, is intelligence a kind of generic commodity at root, something that can be transplanted from, and implanted in, different kinds of entities, both biological and electronic? Is the brain, as has been suggested, merely a meat machine? Since the development of the digital computer, it has become possible to argue with some conviction that this may be the case. It's an argument that becomes increasingly difficult to deny with each leap in the capabilities of the computer and networks of computers.

Here is one way to understand how computer intelligence and human intelligence might be similar. The earliest computers like Colossus and ENIAC were programmed by connecting wires together—by "hard wiring" the program into the circuitry. But before long, programmers had devised a way to feed instructions to the computer's processing unit in machine language. These instructions were of course coded digitally, in 1s and 0s. A typical piece of machine language code might look like this: 0010001011010101. It contains the word or words in the computer's memory that are to be acted upon, instructions as to where to find them, plus instructions as to what action to take when the words are located—instructions such as "add the word in the code to the word in memory." Even a very simple program involves pages and pages of such code.

It's not hard to see why machine language, though a big improvement over hard wiring, presented problems—particularly where long and complex sets of instructions are involved. The possibility of error is enormous; all it takes is transposition of one 0 and one 1 to foul up a program.

Computer programmers thus created the notion of a higher-order language called "assembly language." In assembly language, the computer is given instructions in a code that is much easier for the human programmer to interpret and remember. For instance, instead of writing something like 100010101011 for a given instruction, the programmer using assembly language would merely type ADD (or SUB or DIV). The computer *automatically translates* the assembly language into machine language, so that the instructions can be carried out.

Assembly language is really just machine language made more intelligible to humans. What is really significant about it is that it demonstrates that higher levels of language are possible. If you can write an assembly language which simplifies machine language—thereby making more and more complex programs more feasible to construct—why not develop a third-level language that will simplify the process even further, and thus make programs of a new order of complexity possible?

That is exactly what was accomplished when, in the 1950s, FORTRAN, Algol, LISP and other languages were developed. These are "compiler" languages, and they allow a programmer to express extremely complex machine-language commands in ordinary mathematical notation, or even in simple English. The instructions so written are automatically translated into assembly language and then into machine language for execution by the computer.

The development of these early compilers, and of their many, increasingly sophisticated successors (including voice-recognition programs), has added enormously to the potential power and flexibility of computers. But what's really interesting is that, to communicate with a computer that is running a compiler, you need know nothing whatsoever about machine language and how it gets manipulated in the central processing unit. You no longer need to know how the machine works in order to get results. Even if you did set out to describe the processes in detail right down to the binary bits, your description would be of little practical value because there is so much activity at so many levels that it becomes impossible to interpret what is going on in the machine's "brain" by reference to 1s and 0s. Not only will it not explain it in any useful sense, it will not always predict what the machine will do; computers routinely surprise programmers with unexpected, unpredicted responses.

Which brings us finally to our important point about computer languages and the human brain. If you can have third-level languages like compilers, why not even more sophisticated fourth- and fifth-level languages and so on? And might this not be the way the brain works: processing sensory inputs on the level we're aware of with a very sophisticated symbolic language we call "ideas," but doing its real work many layers down in a "machine" language analagous to that used by computers? Many researchers into artificial intelligence believe this to be the case.

Certainly, at some high level of language sophistication and complexity, it seems simply unreasonable to deny intelligence to a machine that does things that would require intelligence of a human. At that level it makes no more sense to say of a computer, "Of course we understand it—we programmed it," than it does to claim to understand how a child learns what we teach it.

Having come this far, why not take the next logical step and grant to the computer what might be called "intentionality"—in other words, admit that it behaves rationally in terms of its beliefs and goals? Too big a step? Stop a minute to consider: how do we know that humans have beliefs and goals? We don't know. We just assume they do, because they tend to behave rationally. (Of course, if you ask, they'll tell you they have beliefs and goals, but that doesn't really prove anything, does it? We have no way of confirming what goes on inside their heads.)

On the other hand, we know that computer operations are deeply embedded in logic and so take place in a rational way, moving step-by-step from problem through to solution. There seems no good reason, then, not to say the computer acts because it wants to reach a goal. And it seems no less reasonable to say that the machine "believes" the data it is programmed with, in a way that is analogous to a human's belief in the input of senses or the reflections of memory.

What are we left with? A machine that apparently acts according to its beliefs and goals, that communicates intelligibly with humans (and other machines) to provide useful, sometimes unpredictable and surprising information. When we encounter all of this in a domesticated animal, or imagine it in a little green creature from Mars, we have no hesitation in attributing it to innate intelligence. This is the point Alan Turing was making when he dismissed epistemological controversy in favor of his simple "Turing test" for machine intelligence—if it acts intelligent, it is intelligent. Turing's test involved placing a computer behind a wall and allowing humans to converse with it via a keyboard and printer. If humans were unable to tell whether they were talking to a human or a machine, the computer would have passed the test and would have to be acknowledged to be "intelligent." Turing predicted that by the year 2000, a five-minute conversation with a computer would fool an average questioner into thinking that it was human 70 percent of the time. Some of the provocative flavor of his seminal essay on machine intelligence can be gleaned from the following excerpt:

> The game [the Turing test] may be criticized on the ground that the odds are weighted too heavily against the machine. If the man were to try to pretend to be the machine, he would clearly make a very poor showing. He would be given away at once by the slowness and inaccuracy of his arithmetic. May not machines carry out something which ought to be described as thinking but which is very different from what man does? This objection is a very strong one, but at least we can say that if, nevertheless, a machine can be constructed to play the imitation game satisfactorily, we need not be troubled by this objection.[244]

Or, in the vernacular, if it walks like a duck and quacks like a duck, then it is a duck. That, after all, is the intelligence test we apply to other humans.

The unanswered question would appear to be not *whether* a machine will ever be as intelligent as a human, but *when?* Here, the skeptics and technophobes can take some comfort, for it would appear that this is not likely to happen for a very long time. While computers right now operate at processing speeds that greatly exceed those of the human brain, the brain more than compensates by being able to store astronomical amounts of information and by being able to process millions of bits of data at the same time, thanks to the ten trillion neural connections between its ten billion neurons.

New-generation computers that employ parallel processors, neural networks running on supercomputers, and vast computer networks like the Internet already mimic the brain's abilities to some degree, and there is no reason to believe that we are anywhere near the end of the era of very rapid development of computer technology. But even if it becomes possible to construct a computer or computer network with as much brute processing power as the human brain, the far more subtle problem of how to gain efficient access to the appropriate, stored information when it is needed would remain. Chess-playing programs have demonstrated that you can't do it through computing horsepower alone: if/then branching very quickly gets out of control and explodes exponentially. It turns out, in fact, that the number of possible chess moves is greater than the number of atoms in the universe. So much for that idea.

Even seemingly very simple problems can resist the computer's brute force in a similar way, as options multiply. We humans search our brains for problem-solving information through a mysterious process we call "using common sense." It turns out that programming "common sense" into a computer is an exceedingly difficult challenge, one that has been and continues to be among the main roadblocks to real breakthroughs in artificial intelligence research. Progress, though, is being made. In commenting on his historic May 1997 defeat by IBM's Deep Blue, world chess champ Garry Kasparov noted that the machine had become so uncannily humanlike that it had been a mistake for him to prepare for the match by studying the computer's technological limitations and idiosyncracies. In future matches, Kasparov said, he would prepare as he would for a highly skilled human competitor.

Initially, at least, it would appear that the areas in which computers excel at exhibiting apparent intelligence are likely to be of a special kind, where the relevant considerations or variables are relatively few and well defined. These include formal games and disciplines with strict rules or operational frameworks such as accounting, mathematics, physics, some aspects of medicine and law, areas of architecture and graphic design and so on. But as artificial intelligence pioneer John Haugeland of MIT observed in 1981: "It is an open question whether the intelligence manifested in everyday life, not to mention art, invention, and discovery, is of essentially the same sort (though presumably more complicated). [Artificial intelligence research], in effect, is betting that it is; but the results are just not in yet."[245] The question remains open today.

Does it really matter? In one sense, it does not. It is perfectly clear that, whatever we choose to call it, machine "intelligence" is destined to play a very large role in our daily lives as the information economy evolves and computers continue to get smarter. That has been foreordained by the evolution of the technology to date. Whether or not we decide in the end that computer intelligence is of the same species as human intelligence, or something altogether different, there is little doubt that Alan Turing was correct when he predicted that by the turn of the century we would all be able to speak of intelligence in machines without fear of contradiction.

Deep Blue's intelligence may be restricted to the arena of chess, but Garry Kasparov does not hesitate to acknowledge it as authentic, and women's world class champion Susan Polgar said of Deep Blue's play in the 1997 match: "Deep Blue made many moves that were based on understanding chess, on feeling the position. We thought computers couldn't do that."

Evolving machine intelligence makes it clear that we have only begun to understand the true potential of the Internet. Scott McNealy, CEO of Sun Microsystems and one of the specialists responsible for the success of TCP/IP, sums it up with an aphorism of deep metaphorical and practical significance: "The network is the computer." If that is indeed the case, we are building in the Net a computer of truly awe-inspiring capacity. Two "laws" of computer development—actually, real-world observations that have proved themselves accurate over the past several years—help to explain the power-magnifying effect of the Net. Moore's law states that the number of transistors that can be placed on a chip doubles every eighteen months, and that with each doubling, computing power increases fourfold. Metcalf's law states that when computers are networked, their power and value increases as the square of the number of computers.

When these observations are combined with the fact that the number of computers linked to the Internet currently doubles each year, some truly astonishing projections of the Net's power and value can be made. What is meant in this context by "value"? When we compare biological brains in terms of the density of their networks of synapses, one of the ways we describe their differences is in terms of intelligence. Intelligence, it would appear, is an emergent property[246] of complexity. At some stage in its growth, we might reasonably expect the Net to exhibit intelligence in some form. At that stage we will know what is meant by value; it would be well if we were not taken entirely by surprise. Much is at stake, as I hope will be made clear in the concluding chapters.

CHAPTER THIRTY

The New Economy

The information economy—a.k.a. the knowledge economy, the digital economy, the wired economy—is many things. It is new, customer-focused approaches to doing business made possible by the rich new communications environment we live in. It is new ways of manufacturing goods based on the custom specifications of individual purchasers, which means the decline of warehousing and the rise of preordered, prepaid goods, along with the decline of mass production and the rise of mass customization. It is the end of the loyal, servile employee and the rise of the freelance paladin, the decline of corporate hierarchies and the growth of the self-managing work group, the decline of the dependent status of workers fostered by the industrial corporation and the rise of the self-reliant ethic of the individual as entrepreneur, practitioner and go-between. It is, potentially, authentic consumer sovereignty and real autonomy for the small, independent producer. It is, we may hope, a light at the end of the long, dark tunnel of the late twentieth-century crisis of corporate capitalism.

Information is always an important component of any economic system, and in that sense the information economy has always been with us. What is different in today's world is the advent of high-bandwidth digital networks enveloping the globe, multiplying many thousandfold our ability to move information from place to place, and the approaching ubiquity of the silicon chips that allow access to those networks. It is as if we had advanced from primitive irrigation systems to the ability to make rain, from dryland farming to hydroponics.

A complete discussion of the economics of information is beyond the scope of this book. However, before proceeding any further, we should perhaps pause here to define some terms, since the ideas we will be examining are new and novel enough to remain somewhat slippery. First of all, what exactly is the "information economy" in concrete terms? For our purposes, we can include at least the following areas:

- packaged information such as scientific and technical literature;
- informal information such as company reports, news digests and consultant's reports;
- transient information such as news services and market reports;
- permanently stored information such library archives and public records;
- skilled judgment such as professional and technical services and the work of consultants;
- education and training, including textbook publishing and other education services;
- entertainment, including broadcasting, film and the creative arts.

While this is a helpful definition, it is not nearly complete, because information in economic terms is closely related to the concept of invention, and invention is a process which is carried on continuously and at every level of economic activity. The farmer who adjusts his combine harvester partway down a swath of wheat in order to maximize yield is in a direct way part of the information economy at that moment: he has simply changed hats to become his own consultant in machine adjustment and "invented" a new combination of variables within the combine. Pity the poor economist who must try to quantify this sort of elusive, heterogeneous commodity.

Information is, in fact, in classical economic theory, roughly equivalent to technology, that is, technical knowledge that is applied to economic production.[247] The link here is through invention, since technological progress comes through innovation or invention, and invention may be defined as the production of new information or knowledge. Research and development (R&D) is thus a part of the information economy: it is a production process whose output is new combinations of information.

A further, interesting complication lies in the fact that both information and knowledge ("knowledge" may be defined as "information + intelligence + experience") have the peculiar property of self-replication, so that a vendor of knowledge or information retains his or her product even though it has been sold to a purchaser. Producing multiple copies of a software program for sale in no way reduces the inventory of the product as does, for example, the production of multiple batches of aluminum ingots: the producer of software retains all of the product regardless of how much is sold. In fact, there is no way for the vendor to rid himself of information once obtained: he can only share it. In this sense, it can be observed that the information economy operates not so much on buying and selling as sharing. In spite of this, for it to function at all, arrangements must be found for the fair compensation of information and knowledge providers (one reason why laws of copyright have become a hot topic in recent years).

Information economics is also a migraine headache for professionals as an academic discipline, because it challenges almost every assumption basic to the neoclassical economic theory that has been the discipline's backbone since World War II. There is a truism that a parrot could become an economist simply by learning to say, "Supply must equal demand." In the information economy, even this bedrock assumption is in doubt, even irrelevant, as we've just seen. And so, just as twentieth-century physics was forced to acknowledge the world of nonlinear, chaotic systems, late twentieth-century economics has had to come to grips with the fact that economic systems are imbued with a commodity called information. Like nonlinear physical systems in nature, information is very difficult to analyze with traditional tools, which accounts for the fact that it was for so long swept under the carpet, until modern information-processing technology made that tactic impossible. Just as classical, Newtonian physics had concentrated on examining those phenomena that were amenable to quantification and calculation, economics has shied away from acknowledging the important role played by information because it is so intimately tied up in economic processes as to be very difficult to isolate and quantify. Economist Peter Monk notes:

> The neoclassical economics paradigm is founded on the assumption that economic decisions and transactions are costless and "perfect" [i.e., undertaken with the benefit of perfect knowledge]. Even theories which take explicit account of costs [associated with getting information required to make knowledgeable decisions] still operate by assuming "perfection" as the default condition of economic processes. However, in the real world "imperfect" conditions prevail.[248]

As a result of being forced to acknowledge this truth, economics has had to begin a search for a dynamic rather than static theory—for a chaos theory of economics. So far, results have been scant, as the dearth of creative ideas in public policy indicates. Politicians and bureaucrats alike are at a loss as to how to solve persistent, institutional unemployment and underemployment in most Western nations, to take one example, and the reason is that they are applying machine-age economic theory and practice to an information-age economy. It is currently estimated that as much as half of all economic activity in industrialized nations is in the category of the production and processing of information, yet economic theory still largely ignores its impact, not through any willful neglect, but because it is so difficult to disaggregate from more traditional goods and services outputs.

At the same time, it is clear that, like other segments of the economy, the information sectors could benefit from rational investment policies. Once again, however, economists have trouble in this area because investment in information sectors is largely a matter of intangibles rather than bricks and mortar and machines. One thing is reasonably clear: investment in information sectors involves financing the production and distribution of knowledge, and education is thus a more crucial priority for public policy than ever. And as Peter Monk observes so trenchantly, it must be education in the broad sense, as well as skills training:

> The history of technological change clearly demonstrates the opportunity costs, if not dangers, of narrow views of what types of information and knowledge are "relevant," "applicable," or "profitable." An essential feature of technological progress is that it not only involves the creation, development, and use of new types of information but also includes new applications of previously unexploited knowledge. This process is, by definition, uncertain and accelerating. . . . To maximize expected future value from those activities, in unavoidable conditions of uncertainty, the broadest possible range of information production must be undertaken. This means expanding rather than contracting the diversity of disciplines and participants in education. Further technological development in the information economy is likely to depend as much on progress in philosophy, epistemology, literacy and communications studies as on the continued promotion of the engineering disciplines.[249]

Since information sector activities involve, by definition, both the production of information and its consumption, it follows that not only producers but consumers of information as well will need to be equipped with broadly based educational backgrounds rather than simple vocational training, for, as we shall see, consumption and production of information are often heterogeneous activities.

1: Information as an Agent of Control

British economist Gary Hamel appears blessed with a sense of humor rare to his calling. Noting that between 1969 and 1991, Britain's manufacturing output grew by 10 percent, while the number of people employed in the British manufacturing sector was nearly halved, he wondered what the so-called achievement really signified. The press dutifully celebrated the fact that during the period, British manufacturing productivity rose faster and higher than in any other country except Japan. But Hamel was perplexed: "One almost expected to pick up the *Financial Times*," he said, "and find that Britain had finally matched Japan's manufacturing productivity—and that the last remaining person at work in British manufacturing was the most productive son of a gun on the planet."[250]

What's wrong with this picture? Productivity (or cost per unit of output) is a figure made up of both a numerator and a denominator, and by focusing so single-mindedly on cost reductions achieved through "downsizings" and "rationalizations" (the denominator) rather than pursuing new revenue through market expansion (the numerator), British industry, while seeming to become more successful, was in fact surrendering global market share to other, more aggressive competitors. Production costs were down, but so were sales. And, of course, the hardship inflicted by the layoffs involved was immense. The social merit of this corporate bulemia is therefore highly dubious. The notion of "productivity," like "progress," is one that needs careful, critical attention: more of it is not always an unalloyed blessing. Like cholesterol, productivity can be both good and bad.

The technologies of the Information Age increase economic productivity in all sorts of well-documented ways, from improving communications within the corporation, between corporations and between the corporation and its customers, to reducing paper and other administrative costs, as well as by the rationalization of processes and the elimination of waste and error. All manufacturing technologies, whether steam-driven or microprocessor-based, displace labor in one way or another; that is what they are designed to do. In displacing labor, they reduce costs to manufacturers, and these savings are eventually passed on to consumers. In the past, it has always been possible to argue, at least in principle, that everyone has benefited from labor-saving technologies, because the workers displaced were able to find new, often better jobs working in a new industry created by, or ancillary to, the technology that eliminated their former job.

For example, the introduction of electricity greatly reduced the need for manual labor throughout the economy, but it also created entire new industries directly related to electrical production (dam building, power distribution, turbine design, domestic and industrial wiring and so on), as well as ancillary industries to manufacture electrical tools and appliances. The advent of the automobile and the consequent revolution in personal transport brought with it not just the car factory with its assembly-line jobs, but also a boom in road construction and tourism and the development of the petroleum manufacturing and distribution industries. Jobs were lost in one sector, but replaced, and then some, in others.

What makes Information Age's computer-based technology unique in this context is that it operates at a different level than the technologies of earlier industrial transformations. Like the mechanical clock, the computer is a technology that embodies an abstract idea—in this case, logic—and is able to operate autonomously. Get either device up and running and it will continue doing what you've asked of it indefinitely, until it runs out of energy. Like the clock, too, the computer operates in production processes at the level of *control:* control of processes and administrative control. But unlike the clock, the computer operates not just as a tool, but as a factor of production as well. It does more than assist humans in controlling processes; it very often replaces humans. We are accustomed to machinery which replaces people in process jobs, e.g., the new robot on the assembly line; we have very little experience with machines that replace people in administrative and control positions. Even the infamous time-and-motion specialist, with clipboard and stopwatch in hand, can be replaced by computer systems which automatically monitor workers' performances, and which can go beyond that to adjust the working environment in accordance with this input to maximize productivity in a dynamic way.

No matter how sophisticated earlier labor-saving technology has been, it has always required human oversight to ensure its efficient operation. No longer. Computers and automated systems of today can often get the job done without human supervision, either on their own or by empowering customers. In the first case, think of telephone operators displaced by voice-recognition technology, which can dispense information and handle long-distance telephone transactions; in the second, think of voice-mail systems which empower (or require, more properly) the caller to find the person being called without the intervention of a telephone receptionist. ("If you know the person's name, press the first three letters . . . ") On-line banking, which allows the customer to fulfill the role of the bank teller, is another example of so-called customer empowerment. As is Federal Express's use of the Internet to allow customers to track their own parcels through the system without having to call FedEx telephone operators. Or on-line shopping services. All of these computer-based technologies displace workers by getting customers or users, with a little help from microprocessors, to do the work instead.

Computers, as we've seen, can control virtually any process for which formal operating procedures (algorithms) can be described and put down in a manual. And their cost goes down month by month. The clock in its time provided a necessary tool for the development of capitalism, permitting the efficient organization of the various factors of production; today's computer-based information technology is the capitalist's ultimate dream. It is a key factor of production whose price is more or less continuously falling in relation to performance.

There is no practical, foreseeable limit to the numbers of jobs that can be displaced by this new metatechnology. This is not to say that computers can handle any human function you can name; that is obviously not the case. What it does mean is that we cannot draw any meaningful boundary around the areas in which we can expect computer technology to be able to take over human job functions. There is no area of human endeavor in which computer-based technologies cannot fulfill at least some of the functions currently being handled by humans. In this latest technological revolution, therefore, there is no reason to believe that there will be "job creation" in the traditional sense, sufficient to take up the slack caused by the displacement of workers.

From a purely capitalist point of view, this is not a concern and the introduction of new technologies should be encouraged to continue, unabated: rationalization of economic processes is at the very root of the historic successes of capitalism and should not be interfered with. "The role of humans in capitalist enterprise is clear," states Noah Kennedy:

> [H]umans fill the roles in productive processes that are uneconomical to mechanize. This should not be a shocking statement, for if any of our jobs could be done at a lower cost by a machine there is no doubt that this would come to pass. Similarly, there is no doubt that each day technology closes in on new intellectual tasks that previously required human intelligence, which is just another way of saying that the task is being rationalized to the point that it can be reduced to formal description and performed by an algorithm. If there is not at this very moment someone formulating a plan for

> displacing all or part of your labor with machinery, then the sad fact is that you make too little money to make it worthwhile.[251]

British sociologist George Spencer, writing in the journal *Futures*, extends this observation into the world of global economic patterns, with sobering effect:

> This provides a perspective on the high-level policy initiatives to combat structural unemployment, and it opens the question of how radical such initiatives must become, if they are to have any real effect. Put crudely, we can try to make our labor as cheap as its competitors' overseas, but there is no hope of making it as efficient as many processes permeated by [computer-assisted technologies] already are, and as many others promise to become in a range of areas that defies quantification. And, in the context of the changing global division of labor, our chances of winning against the emerging combination of cheap labor *with* [computer-assisted technologies] are not good.[252]

Kennedy, too, finds cold comfort in the observation that in past technological revolutions, displaced workers have always been reabsorbed into the economy:

> Though admittedly it has always been difficult to imagine new types of labor that would rescue society from the technological displacement of the age, does anyone really have any reassuring concept of what meaningful role in production hundreds of millions of laborers can serve if the very pace and capacity of their thought is obsolete?[253]

Perhaps we should take some solace in the fact that, in theory at least, there is no need for economic output to fall behind the point where it is adequate to sustain existing populations at acceptable standards of living. In other words, the increasingly efficient economic processes we are devising should be able to produce enough of everything to go around, despite the fact that they will employ fewer people in creating all of this wealth. The history of agriculture over the past hundred years provides persuasive evidence of this. Despite what common sense would have foreseen as an utterly catastrophic decline in the numbers of people employed in farming since the early years of this century, there is still plenty of food. New agricultural, transportation, biological and chemical technologies have made the difference. And in manufacturing, tracking the *Fortune* 500 list of America's biggest companies shows that their productivity increased steadily by 2 percent a year between 1954, when figures were first available, and 1993, even though for the second half of that period, the number of people employed by those companies was steadily declining.

The problem we now appear to face is how to put money into the hands of enough people to purchase all those goods and services. Where do the unemployed and underemployed, in their burgeoning numbers, get the cash? Until now, having a job was the way for most people to collect their share of society's wealth. What if computers make it impossible for enough jobs to be created? What will the future hold for such a society?

It is impossible to conduct this kind of an inquiry without at least cursory reference to the greatest and most profound of all theorists of market capitalism, Karl Marx. Not Marx the social theorist, but Marx the economic theorist, to whom all succeeding generations of economic thinkers owe indisputable debts. We will find that, with a little judicious stretching, we can extend his thought into the future of information-age society with results that are, if not earthshaking, then at least tantalizing.

Marx predicted that capitalism would collapse in a series of downward-spiraling cycles of economic disruption caused by the fact that the displacement of labor by labor-saving machinery would continually cut profits to unsustainable levels. How can cutting costs reduce profits? Because in a competitive market, the company that cuts costs will make commensurate cuts in prices in order to gain market share. This will force the competition to do the same, with prices being forced lower and lower with each new wave of cost-saving technology.[254] As more and more labor is displaced, however, and unemployment rises, there are fewer and fewer customers for the products being produced. Demand falls behind supply, forcing yet further price reductions. And so on, until businesses become unsustainable and implode in bankruptcy.

After each cyclical collapse, Marx said, a few companies would benefit from cheap machinery thrown onto the market and from the demise (or takeover) of former competitors, but the cycle would inevitably repeat. Corporations would get bigger and bigger, and fewer in number, and all the while there would be more and more displacement of workers and unemployment. Eventually, and inevitably, capitalism would collapse, with a little nudging from a well-organized and politicized industrial proletariat, which is by now thoroughly fed up with a system which abuses its members so badly.

Marx's logic is, in its own terms, hard to refute, but historical evidence has seemed to indicate that, logic or no logic, he was wrong. Capitalist ingenuity opened more new market frontiers than he could have predicted. Nor was government quite so solidly in the grip of the entrepreneurial class as he had supposed, nor so ill-equipped to root out abuses of the system. And finally, new management practices enabled huge gains in productivity, which permitted enormous increases in the real wages paid to workers. The logic of the Marxist endgame remains, but capitalism keeps moving the goalposts, finding new sources of expansion, new fields for job creation, new efficiencies, new ways for governments to smooth some of the rough edges with creative social policy. And Marx made a fundamental psychological miscalculation: the underclasses, it turns out, do not want to stamp out the leisure class; they aspire to join it.

Still, a lot of what he had to say remains remarkably relevant. There is no need to document the cyclical boom-and-crash nature of the capitalist economies, though that was never enough in itself to give credence to Marx's apocalyptic vision. Modern market economists can point to some authentic success in keeping business cycles from spiraling entirely out of control as Marx had predicted they would.

On the other hand, hardly a day goes by without news of the kinds of corporate mergers and acquisitions that have led to industry-dominating monster corporations like Time Warner, British Telecom/MCI, News Corp., BCE, Microsoft, American Airlines, Intel, Shell, Sony and so on.

This may seem, from the vantage point of the 1990s, an easy call for Marx to have made. But, as the American economist Robert Helibroner points out, "When *Das Kapital* appeared, bigness was the exception rather than the rule and small enterprise ruled the roost. To claim that huge firms would come to dominate the business scene was as startling a prediction in 1867 as would be a statement today that fifty years hence America will be a land in which small-scale proprietorships will have displaced giant corporations."[255]

What about Marx's prediction that business profits would continuously decline over time? As recently as 1958, John Kenneth Galbraith was able to assert, in *The Affluent Society:* "In this century *profits have shown no tendency to fall,* and capital accumulation has continued apace. As a result, the declining rate of profit cannot be taken seriously as a cause of depression [emphasis added]." But more recent evidence collected by economist Spyros Makridakis on the economies of Great Britain and the United States tells a different story. It shows that, for *Fortune* 500 companies, real profits, profits as a percentage of sales and profits as a percentage of GNP, have all declined more or less steadily since about 1978 (in the case of profits as a percent of sales, since 1956, two years before Galbraith's book was published). At the same time, the number of people employed by these companies also declined, and even more dramatically. Makridakis concludes:

> Thus, it seems that in a short period of time, economies of scale and scope seemed to provide no advantages to large firms, reversing a trend that had established itself since the beginning of the industrial revolution. . . . The largest U.S. industrial firms not only saw their sales and profits decline, but they also downsized, firing rather than hiring employees, while becoming less important in terms of the sales, profits, or employment opportunities in comparison to other firms in the U.S. economy.[256]

This brings us full circle to our original point, that productivity is a suspect concept, that it cannot automatically be assumed that more of it is better for society in general. It often means that those workers who are able to hang on to their jobs in the face of automation are simply "sweated" that much more by their employers because they remain the capitalist's only source of profit.[257] And for society at large it means having to deal with the effects of increasing, chronic unemployment. In purely economic terms, there is the problem of how to put enough purchasing power in the hands of the population, a diminishing percentage of which remains employed, to buy up the goods being produced by increasingly productive industries. It is often observed that a robot can build a car, but it can't buy one.

A dyed-in-the-wool capitalist like American management guru Peter F. Drucker would of course not accept Marx's labor-value theory in its pure form, but Drucker nonetheless sees a crisis looming. Drucker has confirmed that the productivity revolution that he credits for having so long kept the dogs of Marxist social upheaval at bay—by ensuring ordinary workers a comfortable living, and by nearly halving working hours between 1914 and today—is over.

> Now, these gains are unraveling, but not because productivity in making and moving things has fallen. Contrary to popular belief, productivity in these activities is still going up at much the same rate. . . . The productivity revolution is over because there are too few people employed in making and moving things for their productivity to be decisive.[258]

In other words, the productivity revolution has been a victim of its own remarkable success in reducing the numbers of workers employed in manufacturing and transportation. And as Marx would argue, those workers have now been "sweated" to the maximum limits of their tolerance. Drucker sees an urgent need, therefore, to boost productivity in the service industries, where he believes there is still slack to be taken up. And what will happen when the limits of productivity are reached in that sector as well? He has not yet addressed that question.

The immediate concern in the service industry is not so much numbers of jobs—job creation has been encouraging—but the character of those jobs. Will they be "good" jobs, meaning, will they pay a living wage? David Snyder makes several salient observations in *The Futurist*:

- The U.S. average wage has fallen more than 20 percent in the past twenty-two years. Nearly a fifth of the 85 million full-time, year-round workers in today's America earn wages below the poverty line. Among the young, between eighteen and twenty-four years of age, 47 percent have jobs that pay below the poverty line.
- An increasing number of young Americans are unable to leave home anymore when they want to and are ready to. In 1980 only eight percent of twenty-five to thirty-two-year-olds lived with their parents; today the figure is about 25 percent. Among those young people who do manage to leave home, more than half return within thirty months, often with a young family.
- Low wages have led to an explosion in the two-income family. In the U.S. thirty years ago, 25 percent of families were two-income. Today, three-quarters of married couples between twenty-five and fifty-five both work full time. As Snyder points out, this has sustained household incomes, but at a terrible cost to children, in a society with hopelessly inadequate nursery and day-care facilities.[259]

Endemic unemployment, penuriously low wages, family breakdown, downward-trending profits, rampant corporate gigantism—the collapse of communism in Europe notwithstanding, we cannot, it seems, be entirely confident in denying Marx the last laugh.

2: Information as a Commodity

On the other hand, Information Age technology shows promise, in principle at least, of being capable of reshaping capitalist market economies in ways Marx could not have envisioned. If, as many predict, the economy of the future will be based to a significant degree on a new and inexhaustible resource called information, then the basic premises of Marxist economic analysis are incomplete, at best. Information, of course, existed in Marx's world. But there is an important difference today. Information today may be treated as a discrete commodity, one which has the very desirable quality from a manufacturing point of view of perfect uniformity in its essential characteristics. It is all bits and bytes (or can be). In Marx's day, there were no means of refining or distilling information out of the content in which it was captured into a raw material appropriate to the manufacturing process, like pig iron or wheat flour or raw lumber, ready for incorporation into new products, the sale of which generate new wealth.

The science-based technologies we have been discussing in this book have transformed information over the past century and a half in exactly this way. First, the telegraph broke the age-old bonds that tied communication (and the information bound up in it) to transportation. No longer did information have to be physically transported in the form of documents from one location to another by train or boat or on horseback. Second, information theory, as developed following World War II, further rationalized the field by breaking the bonds that tie information to communication. Norbert Weiner, Claude Shannon and the other seminal information scientists identified the molecules of the universe of communication—bits of information.[260] Information was defined as "the elimination of uncertainty." Having identified the basic constituents, information theory then taught us how to manipulate them, in the same way as the chemist or physicist learns to manipulate molecules or the geneticist to tinker with DNA.

The formal economic relationships characterizing information's role in the economy closely resemble the information theorist's definition of information, and can be described this way:

> [U]nder conditions of uncertainty, information which affects individuals' actions (or expectations) will have a scarcity value and hence become a commodity. For example, an entrepreneur with imperfect knowledge of production techniques has an incentive to acquire information about more profitable means of production.[261]

The tool that employs information as a raw material and assists in its transformation into knowledge and other useful products is the digital computer: it is the steam engine of the Information Age. In the Industrial Revolution, the steam engine was developed to convert raw chemical energy into useful work. In the current postindustrial revolution, the computer takes raw information and manipulates it into useful aggregates, combinations and relationships. But it does more than this. It is as if the steam engine had been capable of taking over the entire production process, in some cases managing it to ensure maximum efficiency, and in others actually producing the final product internally. We've already seen how computers can be valuable in management roles of various kinds. But they can also manufacture the commodity called knowledge, by providing synthetic human experience and artificial intelligence to combine with raw information inputs. Computers can be observed at work in this way in most modern telephone systems using voice-recognition technology to deal with routine customer inquiries. Internet search engines which return suggested information sites based on users' inquiries in ordinary English are another example. Automated banking, either via the Internet or at an ATM, is a third. So-called expert systems which fulfill the criterion of producing a finished product abound. (These are computer programs which incorporate human knowledge and heuristic or "rule-of-thumb" logarithms that enable them to answer complex questions about, for instance, medical diagnosis or geological profiling.)

The question of determining economic value in an information-based economy can be elusive. In traditional economics, value is a function of scarcity and utility. How much we pay for a good or service depends on how much we want or need it on the one hand, and the abundance of supply on the other. The scarcity side of this equation is moot in information economies, leaving traditional economists scrambling for new paradigms. In the information economy, so-called knowledge workers create value out of information by taking the (essentially free) raw material and fashioning it into copyrighted text or images or software, or proprietary studies and reports, and patented products. The controlled access to these products gives them an element of scarcity which makes them valuable. Raw information, being abundant, has very little value on its own: its relationship to knowledge-based end products is analogous to that of beach sand to integrated circuits of silicon. It becomes valuable when it is placed in a specific location and time in relation to real needs. Much of the information-based product bought and sold over computer networks today has to do with commodity prices or financial transactions or news of other kinds. Material of this sort has a short half-life: its value decreases quickly as a function of time. Most of it, as well, is public information, available free at its point of origin; its value as a commodity is therefore wrapped up in speedy and convenient distribution, as well as the context in which it is placed. (The value of the information in a book or a video-cassette movie is also linked to their location and how they may be moved to another location where they are wanted: no information has value if it is not accessible.)

This idea of information's value being closely related to its being in the right place at the right time provides an important clue to an understanding of the economic value of the Internet. The Net is a means of moving undifferentiated bits of information from one location to another very quickly. Those bits become valuable if they arrive at the right place at the right time. Whether or not they do depends in a practical sense on the software used on the sending and receiving computers, the bandwidth of the intervening network lines and, frequently (if the address is not known), on the efficiency of a search engine resident on a third computer elsewhere on the Net. There is opportunity to create value at each of these points along the line in the process of turning information into knowledge. Hence, Microsoft and Netscape fought their epic battle for control of the Internet browser market in the knowledge that the victor stood to reap billions of dollars in rewards from purchases of software for both client and server computers. Each knew that, soon, the idea of a Net browser will be obsolete as the market moves on to full-blown intelligent agents, software surrogates that will learn our tastes, requirements and preferences and do much of our information-handling for us. Both companies want domination of that very lucrative field. Meanwhile, Yahoo!, Lycos, Alta Vista, Excite and other Internet search engines become either hot stock market properties or sought-after corporate acquisitions. Telephone companies and cable television distributors rush to upgrade their facilities to accommodate increased demands for bandwidth and thereby earn even greater profits.

With everything from personal communication to shopping to health and financial services and education and government services hustling on-line in a land-rush frenzy, the Net is already an imposing feature of the economic landscape as an increasingly indispensable source, conveyance and organizer of information of all kinds. Already, it has created a sizable industry in fields related to establishing and maintaining commercial presence on the World Wide Web, such as Web site development, custom programming, graphic design and on-line marketing.

As a vehicle for commerce, the Net radically alters the rules of the game by flattening the playing field. On the Net, the effectiveness of a business is determined by the quality of its Web presence, and there is no practical reason why a small business cannot outshine the even largest corporations. The real costs of Web marketing are low, the audience is vast and nimbleness and creative ingenuity have been shown to succeed where sheer economic clout can strike out. The conditions are thus a closer approximation to the economist's theoretical model of "pure competition" in which the economy is populated by a potentially unlimited number of producers who are free to jump into the market.

In the conventional economy, which is infested by oligopoly—that is, small groups of players controlling most market share—it is common for prices to be artificially manipulated. This can be done in the good old-fashioned way of holding back supply of a scarce good for which there is lots of demand: that forces up the price. Commodity traders are past masters. More often, these days, price manipulation is done in an entirely synthetic way, by identifying—or creating—demand for a product that was previously nonexistent, and cornering the supply. Think of the Cabbage Patch doll or Tickle-Me Elmo or pump sneakers or the Mazda Miata or microwave popcorn, or the fashion and cosmetics industries.

This sort of thing is more difficult to do when it is easy for competition to enter the market and supply the pent-up demand, so to the extent that the Net lowers entry barriers by leveling the playing field and keeping marketing costs low, it assists the market in regulating prices. The impact of this can only increase as consumers and corporations turn more and more to cyberspace as a venue for making purchases.

The supermarket might be used as a metaphor for the conventional marketplace: shelf space is limited, competition is fierce and the marketing muscle that comes with size in the real world ensures success. Tide and Sunlight dominate the detergent aisle. Kellogg and Post rule in cereals. Small companies must be satisfied with the crumbs, the niches not yet filled by the big players. But on the Net, shelf space in the form of Web pages and domains is cheap, unlimited and instantly accessible—and all of it is at eye level!

The Net also tends to greatly accelerate the speed of market processes, eliminating distorting and potentially misleading lags in information feedback. On the Web, a superior product can rake in market share with breathtaking speed. The familiar story of the success of the Netscape Navigator Web browser, which went from 0 to 80 percent of the market in less than a month, is a case in point. Conversely, the rapid spread through the Internet of damaging information about a flaw in certain Intel chips in 1995 shredded the company's attempts at damage control, which were based on successful experience in the conventional marketplace. Intel's effort to stonewall collapsed within days, and the chips were withdrawn or replaced, with apologies.

This points to another area in which the Net can be expected to assist the smooth functioning of the market. Economic theory, in propounding the self-regulating market system, assumes that consumers are always well informed, that they have the knowledge they need to make intelligent choices among competing products. That is what, in theory, ensures that the best product will win out, and force the competition to improve or die. Everybody benefits. Of course, this is not the case in the real world, where few of us have the time to do adequate research prior to making a purchase, and where much of the formidable energy of marketers is directed at gently misleading consumers with exaggerated advertising claims or creating wholly spurious needs.

To the extent that it makes it easier for consumers to find the information they need to make smart purchases, the Net once again can be said to be oiling the gears of commerce to everyone's benefit. A well-publicized example is the fact that it is now possible to purchase a new car on the World Wide Web. In doing so, the purchaser posts his or her requirements at one of several sites involved in new car sales; these requirements are assessed by car dealers all over North America, who offer to supply the required vehicle at a firm price. The purchaser presumably accepts the lowest bid and the car is either shipped or picked up from the successful dealership.

3: The Net as a Marketspace

The "marketspace" created by the Net is a historically unique commercial environment. It exists nowhere and everywhere at the same time, and can be accessed from anywhere in the world. It provides a rich, multimedia communication environment exceptionally well suited to commercial transactions of all kinds, from initial contact, through sales, to service and customer-retention. The closeness with customers which it affords by making communication cheap and easy allows for marketing of an entirely new kind, in which the customer gets involved in the product cycle as early as in the design stage. The Net is radically superior to any other medium for serving niche markets around the world. And for all of its advantages, it is remarkably inexpensive to set up shop there.[262] Virtually anyone with the skill to produce something of value to others now has the opportunity to market it worldwide. While it is now possible to purchase virtually anything on the World Wide Web, some of the most encouraging success stories involve niche products and services for which the market is substantial, but geographically dispersed. Specialty food products, art and crafts, designer and special-purpose clothing, rare cars and antiques, specialized books and instruction, unique resorts and accommodation and specialized consulting services all find a much-enlarged and constantly expanding market on the Net.

Marx believed that workers put out of their jobs by technical innovation would remain unemployed and in desperate straits, since they owned no land and the capitalist class enjoyed a monopoly on the means of production. The capitalists owned the factories, and that is where the jobs were. But Information Age technologies, to a significant degree, have put the means of production back in the hands of ordinary people. They may not be able to find jobs, but they are increasingly able to find productive employment. Jobs, in fact, are less and less the issue.

William Bridges, in a 1994 cover article for *Fortune* called "The End of the Job," wrote: "What is disappearing is not just a certain number of jobs—or jobs in certain industries or jobs in some of the country or even jobs in America as a whole. What is disappearing is the very thing itself: the job." The job, as Bridges points out, is a recent invention, an artifact of the nineteenth-century Industrial Revolution, made necessary to package the work that needed to be done in the factories and the bureaucracies of industrializing nations.

> Before people had jobs, they worked just as hard but on shifting clusters of tasks, in a variety of locations, on a schedule set by the sun and the weather and the needs of the day. The modern job was a startling new idea—and to many an unpleasant and perhaps socially dangerous one. Critics claimed it was an unnatural and even inhuman way to work. They believed most people couldn't live with its demands. It is ironic that what started as a controversial concept ended up becoming the ultimate in orthodoxy—and that we're hooked on jobs.[263]

In the "post-job organization," people will be hired as members of project teams, in which there is no visible hierarchy. Each will be, in effect, a self-employed consultant focused on the task at hand. They will keep their own schedules and work in different places. Intel and Microsoft are among the many companies already operating this way.

Peter Drucker takes us one step further in the context of his argument that the knowledge and service industries can and must be the subjects of huge gains in productivity in coming years. Drucker argues that these productivity gains will come mainly through contracting out service work to a company that

> has no other business, understands this work, respects it, and offers opportunities for low-skilled workers to advance (for example, to become local or regional managers). The organizations in which this work is being done, the hospitals that own the beds, for instance, or the colleges whose students need to be fed, neither understand nor respect it enough to devote the time and hard work that are required to make it more productive.[264]

The opportunities for small businesses will proliferate.

So-called knowledge work is amenable to the same productivity-enhancing approaches, and it is not, like much service work, dependent on locale. In other words, many if not most knowledge workers can telecommute, because they deal in information which can be digitized and transmitted over the Net.

Thus, self-employment in the area of both service and knowledge work appears to be an emerging trend of real significance, at the same time as information technologies have evolved to a stage where they can support such a trend by providing access to information and knowledge, facilitate telecommuting and providing an entrée to worldwide commerce for freelance makers and doers.

Marshall McLuhan published his greatest works long before the Internet existed in anything like its present form: he died the year Apple went public, in 1980. Nevertheless, he foresaw that modern communications technologies would change the economic structures of nations. Computer-based communications technologies, he predicted, would lead to a revival of the cottage industry of preindustrial Europe.

The idea of cottage industries and piecework has traditionally been an anathema to the organized labor movement, for both self-serving and altruistic reasons. In the first case, cottagers and pieceworkers are difficult to organize, if only because they are dispersed geographically. Much better for labor militancy if the workers are employed in close proximity to one another, so that peer pressure can be used more effectively. On the other hand, because they are usually unorganized, cottagers and pieceworkers have historically tended to be among the most ruthlessly exploited workers in the capitalist economies.

But as we've seen, past experience does not provide a good template for what to expect with the Information Age economy. The Net has changed the ground rules by making it feasible to create and market knowledge-based products and services with minimal capital investment. Pieceworkers and cottage workers in the past have been exploited because they had no access to the means of production or to the means of marketing their produce on an economic scale. They were merely geographically dispersed components of a larger production machine over which they had no control and on which they nevertheless depended for their livelihood. In the era of the Net, and in an economy in which information is the principal raw material, and in which marketing can be effectively done at low cost, that has all changed.

The traditional trade-off between the job and the freelance life has been security versus freedom, and depending on which commodity you valued most, you chose one over the other. That, at least, was the freelancer's point of view. The job-seeker argued that freedom arises out of economic security. Most people chose jobs and the security they provided, and many believed that in doing so they had achieved a kind of freedom. But jobs do not provide security anymore—only, at best, the illusion. And if there was ever a sense in which they provided freedom, it exists no longer.

George Gilder is an outspoken, optimistic American technophile whose prognostications are often on target. He calls the Internet "the most egalitarian force in the history of the world economy," because, he says:

> The Internet creates jobs by making workers more productive, and thus more employable, regardless of where they live. By engendering more investable wealth, it endows new work, providing the key remedy for the job displacement entailed by all human progress. By aggregating distant markets, the Internet enables more specialization, and more productivity and excellence. It will help all people, but most particularly the poor, who always comprise the largest untapped market for enterprise. And the Internet will continue to grow, transforming the global economy with its power and building a new industry even larger than the PCs.[265]

Similarly euphoric is Canadian Don Tapscott, author of *Paradigm Shift* and *The Digital Economy:*

> The Age of Networked Intelligence is an age of promise. It is not simply about the networking of technology, but about the networking of humans through technology. It is not an age of smart machines but of humans who through networks can combine their intelligence, knowledge and creativity for breakthroughs in the creation of wealth and social development. . . . It is an age of vast new promise and unimaginable opportunity.[266]

The Stanford University Center for Economic Policy in California has done intriguing research into the economic impact of the electric motor as a kind of template for what we might expect with the microprocessor. In examining the transition from steam to electric motive power in the United States, the study concludes that it takes fifty to seventy-five years to thoroughly exploit the productive potential of a new technology. During the first half of the transition, economic performance and prosperity collapse before they go up. The really substantial benefits from new fundamental technologies do not show up until two-thirds of the way through the transition.

Projecting this onto the integration of information technologies, which has been underway for some time now, the Stanford study suggests,

> Since the early 1970s U.S. productivity has improved very slowly and real wages have been falling as the nation's employers have struggled to get undertrained workers to generate improved performance with immature technologies. But by 2010 to 2015, the U.S. will become a mature information-intensive economy and surpass the levels of general prosperity and upward mobility experienced during the 1950s and 1960s.[267]

Perhaps Gilder and Tapscott and the Stanford researchers are right. One can only hope so. But for the average working stiff to buy into their rosy vision of a future of economic liberation, some long-standing paradigms and prejudices are going to have to be changed. We've already seen how the phenomenon of "the job" is vanishing, and with it, traditional expectations of economic security. In future, more and more people are going to have to invent their own work, create their own means of support. The Net will provide a powerful tool for doing that, but more of us will nevertheless need to be convinced that, in the words of William James, "Lives based on having are less free than lives based either on doing or being."

Without the economic security provided by lifelong employers and benefits packages and union membership, people will have little choice but to live more frugally, to choose their pleasures more carefully, to reexamine their material needs. (This process should be made easier as the volume is progressively turned down on the broadcast technologies and their advertising messages.) The economic equations which gave us the post-World War II consumer society and unprecedented material standards of life are going to change—which cannot be an entirely bad thing, since the same equations gave us the breakdown of the family, a worldwide environmental crises, vanishing natural resources and urban behavioral sinks.

Economist Fritz Schumacher, in commenting on the late twentieth-century economic landscape, observed: "The cultivation and expansion of needs is the antithesis of wisdom. It is also the antithesis of freedom and peace. Every increase of needs tends to increase one's dependence on outside forces over which one cannot have control, and therefore increases existential fear."

The Age of Information, in taking away traditional material securities with one hand, seems to be offering a new, existential freedom with the other.[268]

CHAPTER THIRTY-ONE

The Promise of the Age of Information

When the first manned, heavier-than-air flights took place over Kitty Hawk in 1903, the popular media greeted the birth of aviation as the dawn of an era "in which planes would conquer distance, abolish national boundaries, and make all men brothers."[269] They spoke of the airplane as guaranteeing equality, democracy and perpetual peace—as a kind of universal or omnipotent technological fix. When household electricity was introduced in the 1890s, similar, even more extravagant hopes had been expressed for the civilizing of the world and the triumph of universal democracy. The telegraph, and in particular the transatlantic cable, were hailed as inventions that would put an end to war by facilitating communication. Radio, as we've seen, was greeted with the same enthusiasms, the same hopes expressed for universal peace and the triumph of democracy.

New technologies, particularly those which dramatically improve transportation and communication, carry with them popular yearnings for universal causes such as peace, justice and freedom. These deep-seated desires are projected onto the technologies in most cases without much thought being given to the appropriateness of the expectations; indeed, in most cases the most fervent expression of hope comes from those who least understand the technology. It has been called "the fallacy of the technological fix," and it is probably unavoidable given humanity's genetic bent toward optimism. It is a sometimes dangerous fallacy, nevertheless, because it can be used as a substitute for politics, for the messy and often tedious business of dealing with the real social dimensions of a problem. The well-intentioned but horribly destructive urban renewal projects of the 1950s, in which whole neighborhoods were razed and replaced with sterile, psychosis-inducing tract housing, is an obvious example. Rural and urban electrification, "living better electrically," as a silver bullet for the decline of the North American family under pressure from urbanization, industrialization and the growth of the capitalist corporation, is another. The excesses of mid-twentieth century agriculture's "Green Revolution" in accelerating desertification, pollution and soil degradation are a third.

The technological determinist belief we hear most often in connection with the rise of the Internet is one we have already alluded to: the idea that the Gutenberg press revolutionized European literacy and ultimately European society. In the same way, the argument asserts, the Internet will revolutionize global communication through a new media literacy based on hypertext. The notion that "the Internet is the most revolutionary communications advance since the printing press" has indeed reached cliché status. The analogy between the printing press and the Internet is an unavoidable one, but it is wise to use caution in directly associating revolution, in the sense of rapid and fundamental change, with either. A closer look at the introduction of the printing press will show why. Scott D. N. Cook points out, in an essay called "The Structure of Technological Revolutions and the Gutenberg Myth,"[270] that for the newly invented printing press to have sparked a revolution in the sense of a fast-moving fundamental change in society, two further conditions would have had to be in place: an abundance of paper or other material on which to print, and a literate population. Neither of these conditions applied in 1500. Printing was done either on hide parchment or rag paper. Parchment sufficient to print a large Bible required the skins of between two and three hundred sheep or calves; rag paper was scarce and expensive right through to the mid-nineteenth century, when, after hundreds of years of fruitless experimentation, a substitute was finally found in wood pulp paper.[271] A single, stunning fact will give some idea of the pressing nature of the scarcity of rags for paper: in the nineteenth century, Egyptian mummies were imported to the United States by the shipload for the sole purpose of recycling the linen in which they were bound to make paper. (The desiccated bodies inside were burned as boiler fuel!)

As for literacy, it is estimated that 90 percent of the European population could not read or write at the time of the introduction of Gutenberg's press. Had more people been able to read, they would still have had trouble getting their hands on books, which were to remain so expensive as to be limited to a market among the aristocracy, clergy and academics for at least another 150 years. The printing of books was for generations pursued as a more efficient means of producing traditional manuscripts for traditional markets than as an entirely new mass-production technology. Pages were printed on presses, of course, but then they were hand-illuminated and hand-bound between handmade embossed leather covers, all of this work being done by the same highly skilled craftsmen who had produced earlier manuscripts. A single volume might be in production for months.

Widespread literacy, when it did arrive, may be more logically attributed to the changes in moral and political values encouraged by the egalitarian teachings of Hobbes, Locke, Rousseau, Jefferson, Kant and other Enlightenment thinkers than to the printing press. Illiteracy in Europe only declined dramatically with the introduction of state-funded elementary schooling in the mid-nineteenth century. In 1800, half the population was still illiterate; by 1900, this had fallen to less than 10 percent. Cook also points out that high levels of literacy were found in some societies centuries *before* the advent of printing with movable type: among both Jews throughout Europe and Moors in Spain, literacy was a prerequisite of formal passage into manhood, and many women knew at least the rudiments of literacy and numeracy due to their involvement in commerce.

If the history of the revolution of movable-type printing teaches anything, it is that potent new technologies tend, initially at least, to reinforce existing social, economic and political power bases. In doing so, they can upset the occasional applecart among institutions of governance and control if the impact is unevenly distributed, as was the case with printing as it relates to religion: putting the vernacular Bible in the hands of thousands and then tens of thousands of lay people proved disastrous for the Roman Catholic hegemony in Europe. But the rise in literacy among the ordinary citizens of Europe came too slowly to meet anybody's definition of a revolution, and when it did come, it came through *political will* as much as through new technology. The lesson, then, is that technologies do not provide solutions, they offer opportunities.

The story of the printing press also teaches that new technologies have maximum impact only when they are compatible with existing practices. Cook notes that the Chinese knew about movable block type—indeed, that is likely where Europeans got the idea—but declined to use it. Instead, they adopted plate blocks containing entire texts because they felt the single-character blocks were too crude to adequately embody their high aesthetic standards. "This suggests," Cook says, "that the initiation of a technological revolution may often depend on *new* technologies being functionally compatible with the *old* craft practices and *traditional* values of the surrounding culture."[272]

The distressing fact is that it is simply impossible to predict in any detail the eventual impact of a given technology. The point is worth reiterating. In the words of Langdon Winner:

> The uncertainty and uncontrollability of the outcomes of action stand as a major problem for all technological planning. If one does not know the full range of results that can spring from an innovation, then the idea of technical rationality—the accommodation of means to ends—becomes entirely problematic. The means are much more productive than our limited intentions for them require. They accomplish results that were neither anticipated nor chosen and accomplish them just as surely as if they had been deliberate goals.[273]

As Winner and others have pointed out, we are effectively committed in our dealings with technology to drifting with the current of unintended and unanticipated consequences which we call "progress." This is not to say that we can have no impact whatever on how technology affects society, far from it. The fact that its effects are so multifarious and so widely dispersed means that there are a great many opportunities for intervention at the level of local impact, at the interface of change, as it were, to optimize benefits and control damage.

But to attempt to manage technological change systematically at the macro level would be profitless.[274] Economics provides an analogy in that history has demonstrated the futility of trying to run a "demand economy" on the Soviet model, even with advanced information processing machinery. The complexity of the enterprise overwhelms. Market economies, on the other hand, though they resist macromanagement, respond well to the tweakings of bureaucrats and persistent local interventions of decisions of millions of individual players. With proper political management these small, individual actions can, by consensus, steer the great hulk of the economy in roughly the agreed-upon direction. In the same way, while it is usually pointless and counterproductive to attempt to manage the course of technological evolution in any top-down or authoritarian sense, it is nonetheless possible to achieve desirable results through micromanagement of local impacts, that is, through bottom-up management.

All of this needs to be kept in mind when thinking ahead to possible futures for the so-called information revolution.

Where do we stand, then, vis-à-vis the question of social change inherent in the technologies of the Age of Information? For one thing, computer-mediated technologies are, demonstrably, compatible with existing technologies, craft and aesthetics: they are in fact *extensions* of them. And for another, the first generation of a truly computer-literate population is already among us. We can therefore anticipate change on an authentically revolutionary time scale, and we are already experiencing it. Questions remain to be answered, however, in the areas of the nature and extent of the change.

Computers and computer-mediated communication technologies are also tools which support in very direct and immediate ways fundamentally human traits such as the desire to communicate, the desire for freedom from arbitrary authority, a resistance to uniformity and a preference for diversity, a love of the unexpected and the serendipitous.[275] They do this, or can in principle, while assisting in the maintenance of the social fabric through richness and fluidity of communication, access to information and the provision of tools to make information useful and meaningful. It could be said that, as a metamachine, the computer and its networks are paradoxically putting into the hands of people the instruments they need to break the bonds of the machine age and regain their threatened humanity. Access to information; the virtual corporation; global, multimedia communication; new tools for creativity of all kinds and in all media—all of these developments seem to militate toward redressing the imbalance between the subjective and the objective, the rational and the intuitive, the yin and the yang that has grown progressively more troubling as we have fallen more and more under the spell of our machines and machine-derived management techniques and social structures.

In the end, it is the *networking* of computers that has made the difference, that has differentiated their impact from that of ordinary machines. Or one might say the difference derives from the fact that computers are the first machines that can actually communicate with one another autonomously, making the construction of networks feasible. Communication being the most human of attributes, it might be said that we have finally invented a machine that operates more in sympathy with us, one that we can adapt to our rhythms and preferences, rather than the reverse, as has been the case with most machine technologies.

The machine age that preceded our own, and lingers still, dictated economic processes and values that meshed with machine processes. It meant humans were to be considered in the undifferentiated aggregate, i.e., not as employees or even workers, but as "human resources," the civilian counterpart to "cannon fodder."[276] The army was the ideal organizational model toward which machine society impelled its masses. The factory, as the hub of production and employment, simplified the logistics of production and permitted the regimentation of workers according to the dictates of the manufacturing process. It demanded mobility of labor—workers had to locate themselves near their place of employment—which instigated a breakdown of traditional family and social relationships and values. It demanded collective effort, uniformity, order and obedience.

Frederick Taylor was the patron saint of machine-age management and his ideas on efficiency and productivity dominated the era. His contributions may be summarized in three principles: (a) dissociation of the work process from the skills of the workers—managers alone should be responsible for organizing the processes of labor; (b) separation of conception from execution, so that workers can be confined to performing a series of management-prescribed actions; and (c) the strict maintenance of management's monopoly on knowledge of each step of the production process and their labor requirements.

As Lewis Mumford observed at the height of the machine age in 1930, the most noticeable feature of the era has been its punctuality. Every facet of the daily rhythm is governed by the clock. The household arises at a set time, no matter how tired or apathetic. If one arises late, there is a frantic rush to "make up time." We dine at intervals set by the clock, regardless of appetite. Our entertainment is dictated by the television time-listings. Millions observe the same schedule, so that only perfunctory provision is made for those who have to perform these functions at different times.[277]

And looking back from his own vantage point to the previous century, he painted a picture which is startling in its relevance to our own time of obsessive concern with the quantitative. Mumford wrote:

> The leaders and enterprisers of the period believed that they had avoided the necessity for introducing values, except those which were automatically recorded in profits and prices. They believed the problem of justly distributing goods could be sidetracked by creating an abundance of them; that the problem of applying one's energies wisely could be cancelled out simply by multiplying them; in short that most of the difficulties that had hitherto vexed mankind had a mathematical or mechanical—that is a quantitative—solution. The belief that values could be dispensed with constituted the new system of values.[278]

The social structures dictated by the machine, in encouraging collective exertion, greatly amplified the impact of human effort, but only at the sacrifice of much that is essentially human. And we have carried on in the misguided belief that there is a useful or workable "value-free" approach to solving human problems, available to us through the use of technology.

The Information Age, due to the distinctive nature of its economic demands and consequences, has much less need for formal structure than the mechanical age with its factory paradigm and characteristic corporate hierarchies. Digital communication networks make it possible, even economically desirable, to disperse workers who no longer need access to machinery that in earlier times could be provided economically only at a central factory location. Most of the tools needed by the information worker are available to him or her via digital networks connecting computers. And "information worker" can be very broadly defined here: Hollywood film editors can and do work at home thanks to high-speed fiber optic networks and digital editing software, as can just about anyone else whose on-the-job raw materials can be reduced to digital format and whose needs for most personal communication can be met by telephone, e-mail and video conferencing. Sales reps need no offices at company headquarters and companies may need no headquarters. Most bureaucratic processes, where they survive, can be dispersed. In a very broad sense, decision making and responsibility can be widely distributed in novel ways throughout society because of the new possibilities for continuous communication between the providers and users of services and products of all kinds.

In the Information Age, in the era of distributed networks, machine-age values are out of place and counterproductive. Creativity is valued over corporate loyalty in workers, who in turn have little or no tolerance for the regimentation of traditional factory systems organized along military lines. Excellence is not achieved by division of labor and isolation of tasks as on an assembly line, but by small groups, sometimes only two or three, working intensively as a team. Significantly, the tools and the raw materials of production, principally information, computers and access to networks, can be possessed or at least tapped by any individual worker: the capitalists have lost their historic monopoly on the means of production, along with the power that came with it to set wages and determine working conditions. Their challenge is now to be able to identify the young men and women who are likely to come up with the next bright idea and to get them under contract. In the Information Age, it's a seller's market for labor, and labor often arrives equipped with its own tools of production, its own "factory." Capitalism is not dead; it is being democratized.

Kevin Kelly writes in *Out of Control* :

> In the last half-century a uniform mass market—the result of the industrial thrust—has collapsed into a network of small niches—the result of the information tide. An aggregation of fragments is the only kind of whole we now have. . . . That's almost the definition of a distributed network.[279]

Or, as Jay David Bolter has observed: "Just as our culture is moving from the printed book to the computer, it is also in the final stages of the transition from a hierarchical social order to what we might call a 'network culture.'"[280]

Hierarchies from the Church to government to education to business may remain in place in principle, but they have lost much of their moral authority. It evaporated along with the destruction of the determinist certainties of Newtonian science and nineteenth-century technology, which treated values as inherent in immutable and divinely ordained (or at least humanely elegant) natural systems. We learned at the beginning of our own century that those systems are anything *but* immutable, are in fact indeterminate and relativistic. The authority of science has been based on its mathematical quantification, but the math is based on measurement, which assumes a concreteness (a "measurability") which we now know to be an illusion. Much as Plato's man saw reality through shadows projected on a cave wall (the brilliance beyond the mouth being too intense to behold), we perceive it in the movements of our test gauges and scrolling symbols on our computer screens.

And so traditional hierarchies are being superseded by control networks, which seem more appropriate to the world as we now understand it. No one pays the old power centers much attention anymore; without moral authority there is no justification for their continuing existence. But the replacement of hierarchies with networks does not mean the collapse of civilization, though it might seem that way to those at the apex of the pyramid. From the perspective of the great mass of ordinary citizens who populate the pyramid's lower expanses, it means, instead, fewer roadblocks in the way of fulfillment of potential, fewer barriers to open communication with peers and, in general, a greater degree of control over social and personal life in the hands of the individual.

A new worldview suggested by digital computers and their networks as a defining technology is emerging, half-formed but nevertheless recognizable. The idea of time as a river, of clockwork orderliness in the universe, of predictability of cause and effect, of beginnings, middles and endings, of closed systems, of top-down organization, of social harmony based on conformity and uniformity—none of these notions born of earlier technological eras fits a digital world. Nor, for that matter, does the machine-age idea of entropy, of the irresistible, fatal drift of all things toward chaos and disorganization and scarcity. The world suggested by digital networks is one of time as discrete segments or bits that are randomly accessible, of order underlying seemingly random complexity, of self-organizing systems and emergent properties. It is a world that finds beauty, meaning, order and life in chaos. It suggests a civilization that promotes and values diversity, even anarchy. It suggests a politics that organizes from the bottom up, valuing nimbleness above persistence, honor above duty, freedom above security, cooperation above competition, consensus above authority, and approach to communication that is bilateral rather than unilateral, valuing the informal conversation over the formal address.

The notion of national sovereignty is also undergoing change, though with what eventual outcome, no one can tell. Traditionally defined in geographical terms, it has now to incorporate the idea of communications or information sovereignty and integrity. No longer can despots seal their borders from the outside world with machine guns and barbed wire. Data percolates like water in an underground aquifer, indifferent to political frontiers. Power still flows from the barrel of a gun, as Mao said, but increasingly, it also comes from information and the appliances that manipulate it. Information, not guns, overthrew the Soviet Empire, liberated Eastern Europe and reunited Germany. National sovereignty, in its traditional form, is in peril, as vulnerable to information technologies as walled cities were to modern artillery. In Canada and France, both early adopters of information technologies, the emphasis has shifted to the defense of what really mattered all along: cultural sovereignty. Cultural imperialism is the adversary, and it advances under corporate and not national banners.

The machine age gave the West a kind of political and economic "democracy," in the same way as the mass broadcast media gave us a kind of "democratic" popular culture. In each case, what was delivered fell somewhat short of the promise inherent in the idea. The traditional mass media exemplified by radio and television and, to a somewhat lesser extent, the cinema and newspapers, because of their "point-to-multipoint" structure and commercial financing, are democratic only in the sense that they seek to satisfy mass taste. They are, in fact, rigidly hierarchical structures, obdurately resistant to outside influence. Moreover, much of their energies are spent on *shaping* mass taste to conform with commercial goals. The modern market economy, it may be argued, is democratic in essentially the same, limited way as the mass media, in that it serves up to the mass market not so much what it needs, but what advertising convinces it it wants. And representative, parliamentary democracy has resulted in hierarchical, bureaucratized structures which, over time, become increasingly isolated from, even hostile to, the citizenry they are meant to represent.

Our time is one in which content in political campaigns is debased or erased by marketing techniques. It is a time of "rationalization" and "rightsizing," when "weed 'n' feed" is considered an astute approach to managing so-called human resources and outsourcing production to unhealthy and inhumane Third World sweatshops is applauded as an imaginative response to global competition, when statistical audience ratings and psychographic profiles determine the success or failure of most cultural enterprises, when education authorities refer to students and their parents as "customers," when governments design public policy on the basis of opinion samplings based on off-the-cuff responses to glib questions about deep issues. This is the place to which machine-age values of scientific rationalism—really, the absence of values—have led us.

At the same time, and taking a longer historical perspective, we cannot ignore the fact that our recent progress in technology of information and communication is on a continuum which emerges in the Middle Ages with the efficient harnessing of the draft horse and the power of the wind and water to do heavy, mind-numbing work that had previously been assigned to human labor. This was essentially humanitarian technology in that it arose out of abstract beliefs as much as any economic necessity. Historian Lynn White, Jr. has written:

> The labor-saving power-machines of the later Middle Ages were produced by the implicit theological assumption of the infinite worth of even the most degraded human personality, by an instinctive repugnance toward subjecting any man to a monotonous drudgery which seems less than human in that it requires the exercise neither of intelligence nor of choice.[281]

One need not ascribe this point of view to organized religion as White does; the rise of humanism in the Middle Ages, as we've already noted, is sufficient cause. White goes on eloquently to observe:

> The chief glory of the later Middle Ages was not its cathedrals or its epics or its scholasticism: it was the building for the first time in history of a complex civilization which rested not on the backs of sweating slaves or coolies but primarily on non-human power. The study of medieval technology is therefore far more than an aspect of economic history: it reveals a chapter in the conquest of freedom.[282]

It is possible to see in our current technological era the writing of another such chapter. Despite the "Gutenberg myth" and the "fallacy of the technological fix," despite Postman and others who decry the "information glut," it is difficult to resist the belief that the Information Age holds the promise of a truer, more authentic political and economic democracy, not only by providing the tools to make the sought-after ideal a functioning reality, but by stimulating a social structure within which it is not just desirable in the abstract, but a practical necessity. Whereas industrial technology for most of the preceding two centuries has, in the words of philosopher Herbert Marcuse, provided "the great rationalization of the unfreedom of man and demonstrate[d] the 'technical' impossibility of being autonomous, of determining ones own life,"[283] Information Age technology appears to militate in the opposite direction, *toward* autonomy. We need to ensure that this inherent tendency is realized, but we will not succeed so long as we cling to the twentieth-century notion that technology is in one way or another beyond our powers to control, or as it is more frequently framed, that "progress is unstoppable."

By "progress," of course, we mean any and all effects of technology, be they welcome or unwelcome. Both technophobes and technophiles allude to this characteristic of out-of-controlledness in their assessments of current social and political prospects: the former assert that we are bound for hell in a hand basket; the latter insist that the logic of technological change is essentially benign and guarantees a Utopian future.

The viewpoint implicit throughout this book has been that while technology has its own evolutionary logic in that it builds on previous developments and tends to incorporate so-called breakthroughs into existing patterns, it does not operate outside human control. Indeed, how could it? Technology is a product of human endeavor. Controlling it, in the sense of minimizing the damage it causes and optimizing its benefits, is a job for representative politics at all levels, from community to national and international. In fact, given the profound impact of technological change on society, it is far too important a job to be left to any other social agency, let alone left entirely unattended.

It is a symptom of the disaffected nature of our times that many will find the idea of social control of technology preposterous or impractical. Here is an analogy that may help to explain why it is neither: a single action taken in isolation in the world (were it possible for such an event to occur) might look like a single raindrop falling upon a still pond. The concentric circles of its impact would spread outward in a widening and (apparently) diminishing wave of influence on the surrounding waters. Actions in the real world, however, are taken in a context more closely resembling a more or less continuous rainfall, in which many raindrops hit the pond's surface virtually simultaneously. As their concentric waves of impact spread, they intersect one another, with disrupting, one might say chaotic, results.

If we imagine the pond under the influence of a rainfall of technological, as well as social and political, change, with their many widening, concentric waves of influence and many more points of intersection of those waves, we have a picture of how the impact of evolving technology is felt.

It is in the points of convergence or intersection of the waves that both the challenge and the opportunity for social intervention and control occur. As we can see in our mind's eye, there are many, many such points of opportunity, places where technological impact can be usefully micromanaged.

The analogy also indicates why it is normally futile to try to control technology in a top-down way, by predicting its likely outcomes and intervening to somehow influence those macroresults. The reason, simply stated, is that the macro results, being composed of the sum of so many microresults, are essentially unknowable. As Hannah Arendt has said in a different context:

> Action, no matter what its specific content, always establishes relationships and therefore has an inherent tendency to force open all limitations and cut across all boundaries. . . . To do and to suffer [to be affected] are like opposites of the same coin, and the story that an act starts is composed of its consequent deeds and sufferings. These consequences are boundless, because action, though it may proceed from nowhere, so to speak, acts into a medium where every reaction becomes a chain reaction and where every process is the cause of new processes. . . . The meaning of the deed and its consequences can only be clear when the chain reaction which springs from it is complete.[284]

Or as Jean-Paul Sartre put it, somewhat more succinctly:

> The consequences of our acts always end up by escaping us since every concerted enterprise as soon as it is realized, enters into relations with the entire universe, and since this infinite multiplicity of relations goes beyond our intentions.[285]

If the results of the introduction or modification of a given technology cannot be known (and yet we can safely assume, based on past experience, that at least some of those unpredictable results will be positive), then it becomes politically very difficult to argue against the technology or try to exert control over its implementation at a macrolevel. Even if we were successful in our efforts, we would have no way of predicting whether our intervention would be for better or worse! Should we have banned development of the automobile a hundred years ago? Who knows? Should we have prevented the development of atomic weaponry when we had the chance? Possibly, but the issue is by no means clear-cut. It can be argued they helped us avoid World War III. Should we have banned human heart transplants, as many were demanding following the successful transplant operations of Dr. Christian Barnard in 1967?

Micromanagement at the level of local impact is the only practical, perhaps the only feasible, approach to social control over technology. Given the unknowable nature of technology's positive as well as negative consequences, micromanagement of impacts is also probably the only *desirable* approach in any case; history has shown that few would wish to forgo a technology's demonstrable benefits on the basis of speculative fears of drawbacks. Like everything else in politics, management of technology's impact can be a messy and often confusing affair, but it *is* both possible and necessary. The appropriate role for national and international authority in this context is to ensure that obstacles to such micromanagement are removed.

It is important, then, to recognize that it is as true when considering the opportunities offered by technology as it is in dealing with political institutions, that eternal vigilance is the price of liberty.[286] Technological processes are amenable to management, as we've seen, but in the absence of continuous monitoring and intervention at appropriate decision points, they will manage themselves in ways that may or may not be to our advantage. The decision not to manage technology has at least as much impact as the decision to manage, but that impact is less predictable. Never has there been a better example of this principle than the Internet: it developed as an open and democratic institution because it was deliberately designed that way, and it will remain so only so long as each of us respects its qualities and works to preserve them.[287]

We may dare to hope that the Age of Information, finding fruition as it does at the end of the bloodiest century in history, will help to usher in an era of more humane civilization. The opportunity is certainly there, as perhaps never before. But we must seize it.

N O T E S

1 The terms "information technology" and "communications technology" are used more or less interchangeably throughout the book. While I recognize that there are important semantic differences between the two, in practical terms they amount to the same thing, for most purposes. I note as well that in academe, Information Theory and Communication Theory have identical roots. Hence, the Age of Information might just as well be called the Age of Communication. Simple convention has persuaded me to use the former.

2 Neil Postman, *Amusing Ourselves to Death: Public Discourse in the Age of Show Business* (Penguin Books, 1986), 161.

3 Neil Postman, *Technopoly: The Surrender of Culture to Technology* (Vintage Books, 1993), xii.

4 Ibid.,189–90.

5 Ibid., 161.

6 G. J. Mulgan, *Communication and Control: Networks and the New Economies of Communications* (Polity Press, 1991), 251.

7 Gene I. Rochlin, *Scientific Technology and Social Change* (W. H. Freeman and Co., 1974), 149.

8 This reached its height with the fall of Byzantium (Constantinople), the surviving Eastern Roman Empire, to the Turkish army in 1453.

9 Arthur Koestler, *The Trail of the Dinosaur* (Macmillan, 1955), 245.

10 J. N. W. Sullivan, *The Limitations of Science* (New York, 1949), 147.

11 Lewis Mumford, *Technics and Civilization* (Harcourt, Brace and Co., 1930).

12 Ibid., 50–51.

13 Richard Lanham, *The Electronic Word: Democracy, Technology and the Arts* (University of Chicago Press, 1993), 252.

14 Ibid., 254.

15 Greek writing can be shown to have evolved from Semitic alphabets, which in turn grew out of Egyptian prototypes, but by incorporating phonetic indicators and vowel signs into the letters themselves, it made a leap so far beyond its predecessors that it is clearly different in kind rather than merely in degree.

16 Eric Havelock, *Preface to Plato* (Basil Blackwell, 1963), 200.

17 See also Lewis Mumford, *The Condition of Man,* ". . . the person is an emergent from society, in much the same fashion that the human species is an emergent from the animal world"; W. G. Greene, *Moira,* ". . . the whole trend of Greek thought is from an external toward an internal conception of life"; Werner Jaeger, *Paideia,* ". . . other nations made gods, kings, spirits: the Greeks alone made men."

18 Marshall McLuhan, *Understanding Media: The Extensions of Man* (Signet, 1954), 85.

19 Quoted in Elizabeth Eisenstein, *The Printing Press as an Agent of Change: Communications and Cultural Transformations in Early-Modern Europe,* vol. 1 (Cambridge University Press, 1979), 66.

20 Jane Austen, *Emma* (Oxford University Press, 1971), 266–67.

21 Paul Johnson, *The Birth of the Modern* (Weidenfeld and Nicolson,1991), 167.

22 Geoffrey Wilson, *The Old Telegraphs* (Philmore and Co., 1976).

23 "Like spirits in the guise of mechanism" is strongly evocative of a later phrase often adduced in writings about communications technologies and the computer in particular: "The ghost in the machine." The latter was coined by British philosopher Gilbert Ryle, who used it in a discussion of claims for a human soul, the "ghost" in the "machine" of the human body. The strong sense that intelligence is somehow operating within complex technologies is especially prevalent where communications technologies are concerned, and might be attributed to the notion that these inventions are, in more than merely a metaphorical sense, "extensions of man," and are therefore closely identified with human attributes such as consciousness and intelligence.

24 Johnson, *The Birth of the Modern*, 166.

25 Bern Dibner, *Ten Founding Fathers of the Electrical Science* (Burndy Library, 1981), 37.

26 *The Book of Knowledge*, vol. 7 (The Grolier Society, 1911), 2117.

27 New York *Herald,* May 30, 1844.

28 It doesn't take a nuclear physicist to recognize that the properties of time and space are so intimate to an understanding of the universe that tinkering with them will require wholesale changes elsewhere in the theoretical construct. Thus, by the time Einstein had finished working through his idea of special relativity, he had come to the breathtaking conclusion that mass and energy were interchangeable, that one could be converted into the other, and that the relationship between them could be summed up in the famous equation: $E = mc^2$.

29 John Gribben, *Schrödinger's Kittens* (Weidenfeld and Nicolson, 1995), 18.

30 Werner Heisenberg, *Physics and Philosophy* (Harper and Row, 1958), 41.

31 Niels Bohr, *Atomic Theory and Human Knowledge* (John Wiley, 1958), 62.

32 J. A. Wheeler, K. S. Thorne and C. Misner, *Gravitation* (Freeman, 1973), 1273.

33 Quoted in M. Capek, *The Philosophical Impact of Contemporary Physics* (Van Nostrand, 1961).

34 W. Thiring, "Urbausteine der Materie," *Almanach der Osterreichischen Akademie der Wissenchaften*, vol. 118 (1968), 160.

35 David Bohm and B. Hiley, *On the Intuitive Understanding of Non-locality as Implied by Quantum Theory* (Birkbeck College, University of London).

36 It will not have escaped the notice of some readers that quantum field theory bears a striking resemblance to the idea of *ch'i*, or life force, in Chinese traditions of Confucianism, Taoism and Buddhism. Chang Tsai said: "When the ch'i condenses, its visibility becomes apparent so that there are then the shapes [of individual things]. When it disperses, its visibility is no longer apparent and there are no shapes. At the time of its condensation, can one say otherwise than that this is but temporary? But at the time of its dispersing, can one hastily say that it is then non-present?" Those wishing to delve further into this area of convergence are directed to Fritjof Capra's *The Tao of Physics* and Gary Zukav's *The Dancing Wu Li Masters*.

37 It is in this connection that the Internet, as a superior medium of communication among scientists and academics, has been called the greatest scientific tool of this century.

38 Letter from Samuel Morse to U.S. Treasury Secretary George M. Bibb, December. 12, 1844. Morse was trying to convince the U.S. government to purchase rights to his invention.

39 Ibid.

40 Letter from Samuel Morse to U.S. Treasury Secretary McLintock Young, June 3, 1844.

41 The San Francisco to New York rate was soon moderated to six dollars for the first ten words (which included place of origin, time and date), seventy-five cents per additional word.

42 An equivalent logic drives the current expansion of business on the World Wide Web: What is possible becomes necessary.

43 Daniel Czitrom, *Media and the American Mind* (University of North Carolina Press, 1982), 8.

44 Charles Briggs and Augustus Maverick, *The Story of the Telegraph and a History of the Great Atlantic Cable* (Rudd and Carleton, 1858), 13.

45 Annie Ellsworth, daughter of the U.S. Commissioner of Patents, and a romantic interest, was first to tell Morse that Congress had passed the bill supporting his test line from Washington to Baltimore.

46 Numbers 23:23

47 Taliaferro P. Shaffner, "The Ancient and Modern Telegraph," *Shaffner's Telegraph Companion* (February 1854), 85.

48 Daniel Czitrom, *Media and the American Mind,* 11.

49 H. L. Wayland, "Results of the Increased Facility and Celerity of Inter-Communication," *New Englander*, November 1858.

50 W. F. Butler, *The Wild North Land* (London, 1874), quoted in Rosemary Neering's *Continental Dash* (Horsedal and Schubart, 1989).

51 *New York Times* (August 6, 1858).

52 Ibid.

53 Field's trials were not at an end. He lost most of his cable fortune speculating in stock of New York's Elevated Railway Company. His wife died soon after, and a son and daughter both succumbed to insanity. He died in his sleep in 1892.

54 With the advent of live broadcast news as promulgated by CNN and its competitors, currency became the paramount defining criterion.

55 Oliver Gramling, *AP: The Story of News* (Farrar and Rinehart, 1940), 34.

56 Postman, *Amusing Ourselves to Death,* 65.

57 It must be admitted, however, that it is very difficult to know with certainty which pieces of random information will end up being of use to an individual, and which will not. What seems irrelevant today may be relevant in the changed circumstances of tomorrow. Maine and Texas, *pace* Thoreau, certainly did have much to talk about on the eve of the American Civil War.

58 Postman, *Amusing Ourselves to Death,* 69.

59 "The Intellectual Effects of Electricity," *The Spectator*, November 9, 1889.

60 W. J. Stillman, "Journalism and Literature," *Atlantic Monthly*, November 1891.

61 Czitrom, *Media and the American Mind,* 21.

62 J. D. Bernal, *Science in History* (MIT Press, 1971), 35.

63 Sir Desmond Lee, "Science, Philosophy and Technology in the Greco-Roman World," Greece and Rome, vol. 20 (1973), 70–71.

64 Arthur Koestler, *The Act of Creation* (Pan Books, 1971), 267.

65 David F. Noble, *America by Design: Science, Technology and the Rise of Corporate Capitalism* (Alfred A. Knopf, 1977), 34.

66 Ibid., 34.

67 Langmuir received the Noble prize for research into molecular films on solid and liquid surfaces. He was also responsible for significant improvements in both the light bulb and the vacuum tube.

68 Noble, *America by Design*, 118.

69 George Basalla, *The Evolution of Technology* (Cambridge University Press, 1988), 128.

70 Theodore Vail, quoted in N. R. Danielien's *AT&T: The Story of Industrial Conquest* (Vanguard Press, 1979), 98.

71 L. H. Baekeland, "The U.S. Patent System, Its Uses and Abuses," *Industrial and Engineering Chemistry* (December 1909), 204.

72 Noble, *America by Design*, 108–9.

73 Gene Rochlin, *Scientific Technology and Social Change* (W. H. Freeman and Co., 1974), 102.

74 Elisha Gray to A. L. Hayes, November 2, 1876; quoted in David A. Hounshell, "Elisha Gray and the Telephone: On the Disadvantages of Being an Expert," *Technology and Culture* (April 1975), 157.

75 Lloyd W. Taylor, "The Untold Story of the Telephone," *The American Physics Teacher*, vol. 5 (1937).

76 The Western Union directors were not the only potential investors suffering from myopia. George Brown, founder of the *Globe* newspaper in Toronto (forerunner to the *Globe and Mail*), passed up an offer from Bell to purchase patent rights outside the United States. And Sir William Preece, the chief engineer to the British Post Office, reported to a House of Commons committee in 1879 that there was little need for telephones in Britain: "Here we have a superabundance of messengers, errand boys and things of that kind . . . if I want to send a message—I use a sounder or employ a boy to take it."

77 The Edison transmitter also used a small induction coil or transformer to step up the current leaving the carbon button before passing it along to the telephone line.

78 Herbert Casson, *The History of the Telephone* (1910; reprint, Books for Libraries Press, 1971), 66.

79 Quoted in Carolyn Marvin, *When Old Technologies Were New: Thinking About Electric Communication in the Late Nineteenth Century*, (Oxford University Press, 1988), 80.

80 Ibid., 226.

81 Francis Jehl, *Menlo Park Reminiscences*, vol. 1 (Dearborn Michigan, 1937), reprinted in *The Telephone, An Historical Anthology* (Arno Press, 1977).

82 Communications from another planet seem to have been a persistent, if discreet, theme in turn-of-the-century engineering circles. Marconi is reported to have listened for signals from Mars aboard his yacht *Elettra*, no doubt provoked by spurious, low-frequency radio signals that had many of the same qualities as telephone-line noise. H. G. Wells, who had begun his science fiction career in 1895 with *The Time Machine*, followed up in 1895 with *The War of the Worlds*, in which Mars invades Earth. It is now understood that the mysterious radio signals are generated by high-energy solar radiation acting on the Earth's outer atmosphere.

83 Casson, *The History of the Telephone*, 121.

84 Much of the work done in clearing up distortion owes its success to early research done by the British scientist Oliver Heaviside, a brilliant but socially inept mathematician whom Norbert Weiner has described in *Invention* as "sincere, courageous, and incorruptible" which no doubt accounts for the fact that he "was born poor, lived poor and died poor" (MIT Press, 1993). Heaviside also made important discoveries in the propagation of radio waves by atmospheric phenomena.

85 As distinct, of course, from the telephone *company,* which does come in for its share of criticism, and more.

86 Ivan Illich, *Tools for Conviviality* (Harper and Row, 1973), 23.

87 Ibid., 24.

88 Ibid., 23.

89 Ibid., 69.

90 *The Age of Access, Information Technology and Social Revolution,* posthumous papers of Colin Cherry, compiled and edited by William Edmondson (Croom Helm, 1985), 64.

91 McLuhan, *Understanding Media,* 238.

92 Steven Lubar, *Infoculture* (Houghton, Mifflin Company, 1993) 132.

93 Casson, *The History of the Telephone,* 159.

94 Quoted in McLuhan, *Understanding Media,* 240.

95 Ann Moyal, "The Feminine Culture of the Telephone: People, Patterns and Policy" in *Information Technology and Society,* Nick Heap, Ray Thomas, Geoff Einon, Robin Mason and Hughie Mackay, eds. (Sage Publications/Open University, 1995), 303.

96 Sir William Crookes, *London Fortnightly Review* (1892).

97 It is a continuing challenge. Lawyers at Bell Laboratories were initially unwilling even to apply for a patent on the laser, on the grounds that it had no possible relevance to the telephone industry. At that time (1966) the best transatlantic phone cable carried 138 simultaneous conversations. The first fiber optic cable (1988) could carry 40,000. Fiber cable of the 1990s can carry 1.5 million simultaneous telephone conversations—thanks to the laser.

98 December 2, 1932. Marconi was then exploring the VHF and microwave reaches of the radio spectrum.

99 Cable interests were to merge with Marconi's company in 1928.

100 Degna Marconi, *My Father, Marconi,* 2nd edition, rev. (Balmuir Book Publishing Ltd., 1982), 95.

101 It operated at a wavelength of about two thousand meters.

102 In 1919, a third Cape Breton station erected at the site of fortress Louisbourg was linked with Letterfrack, Ireland, for the first transatlantic wireless voice—wireless telephony—connection.

103 Gleason L. Archer, *History of Radio* (American Historical Society, 1938), 64.

104 It was at this conference that the word "radio" was formally adopted to describe communication through space without wires. The word took some time to catch on in North America, and the British continued to use the word "wireless" until only very recently. The conference also adopted SOS as the international distress signal: it had no literal meaning, but was easy to remember and send in Morse (· · · – – – · · ·).

105 To some extent this fallacy of limited spectrum has carried over into the present era's attempts to restrict competition among cable specialty channels by arguing

space limitations on cable systems. Off-the-shelf digital technologies and distribution systems can deliver many more channels than are currently available, but installing this equipment would cost cable operators handsomely in the short run.

106 Hermann Weyl, *Philosophy of Mathematics and Natural Science* (Princeton University Press, 1949).

107 As usual, Marconi was making an inspired adaptation of someone else's discovery: the coherer had been invented by French physicist Edouard Branly in 1890, and improved by Sir Oliver Lodge. On his ninety-fifth birthday, in 1939, irked by wartime propaganda on the radio, Branly said, "It bothers me to think I had something to do with inventing it."

108 George Shiers, in *Scientific Technology and Social Change,* ed. Gene I. Rochlin (W. H. Freeman and Co., 1974), 141.

109 Quoted in Archer, *History of Radio,* 141.

110 Quoted in A. F. Harlow, *Old Wires and New Waves,* Appleton (1928).

111 Ibid.

112 Carneal, *Life of De Forest,* quoted in Archer, *History of Radio,* 100.

113 Owen D. Young, former undersecretary of the navy, a vice-president at General Electric, later to become founding chairman of RCA, quoted in Archer, *History of Radio.* 164.

114 Within less than five years the alternators, in their turn, would be made obsolete by new high-power vacuum tubes for transmitters, but that, of course, was not anticipated at the time.

115 Details of the government's role in the birth of RCA were made public during U.S. government antitrust hearings which reported December 1, 1923.

116 They were, after all, no strangers to the notion of oligopoly. General Electric, formed initially in the merger of Edison's electric light interests with the Thompson-Houston electric company in 1892, had been created as a means of pooling the two companies' patents to place the new corporate entity in an unassailable market position vis-à-vis smaller competitors. And in 1896 General Electric and Westinghouse had in turn called a truce in their war for dominance in the electrical equipment market. With more than three hundred patent suits outstanding between them, they agreed to pool their resources, with GE assigned 62.5 percent of the business flowing from the shared patents. Not only did the deal allow the two companies to get on with the business of making money rather than spending it on lawyers, it served once again to limit competition, giving GE and Westinghouse a shared monopoly that was almost impossible to penetrate. Public interest, of course, took a backseat to corporate comfort in these competition-limiting sweetheart deals, and government, prodded by public protest, was to become increasingly engaged in what was referred to as "trust-busting."

117 Amateur radio would be officially sanctioned in international law for the first time at the international radio conference of 1927, at which time amateurs were given exclusive use of several small slices of the high- and very high-frequency spectrum.

118 Just prior to and following World War II, the high-frequency spectrum began to be populated by international shortwave stations, most of them operated by national governments, and most of them maintained primarily for propaganda purposes. Shortwave outlets of the major powers, Voice of America and Radio Moscow, broadcast in scores of languages, carrying their government's message to the far corners of the world. Often, high-quality educational, musical and dramatic programming was served up as well. Other, less bellicose voices were heard from the countries of Western Europe, Canada, the Antipodes and the Spanish-speaking world, and from the standard-setter, the BBC World Service. Of course, this kind of cross-frontier communication was decidedly unwelcome with autocratic regimes, and a major feature of the shortwave bands, until the end of the Cold War brought blessed relief, was the idiot roar of the high-powered "jamming" station, whose signals blanketed unwanted shortwave broadcasts and anything else in their vicinity.

119 Despite their compact of 1896 to share the market in heavy electrical equipment, there was no love lost between General Electric and Westinghouse. In the 1880s when Edison was introducing his electric lighting systems, George Westinghouse was perfecting a superior system based on European technology. It used alternating current (ac), as opposed to the direct current (dc) employed by Edison. The battle that ensued between the two systems was one of the most vicious in American corporate history, with Edison's side claiming that ac was a danger to public safety. As part of their campaign, the dc proponents went so far as to have a Westinghouse ac system adopted as the official means of public execution in New York State. Edison himself, in a series of gruesome late-night experiments at Menlo Park, had several animals of varying sizes killed with alternating-current shocks to demonstrate the hazard. Word of the "top secret" experiments was carefully leaked to the press. In the end, Westinghouse won the war because it was a superior technology, and the company was given the contract to develop Niagara Falls for hydroelectric power. Edison's GE had to sue for peace.

120 The great radio trust would break down through internal bickering by 1924. Two years of secret negotiations among the players resulted in a series of deals that had historic implications. AT&T withdrew from broadcasting to stick to its knitting—long-distance telephony—selling radio station WEAF to RCA for $1 million. The telephone company retained exclusive patent rights to two-way telephone services by both wire and radio, while GE and Westinghouse received rights to all AT&T and RCA patents related to radio. Once again, the positions of these corporate giants had been consolidated and rationalized into an effective oligopoly, and it had become more difficult than ever for competing interests to enter their markets. Following the agreement, RCA, Westinghouse and General Electric together formed a new broadcasting venture which would network their respective stations and supply them with daily programming: it was to be called the National Broadcasting Company, or NBC. In 1932 the oligopoly was somewhat diluted as a result of a U.S. federal antitrust consent decree.

121 Archer, *History of Radio,* 201-2.

122 Samuel Kintner, *Proceedings of the Institute of Radio Engineers,* December 1932.

123 KDKA may not have been, strictly speaking, the first U.S. radio broadcasting station. Other contenders for that distinction are WWJ in Detroit, KCBS (which was born as KQY) in San Francisco and WHA (9XN) in Madison, Wisconsin. But, as has often been observed, when KDKA started up, broadcasting started up.

124 George H. Clark, quoted in Archer, *History of Radio,* 174. Shades of the Macintosh versus IBM/DOS debate!

125 Quoted in several sources, including Tom Lewis, *Empire of the Air: The Men Who Made Radio* (Harper Perennial, 1991), 115–17.

126 The designer Sarnoff went to for his prototype receiver was former Marconi and now RCA scientist Dr. Alfred Goldsmith. A man no less visionary than Sarnoff himself, Goldsmith had in 1918 written a book about radio, in which he characterized the new technology as "the ultimate extension of personality in time and space." This predated Marshall McLuhan's publication of *Understanding Media: The Extensions of Man* by nearly half a century.

127 *Radio Broadcast,* May 1922.

128 Ibid.

129 Reproduced in Archer, *History of Radio,* 257–58.

130 AT&T salesman H. Clinton Smith, long deceased, owns the dubious distinction of having sold this, the first radio commercial. His descendants, if they are aware of this fact, no doubt take comfort in the anonymity afforded by their surname.

131 Gleason Archer, *Big Business and Radio* (American Historical Company Inc., 1939), 305.

132 Frank P. Arnold (director of development for NBC), *Broadcast Advertising: The Fourth Dimension* (John Wiley and Sons, 1931), 41–42.

133 Of course, there were exceptions, in which high-minded sponsors practiced noninterference in program decisions, but instances of this were rare in the extreme and generally short-lived. As well, certain performers have attracted such a wide audience that they have been able to, as it were, speak to their audience over the heads of their sponsors. Once again, the relationship between the sponsor and performer in such cases is so unstable as to be short-lived.

134 G. E. C. Wedlake, *SOS: The Story of Radio-Communication* (David and Charles, 1973), 189.

135 Current North American and European standards use 535 and 625 lines per second, respectively. The results obtained are very similar from a viewer's point of view, with the European picture being of slightly better definition and greater subtlety of color. "High-definition" television as currently proposed is an all-digital system to which these analogue standards are no longer relevant.

136 Quoted in Tom Lewis, *Empire of the Air: The Men Who Made Radio,* 106–7.

137 Some see in Microsoft Corp. the current inheritor of these time-honored tactics. Microsoft describes its approach to dealing with individual software developers as "embrace and extend." For some, it is an unwelcome embrace.

138 Radio is also, of course, saddled with linear scheduling. However, radio is highly portable, and it does not demand complete attention in the way that television does; one can do other things while listening, e.g., drive a car. It is sometimes said that radio is a companion; television is a tyrant.

139 Quoted in Derrick De Kerckhove, *The Skin of Culture* (Somerville House Publishing, 1995), 14.

140 Postman, *Amusing Ourselves to Death,* 128.

141 I freely acknowledge that the BBC has had its share of censorship scandals and other programming outrages. I am speaking here in generalities, hoping to make a broader point. Public financing of television is certainly no guarantee against stupidity or cowardice in management, but then neither is commercial sponsorship and it has a great many additional problems.

142 For more on the networks' convergence strategies, see Chapters 27 and 28.

143 Roper-Starch Worldwide, December 1998.

144 As reported by Peter Schwartz in *Wired* (November 1993), 64.

145 The new frequencies opened up were from 135.5 kHz to 550 kHz, which is the bottom of today's AM broadcast band. Much more bandwidth was clearly available, but its use would have required building and marketing a new generation of receivers capable of tuning the higher-frequency ranges.

146 It was a thoroughly modern radio design. Armstrong sold his patents for the "superhet" receiver to Westinghouse in 1920, but it was not until 1924 that the first home receivers were put on the market, and they were manufactured by RCA under the newly minted cross-licensing agreement with Westinghouse. Their more complex circuitry made them more expensive than the crystal sets and regenerative receivers then widely in use, but they would eventually take over the entire market because of their superior performance. Armstrong didn't stop there: he went on to invent and perfect FM radio.

147 R. Bierstedt, *Emile Durkheim (Life and Thought)* (Weidenfeld and Nicholson, 1969); Emile Durkheim, *The Division of Labour in Society* (Macmillan, 1933).

148 Hannah Arendt, *Between Past and Future: Eight Exercises in Political Thought* (The Viking Press, 1968), 235.

149 Former U.S. Federal Communications Commission chairman, who viewed television with a jaundiced eye on his appointment to the post in 1961 by President John F. Kennedy.

150 Money is another digital system, as Karl Marx observed: "Since money does not disclose what has been transformed into it, everything, whether a commodity or not, is convertible into gold. Everything becomes salable and purchasable. Circulation is the great social retort into which everything is thrown and out of which everything is recovered as crystallized money. . . . Just as all qualitative differences between

commodities are effaced in money, so money, a radical leveller, effaces all distinctions." It is interesting to note the similarities between the conversion of economic value into money as Marx describes it and the conversion by digital technology of media content into an undifferentiated commodity called information, which can be stored, transferred and manipulated as symbols. Marx notes: "Money itself is a commodity, an external object, capable of becoming the private property of an individual." He goes further to observe: "Thus, social power becomes private power in the hands of a private person." The same could be said of information.

151 Thomas Hobbes, *Leviathan* (Cambridge University Press, 1991), 31.

152 René Descartes, *Meditations on First Philosophy* in *Philosophical Works of Descartes,* eds. Elizabeth S. Haldane and G. R. T.Ross, vol. 1 (Cambridge University Press, 1967).

153 See Chapter 29.

154 A none-too-subtle slap at both Aristotle and the philosophy of rhetoric in this last phrase!

155 Noah Kennedy, *The Industrialization of Intelligence* (Unwin Hyman, 1989), 75.

156 Formal systems are touched on in more detail in Chapter 23.

157 He had an early and instrumental connection with Leibniz. As a young student at Cambridge University, Babbage was part of a movement to shake British mathematics out of a stultifying insularity and force it to recognize the stimulating work being done on the Continent and in particular in France. He and his fellow radicals were especially insistent on recognizing that the differential calculus notation of Leibniz, used in most of Europe, was much superior to that of Newton, stubbornly clung to in England for mainly patriotic reasons.

158 Johnson, *The Birth of the Modern,* 543.

159 Charles Babbage, *On the Economy of Machinery and Manufactures* (John Murray, 1846), 191–92

160 The formula is a famous one, and speaks of Peel's erudition. It is a prolific generator of prime numbers (numbers that can be divided only by themselves), discovered by Leonhard Euler, eighteenth-century Swiss mathematician and pupil of Jakob Bernoulli.

161 Ada was only a few months old when her parents separated in one of the century's most famous scandals and Byron left England forever. The poet was angry and bitter to his death over losing his daughter. He opened the third canto to *Childe Harold's Pilgrimmage* with the lines:

Is thy face like thy mother's, my fair child!
Ada! sole daughter of my house and heart?
When last I saw thy young blue eyes they smiled,
And then we parted—not as now we part
But with hope.

As he lay dying in exile in Missolionghi, he said to his valet: "Oh my poor dear child!—my dear Ada! my God, could I but have seen her!" Ada specified in her will that she was to be interred in the family crypt beside her father.

162 Correspondence quoted in Maboth Moseley, *Irascible Genius: A Life of Charles Babbage, Inventor* (Hutchinson of London, 1964), 221.

163 Lady A. A. Lovelace, in "Sketch of the Analytical Engine Invented by Charles Babbage," *Notes upon the Memoir* by I. F. Manabrea (Geneva, 1842), reprinted in P. and E. Morrison, *Charles Babbage and His Calculating Engines*.

164 Ibid., 35.

165 Completed portions of the original Difference Engine can be seen in the Kensington Science Museum in London.

166 Babbage's son Henry completed parts of the Analytical Engine from his father's drawings in 1906 and used it to print multiples of pi to twenty-nine places as a test. It performed flawlessly.

167 Among the best known of these is the ASCII protocol, which lists 256 characters as completely representing the alphabet and its punctuation marks and assigns a binary code or string of 1s and 0s to each character.

168 Andrew Hodges, *Alan Turing, the Enigma,* (Burnett Books, 1983), 496.

169 The world first learned of this in the 1970s when secret information was declassified under Britain's thirty-year secrecy rule. Colossus was designed by Thomas Flowers, S. W. Broadhurst and W. W. Chandler with assistance from Alan Turing and M. H. A. Newman.

170 The Macintosh Powerbook this book was written on weighs five pounds, is about a million times faster than ENIAC and has about a million times as much storage. It cost less than a used car or a large kitchen appliance and can run for about two hours on a rechargeable battery which I can slip into a shirt pocket. A six-year-old child would have no difficulty handling most of its operations.

171 Herman Goldstine, *The Computer from Pascal to von Neumann* (Princeton University Press, 1972), 86.

172 Details of Mauchly's security ordeal were first made public in Stan Augarten's *Bit by Bit* (Ticknor and Fields, 1984), Appendix 1, 289.

173 *Life*, February 25, 1957.

174 A UNIVAC machine was used by the CBS television network as part of its coverage of the 1952 Stevenson–Eisenhower presidential election, the first time a computer had been used in journalistic coverage of a major event. With only 7 percent of the vote tallied, the UNIVAC predicted a landslide victory for Eisenhower. CBS producers, however, were reluctant to believe it and asked that it be reprogrammed. This time it predicted a result that was too close to call. Of course, the initial prediction was proved to have been accurate.

175 Cherry, *The Age of Access*, 66.

176 Paul Edwards, "Close World: Systems Discourse, Military Policy and Post World War II Historical Consciousness," in *Cyborg Worlds: The Military Information Society,* eds. Les Levidow and Kevin Robins (1989), 149.

177 Simon Lavington, *Early British Computers* (Digital Press, 1980), 104.

178 Norbert Weiner, *Cybernetics, or Control and Communication in the Animal and the Machine* (MIT Press, 1961), 26.

179 Pulse code modulation was one of them. See Chapter 8.

180 Information can thus be measured or quantified statistically by the number of yes/no questions that need to be answered to resolve uncertainty.

181 John R. Pierce, "Communication," *Scientific American,* September 1972.

182 J. M. Jauch, *Are Quanta Real?* (Indiana University Press, 1973), 63–65.

183 Jauch suggests there may in fact be several meanings, depending on the particular filters employed. Each would have equal claim to representing reality, and each would be complementary to the others.

184 Readers familiar with the quantum mechanical description of electrons in Chapter 4 will share with me a feeling that by this point in the description of the transistor, the planetary/Newtonian/Maxwell analogy for atoms and their relationships is strained to the breaking point. I will therefore drop the analogy and proceed simply by describing results achieved. (What is a "hole," really?)

185 It is relatively easy to understand the way a transistor amplifies if it is thought of as two diodes connected back-to-back, i.e., in series. The three zones of doped material that result are called the emitter (p), base (n) and collector (p). In a pnp transistor (pn + np diodes), a current of holes can be made to flow from the emitter to the base by the application of a small *positive* voltage or "bias," as in the case of the diode described above. If, as well, a small *negative* voltage is applied to the base-to-collector section of the transistor, the depletion zone in that region becomes an effective barrier to electron flow, but a powerful attractor of holes. Therefore, holes from the base join holes from the emitter in flowing into the collector, as a tributary would join a river. In other words, a current beginning in the emitter is added to by the current arising in the base and the two merge to flow out of the base to the collector: the original emitter current has been *amplified* by the time it reaches the collector. The extra energy has come from the batteries supplying the bias voltages. A weak radio or audio signal applied to the emitter can thus be stepped up several orders of magnitude by the time it leaves the transistor at the collector.

186 Noyce was actually preceded by six months by Jack Kirby of Texas Instruments, who created an integrated circuit based on a germanium chip. Noyce's version was better adapted to manufacturing processes, however, and set the industry standard.

187 Jack Kirby of Texas Instruments filed for a patent on an integrated circuit device of his invention several months earlier than Noyce; however, the U.S. Patent Office ruled that his description of the method for providing interconnections between components on the chip was not adequate and gave the patent to Noyce. A ten-year lawsuit ensued, which Noyce won. The idea of an integrated circuit within a semiconductor chip was first proposed in public by the English engineer G. W. A. Drummer, in 1952, who worked on the idea on a contract from the Royal Radar Establishment.

188 It was a four-bit chip running at about sixty thousand operations per second. RAM, ROM and the input-output chip were external to the CPU chip.

189 Stan Augarten, *Bit by Bit—An Illustrated History of Computers* (George Allen and Unwin, 1984), 265.

190 The first eight-bit chip, capable of processing data eight bits (binary digits) at a time. Intel also produced the first commercial sixteen-bit microchip, the 8086.

191 Tom Wolfe, "The Tinkerings of Robert Noyce," *Esquire* (December 1983), 346–74.

192 Stephen A. Campbell, *The Science and Engineering of Microelectronic Fabrication*, (Oxford University Press, 1996), 3

193 Ibid., 491.

194 Hoo-min D. Toong and Amar Gupta, "Personal Computers," *Scientific American* (December 1982), 87.

195 "The Computer Moves In," *Time* (January 3, 1983), 10.

196 Michael Shallis, *The Silicon Idol: The Micro Revolution and Its Social Implications* (Oxford University Press, 1984), 95.

197 The dwarves being Sperry Rand, Control Data Corp., Honeywell, RCA, NCR, General Electric and Burroughs.

198 Steven Levy, *Hackers: Heroes of the Computer Revolution,* revised edition, (Penguin Books, 1994), 40–45.

199 Communism, we may note from the perspective of the century's end, appears to have done its worst and expired; corporate capitalism has shown a sobering and unacceptable ruthlessness through a decade of downsizing and other forms of active disengagement from its responsibilities to the labor force and the broader community.

200 Quoted in Steven Levy, *Hackers: Heroes of the Computer Revolution,* 431.

201 Roberts was assisted in the design by two former Air Force colleagues, engineers William Yates and Jim Bybee.

202 Rick Prelinger in a posting to Computer Memory forum.

203 Jean-Louis Gassé, *The Third Apple,* trans. from the French by Isabel L. Leonard (Harcourt Brace Jovanovich, 1987), 25, 27.

204 The designers had the benefit of having seen leading-edge research into interface design being done at Xerox's Palo Alto Research Center (PARC).

205 Internet posting by Rod Perkins, Lisa interface designer, CPSR (Computer Programmers for Social Responsibility) Mail Digest 17. Perkins also notes that in June 1996, Xerox Parc visits by Apple employees created enthusiasm for the mouse as interface device.

206 Ibid.

207 Bruce Sterling, in a speech to National Academy of Sciences, Convocation on Technology and Education, Washington, D.C., May 10, 1993.

208 Howard Rheingold, "The Future of Democracy and the Four Fundamentals of Computer-Mediated Communication," paper delivered to *Ars Electronica*, Linz, Austria (1994).

209 Baran seems to have been among the earliest advocates of an "information highway." One of his recommendations in "On Distributed Communications" is for a national public utility to transport computer data, much in the way as the telephone system handles voice data. "Is it time now to start thinking about a new . . . public utility," Baran asks, "a common user digital data communication plant designed specifically for the transmission of digital data among a large set of subscribers?"

210 ARPANET Completion Report Draft, unpublished (September 1977): 111–24. Quoted in Michael Hauben, "Behind the Net: The Untold Story of the ARPANET" WWW document (1995).

211 Ibid., 111–21. The quotation is from Robert Taylor, Licklider's successor at ARPA.

212 Work was also being done in Britain and the world's first operational packet-switching network went on stream at the National Physics Laboratories there in 1968. Another early packet-switching experiment was conducted by the Société Internationale de Télécommunications Aeronautiques in 1968–70.

213 The initial plan for the ARPANET was distributed at the October 1967 Association for Computing Machinery (ACM) Symposium on Operating Principles in Gatlingberg, Tennessee. The initial design called for networking four sites. The first ARPANET Information Message Processor (IMP) or gateway computer was installed at UCLA on September 1, 1969.

214 Licklider and Vezza, "Applications of Information Technology," *Proceedings of the IEEE*, 66 (II) 1330 (1978).

215 Brian Reid, Usenet posting, 1993.

216 Ibid.

217 Steve Crocker, RFC 3 (1969).

218 Robert Braden, RFC 1336.

219 Vinton Cerf, Internet newsgroup posting (1995).

220 Ibid.

221 "Chat" may be defined as on-line communication by text, in real time, as opposed to e-mail and newsgroups, which operate on the store-and-forward principle. Chat is, in other words, akin to a telephone call using keyboards and text instead of a telephone handset. It can be a conference call, involving any number of participants, or it can be private. Chat areas of BBSs are typically open forums in which participants can meet and exchange messages in public before moving "off-line" to private communication accessible only to themselves. As might be expected, the anonymity provided by the environment tends to make conversation in chat areas somewhat less inhibited than in real-life meetings between strangers. It might be compared to the masked balls of bygone days. Ogden Nash said in a short poem on getting acquainted: "Candy is dandy/But liquor is quicker." Had he known about on-line chat, he might have added a reference.

222 The first transatlantic telephone cable, installed in 1966, was capable of carrying 138 simultaneous telephone conversations. The fiber optic transatlantic cable opened in 1990 carries 1.5 million simultaneous conversations.

223 Interview with Gary Wolf in *Wired* (October 1994), 150.

224 Kevin Kelly, *New Rules for the New Economy: 10 Radical Strategies for a Connected World* (Viking, 1998), 60.

225 A portal site is the default opening site selected by the user's browser whenever it is launched. Microsoft, of course, pre-sets its Internet Explorer web browser to select the company's own portal (as does Netscape). This setting can be changed with a few keystrokes, but a majority of users don't bother, or don't wish to go to the trouble. Microsoft, Netscape, America Online, Yahoo! and a handful of search engines dominate this increasingly important market niche. (In Britain, interestingly, the BBC Website has become one of the most-used portals.) In a wave of "portal mania" in early 1999, Disney and Infoseek (a search engine) combined to form *SEEK*; NBC and Cnet formed *Snap*; Netscape added Excite (another search engine) to its Netcenter portal (and then merged with AOL); and Yahoo! purchased Geocities in a share deal worth $4.8 billion.

226 Quoted in *Wired* (December 1998), 83.

227 Of course, there is always the risk that advertisers will attempt to integrate their messages directly into sites that are designed to appear to be purely informational, creating in the process the Internet equivalent of the television infomercial. (An example might be an allergy information site operated by an antihistamine maker.) It is a strategy that seems doomed to meager successes, however, thanks once again to the Net's vast resources. Information, no matter how obscure or arcane, is almost always available from several different Net sources, at least some of which are certain to be noncommercial.

228 A novel, if distastefully Orwellian strategy to surmount the "problem" of user sovereignty, was adopted by a U.S. company called Free-PC, lauched in 1999 with the support of USA Networks. Its aim is to give free computers and free Internet access to clients willing to supply personal information to advertisers and to put up with advertising on their screens whether or not they are using the Net.

229 One of the most sophisticated search engines, Excite, for example, works from a single assumption: words that appear frequently together are somehow related. This means that the categories of meaning of words and phrases arise out of the corpus of Web data itself, rather than being imposed by Web librarians and computer scientists. See Steve G. Steinberg's "Seek and Ye Shall Find (Maybe)," *Wired* (May 1996), 108.

230 Business and financial enterprises frequently employ privately leased lines for security reasons. More and more, however, business is finding that adequate security is available on the Internet itself through encryption and the use of "firewalls" to isolate internal networks from the worldwide Net.

231 *tele.com* (January 1997), 17–18.

232 These Ka band satellite systems move data at speeds as high as 9.2 mbps (megabits per second). Here is a chart comparing the length of time it takes to transmit an entire sixty-four-page newspaper, including images, over various carriers:

Carrier	*Bit rate*	*Time*
analogue phone line	9.6 kbps	26.6 hours
consumer modem	14.4 to 28.8 kbps	17.8 to 8.7 hours
ISDN (digital) phone line	56 kbps	4.6 hours
TI (digital) phone line	1.5 mbps	10.2 minutes
cable (TV) modems	56 to 500 kbps	4.6 hours to 30 minutes
Ka band satellite system:		
small satellite terminal	384 kbps	39.9 minutes
large satellite terminal	2.3 to 9.2 mbps	6.7 to 1.7 minutes

233 By "disposable" I mean, of course, capable of being recycled. It should also be noted that Iridium and Globestar have the potential, in principle, to make Net access truly portable via worldwide cellular telephony.

234 Forrester Research (1999). Cisco Systems CEO, John Chambers, predicted in 1999 that U.S. e-commerce would be worth $1 trillion by 2002. Industry analyst Nicholas Lippis predicted $1.5 trillion in a speech at Networld 98.

235 Forrester Research (1999).

236 Jupiter Communications (1999).

237 Amazon keeps many titles on hand in its warehouse, but orders others directly from their publishers.

238 From a business point of view, the Net makes setting up a catalogue operation remarkably inexpensive, since only one catalogue need be prepared: the Net gives it automatic worldwide distribution.

239 With adequate bandwidth, digital or information-based products can be delivered via the Net.

240 Current estimates (1999) put the number of Net users worldwide at about 160 million.

241 I am indebted to Derrick de Kerckhove, head of the McLuhan Program on Culture and Technology and the University of Toronto for bringing the idea of public space to my attention, in conversation and in his fascinating book, *The Skin of Culture* (Somerville House Publishing, 1996).

242 One of the very few works of early science fiction to speculate on machine intelligence is Samuel Butler's *Erewhon,* first published in 1872. In the utopia of Erewhon, all machines had been banished because the people had seen that they were bound to develop intelligence one day, given their very rapid rate of evolution. When that day arrived, machines would be the masters of humans.

243 Searle, 1981, p. 302, quoted in Jay David Bolter, *Writing Space, The Computer, Hypertext and the History of Writing* (Lawrence Earlbaum Associates, 1991), 174.

244 Alan Turing, "Computing Machinery and Intelligence", *Mind,* vol. 59, no. 236 (1950), 6.

245 John Haugeland, "Semantic Engines: An Introduction to Mind Design," in *Mind Design* (MIT Press, 1981).

246 An "emergent property" is one that emerges unpredictably from a complex, nonliner or chaotic system. Life is one such property. Consciousness appears to be another. One could speculate that there is a threshold of complexity in information processing beyond which intelligence is likely to emerge.

247 A useful definition of technology is, simply, the way people do things.

248 Peter Monk, *Technological Change in the Information Economy* (Pinter Publishers, 1989), 50.

249 Ibid., 80

250 *Harvard Business Review* (July–August 1994), with C. K. Prahalad.

251 Noah Kennedy, *The Industrialization of Intelligence* (Unwin Hyman, 1989), 156.

252 *Futures* (August–September 1995).

253 Kennedy, *The Industrialization of Intelligence*, 170.

254 Marx's argument was based on the idea that only labor contributed real value to production, a notion that has been challenged by later economists. The effect, however, remains the same.

255 Robert Heilbroner, *The Worldly Philosophers* (Simon and Schuster, 1961), 138–39.

256 *Futures* (November–December 1995).

257 Remember that for Marx, only human labor could produce a profit, the reason being that labor can be "sweated"—it can be coerced into working extra-long hours for little or no extra pay, thus contributing more to the production process than it takes out in wages.

258 *Harvard Business Review* (November–December 1991). Drucker argues that technology in the knowledge and service industries is a *tool* of production rather than a *factor* of production as it is in manufacturing. "The difference is that a factor can replace labor, while a tool may or may not." In the short run, productivity improvements brought about by computer technologies may or may not lead to increased unemployment in these sectors. In the long run, as we saw earlier, computers can be expected to replace human workers on a large scale. However, as Keynes observed, "In the long run we shall all be dead."

259 David Pearce Snyder, "What's Happening to our Jobs?" *The Futurist* (March–April 1996).

260 See Chapter 26.

261 Monk, *Technological Change in the Information Economy,* 50.

262 Admittedly, a degree of technical competence is required. But the necessary design and programming skills are available at relatively low cost in the new, highly-competitive Web site development industry.

263 William Bridges, "The End of the Job," *Fortune* (September 19, 1994), 64.

264 ibid. Many of the service jobs Drucker cites are not immediately amenable to automation and can be expected to be significant continuing sources of jobs.

265 *Forbes ASAP*, 1995.

266 Don Tapscott, *The Digital Economy: Promise and Peril in the Age of Networked Intelligence* (McGraw-Hill, 1996), xiv.

267 Stanford University Center for Economic Policy Research Report: "Computer and Dynamo: The Modern Productivity Paradox in a Not Too Distant Mirror" (1996).

268 This would not have surprised Marx, who foresaw a time when automation of industrial processes would reach a stage of such completeness that the chain between labor and value would be broken, the labor content of commodities now being provided by virtual labor embodied in the machinery. At that stage, humans would be truly free to pursue their higher interests.

269 Joseph J. Corn, ed., *Imagining Tomorrow: History, Technology and the American Future* (MIT Press, 1986), 221.

270 Scott D. N. Cook, "The Structure of Technological Revolutions and the Gutenberg Myth," in Joseph C. Pitt, ed., *Philosophy and Technology, Vol. 2: New Directions in the Philosophy of Technology* (Kluwer Academic Publishers, 1995), 63.

271 It is interesting to note, however, that rag paper produced from linen was low in price relative to parchment and vellum so that the labor in producing a book on linen paper became a relatively much bigger component of the overall price. That provided an excellent economic rationale for Gutenberg's experiments with movable block printing, since its principal impact was on labor requirements in reproducing texts.

272 Cook, in *Philosophy and Technology*, 69

273 Langdon Winner, *Autonomous Technology: Technics-out-of-Control as a Theme in Political Thought* (MIT Press, 1977), 96.

274 To quote Hannah Arendt: "The reason why we are never able to foretell with certainty the outcome and end of any action, is simply that action has no end. The process of a single deed can quite literally endure throughout time until mankind itself has come to an end." In *The Human Condition* (University of Chicago Press, 1958), 233. This, of course, is not to be taken as an argument against attempts at what we nowadays call technology assessment; only that such assessment must cast a very wide net if it is to be at all useful.

275 Whether these traits are truly "fundamental" in the sense of genetic inheritance, or whether they are merely among the most desirable or admirable of human traits, it seems to me, matters not for the purposes of this argument. They may be either, or both.

276 Philosopher François Perroux said: "Slavery is determined neither by obedience nor by hardness of labor but by the status of being a mere instrument, and the reduction of man to the state of a thing." In *La Coexistence pacifique*, vol. 3 (Presses Universitaires, 1958), 600.

277 In more recent times, work schedules expanded around the clock, as manufacturing processes took advantage of electrical power, and as utilities encouraged industrial customers to minimize load fluctuations (which are inefficient) by offering discounts during minimum-load periods, i.e., overnight. Nevertheless, employees, whether shift or day workers, are forced to adhere to strict schedules.

278 Mumford, *Technics and Civilization,* 269.

279 Kevin Kelly, *Out of Control* (Addison-Wesley, 1994).

280 Bolter, *Writing Space,* 232.

281 Lynn White, Jr., *Medieval Religion and Technology, Collected Essays* (University of California Press, 1978), 22.

282 ibid.

283 Herbert Marcuse, *One-Dimensional Man* (Beacon Press, 1964), 158.

284 Arendt, *The Human Condition,* 190.

285 Jean-Paul Sartre, *Search for a Method* (Alfred A. Knopf, 1963), 47.

286 Actually: "The condition upon which God hath given liberty to man is eternal vigilance; which condition if he break, servitude is at once the consequence of his crime and the punishment of his guilt." Irish statesman John Philpot Curran (1759–1815) in a speech on the right to vote, 1809.

287 If only by resisting abuses to the system in the form of breaches of so-called Net etiquette, which amounts to a digital translation of the ancient Golden Rule.

INDEX